P9-CFJ-333

ELEMENTARY SCIENCE METHODS

ELEMENTARY SCIENCE METHODS

KENNETH T. HENSON
Curriculum and Instruction
University of Alabama, Tuscaloosa

DELMAR JANKE
Texas A & M University

McGRAW-HILL BOOK COMPANY
New York St. Louis San Francisco Auckland Bogotá Hamburg Johannesburg London Madrid Mexico Montreal New Delhi Panama Paris São Paulo Singapore Sydney Tokyo Toronto

ELEMENTARY SCIENCE METHODS

1 2 3 4 5 6 7 8 9 0 DOCDOC 8 9 8 7 6 5 4

ISBN 0-07-028265-X

This book was set in Trump Medieval by Intergraphic Technology, Inc.
The editors were Christina Mediate and Susan Gamer;
the designer was Robin Hessel;
the production supervisor was Diane Renda.
The drawings were done by Fine Line Illustrations, Inc.
R. R. Donnelley & Sons Company was printer and binder.

Photograph Credits

David Janke: Chapter 1; Chapter 3; Chapters 7 through 12; page 298; Chapters 14 through 22.
Sharon Henson: Chapter 2; Chapter 4; Chapter 5; page 110.
Virginia Rayner, Delta State University Audio-Visual Center: Pages 105, 106, 118, and 121.
National Aeronautics and Space Administration (NASA): Pages 5 and 282.
John West: Pages 278 and 293.
Indiana State University Audio-Visual Center: Page 102.
Charles Sappington: Page 80. From Amy Broussard, "Marine Education for Inlanders," *American Education*, vol. 17, no. 1, January-February 1981.
Aylmer Thompson: Page 350.

Library of Congress Cataloging in Publication Data

Henson, Kenneth T.
Elementary science methods.

Includes bibliographies and indexes.
1. Science—Study and teaching (Elementary)
I. Janke, Delmar. II. Title.
LB1585.H46 1984 372.3'5044 83-13615
ISBN 0-07-028265-X

In memory of
Dr. Caspar N. Rappenecker

CONTENTS

PART THREE
TEACHING THE EARTH SCIENCES

PREFACE

This book is about a profoundly important career—teaching science to elementary school children. We believe that any approach to teaching should begin with children. Our own teaching experience has taught us that children are naturally curious, and we believe that teachers should capitalize on this natural curiosity. We also have observed that children are naturally active, and we believe that good teaching is the art of introducing students to meaningful activities. John Dewey was right when he said, "Learning is doing."

No single description fits every child. Each child is unique—with his or her own interests, abilities, and limitations resulting from physical, emotional, mental, and social development. The activities that we have selected for this book are those which involve the "whole" child, physically, mentally, and emotionally.

For our readers' convenience we have grouped the activities in each chapter as "activities for primary grades" and "activities for intermediate grades." However, many of the activities are equally appropriate for both groups; we encourage you to study both groups and select the activities which appear suited to the children you are teaching.

We believe that maximum learning occurs when the learner is enjoying the experience. The activities in this book have been selected on the basis of actual experience in elementary classrooms, where they have been found enjoyable and effective.

We hope that once you have become acquainted with the activities, you will want to keep this book as a teaching resource. To make it more useful as a reference, we have included sections to help you build your own professional library and sections to help you build a classroom library for your students.

We believe that the study of science should be integrated with other subjects; therefore, we have shown how science can be used in teaching many subjects. We have also emphasized everyday precautions that science teachers should take; we believe that safety should always come first in the science classroom.

We wish to thank our own professors of science: Milton Pella, Dave Powell, and the late Caspar Rappenecker. Special thanks go to our former students and student teachers who have helped us test the activities in this book. We also appreciate the typing assistance given by Loretta Johnson, Jean Stoeckel, and Donna Moore. Finally, for their help with our photograph program, we want to thank Tommy Waldrup, Principal, Pearman Elementary School; Johnny Arnold, Superintendent, Bolivar County School System, District 4; Sarah Jordan, Chairperson, Home Economics Department, Delta State University; and Marjorie Haislet, Principal, and Claudette Hollie and Dana Marable, teachers, Sul Ross Elementary School.

Kenneth T. Henson

Delmar Janke

ELEMENTARY SCIENCE METHODS

PART 1

TEACHING SCIENCE IN THE ELEMENTARY SCHOOL

Most prospective teachers enter their first elementary science methods course with many questions. What is teaching science in the elementary school really like? What do I need to know about this age group to get my ideas across and to stimulate pupils to want to know more about science? These are more than rhetorical questions, for the elementary teacher's success hinges on this knowledge. It would not be an overstatement to suggest that success will be directly correlated with how well a teacher understands the elementary school child and the role of science in the elementary school. Part One of this book will help you answer these questions.

But other skills are essential to becoming an expert science teacher. You must learn how to plan and organize daily lessons as well as plan for longer periods of time such as 6-week periods, semesters, and even years. You must know how to write objectives and evaluate students' performance on these objectives. Part One contains a chapter dedicated to each of these important roles of the science teacher. As you begin this study, we suggest that you think of yourself as a teacher and stay alert to these and other questions that may arise as you explore the exciting field of elementary science.

CHAPTER 1

THE ROLE OF SCIENCE IN THE ELEMENTARY SCHOOL

PREASSESSMENT

	AGREE	DISAGREE	UNCERTAIN
1. Science and technology have had a great influence upon our society.	____	____	____
2. Science is a body of knowledge.	____	____	____
3. Children of elementary school age have a natural interest in science.	____	____	____
4. Concept formation is an important part of learning science.	____	____	____
5. Science curricula may be organized around either knowledge or processes.	____	____	____
6. Scientific attitudes and ideals are applicable to everyday situations.	____	____	____
7. I am well prepared to be a teacher of science in the elementary school.	____	____	____
8. I have knowledge of a majority of the topics taught in elementary school science.	____	____	____
9. There are many resources available which I can reasonably use to strengthen my background in science topics.	____	____	____
10. It is a mistake to rely on a textbook in teaching elementary school science.	____	____	____

OBJECTIVES

Upon completion of this chapter you should be able to:

1. Name examples of the influence of science and technology upon our lives.
2. Describe the natural appeal of science to children.
3. Identify two important components of science.
4. Discuss the relationships of science and technology.
5. Name three components of science which are important for the elementary school science curriculum.
6. Explain the difference between a knowledge-centered and a process-centered science curriculum.

7. Justify the statement: "A person who is dedicated to becoming a good elementary school teacher has the qualifications for becoming a good teacher of elementary school science."
8. Describe a plan for strengthening your knowledge of a science topic for which you consider your present background inadequate.

THE ROLE OF SCIENCE IN SOCIETY

Our world is both understandable and perplexing, enjoyable and frightening, simple and complex, supporting and challenging, gentle and powerful. It is in and for this marvelous and awe-inspiring world that we are educating our elementary school children. As a result of teaching science we want our students to enjoy more of their natural environment. We want them to appreciate and deal with the complexities of the natural world and learn to interact intelligently with that world.

To recognize the role of science and technology in our society we need only examine what we do on a given day. Many of the things we do each day are affected by developments in science and technology. We need only reflect upon transportation, communication, food production, and leisure activities to sense the impact of science and technology upon our lives.

To say the least, science and technology occupy a prominent place in our society, and accomplishments in those areas are a matter of national pride. The many committees and agencies at various levels of government which deal with science and technology are an indication of their importance to and impact upon our society. The importance of science

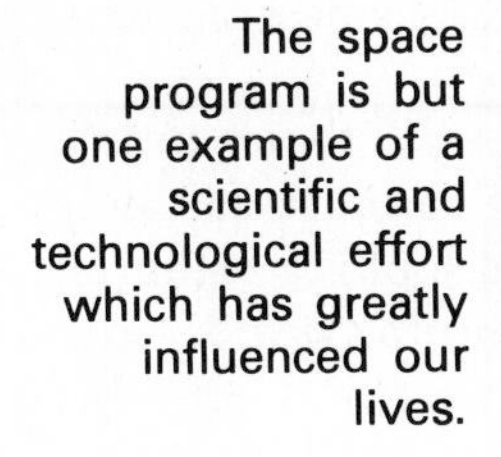
The space program is but one example of a scientific and technological effort which has greatly influenced our lives.

This scientist is working to improve agricultural plants.

and technology to our society has led Bronowski to say, "The world is made, it is powered by science; and for any man to abdicate an interest in science is to walk with open eyes toward slavery."*

Think of the problems and challenges with which we personally are confronted every day. Scientists are also confronted with challenges in their professional work. Certain attitudes and procedures which help them meet those challenges have been identified. Could it be possible that if our students adopt those attitudes and become proficient in using those procedures, they will also become more successful in dealing with everyday challenges? We believe so.

What interests do your students have? What do they talk about and do after school and on weekends? Knowing these areas of interest can help make us better teachers. Our students' interests include looking at, interacting with, and asking questions about things in the natural world. Children can often be seen around bodies of water observing and probably trying to catch living things. If there is a small woods nearby, they can be found there either looking at something specific or just walking around enjoying the environment. Have you ever had the desire to be in the mountains or at the beach? A part of science includes the study of such places.

Some children can spend hours collecting rocks and fossils. Other children immediately become absorbed in playing with a simple magnet. Receiving a microscope or a telescope would excite most children. All these areas of interest are within the realm of science. Can teachers capitalize upon these interests of children and make school an exciting place to be?

*Jacob Bronowski, *Science and Human Values,* Harper and Row, New York, 1965.

Can some of this excitement be carried over into other curriculum areas by relating science to them? In this book you will find many suggestions and activities which capitalize on students' and teachers' interests and which can help make your classroom an exciting place to be.

ELEMENTARY SCHOOL SCIENCE: WHAT IS IT?

To realize that there is not universal agreement upon what studies should be included in elementary school science, one need only review a few actual elementary science programs. On the other hand, such a review also indicates many similarities.

What is science?

Almost everyone who defines science or indicates the purpose of science has a more or less unique understanding of what science is. However, as we look at definitions and descriptions of science, we find certain similarities:

Lachman
Science refers not merely to the data acquired by the scientists and the generalizations derived, but to the fundamental objectives, basic assumptions, principal operating conceptions, and general methodology typically subscribed to by the scientist in connection with his professional activities.

Conant
Science is a process of fabricating a web of interconnected concepts and conceptual schemes arising from experiments and observations and fruitful of further experiments and observations.

Kemeny
Science is all knowledge collected by means of the scientific method.

Clagett
Science comprises, first, the orderly and systematic comprehension, description, and/or explanation of natural phenomena and, second, the tools necessary for that understanding.*

There are two common components of these and other definitions: science includes (1) a body of knowledge and (2) a set of methods by which this knowledge is established.

*S. J. Lachman, *The Foundations of Science,* Vantage, New York, 1956; J. B. Conant, *Modern Science and Modern Man,* Doubleday, Garden City, N.Y., 1959; J. C. Kemeny, *A Philosopher Looks at Science,* Van Nostrand, Princeton, N.J., 1959; C. Claggett, *Greek Science in Antiquity,* Collier, New York, 1955.

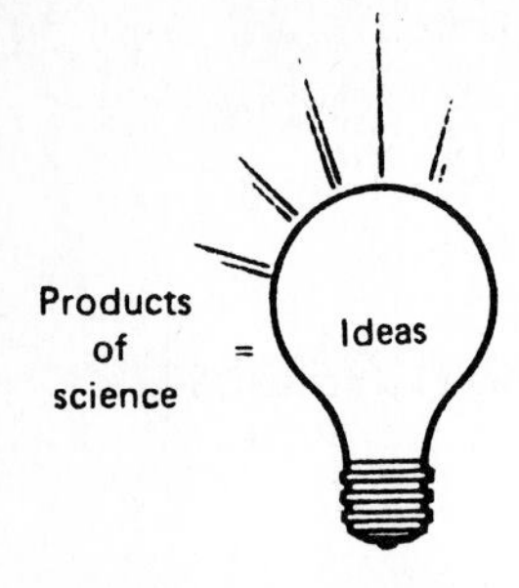

Figure 1-1. Ideas are the products of science.

Knowledge is described by terms such as "concepts," "understanding," "generalizations," and "facts." The generalizations, concepts, etc., of science may be referred to as the "products" of science. *Science produces ideas.* (See Figure 1-1.)

The methods by which scientists generate knowledge are often referred to as the "processes" of science. These processes include such activities as observing, classifying, measuring, and hypothesizing.

We believe that there is a third factor which should be included in all definitions of science: the attitude of the investigator. Scientists and children in science classes share a thirst for knowledge, a love for exploring the unknown, and an appreciation of the natural beauty of the world around them.

How do our students look at the world?

Perhaps of greatest importance to us is how our students look at the natural world. As teachers we also want to know how we may help our students look in additional ways so they can better understand their natural environment and learn to live wisely in it.

Young children look at the world in a holistic manner. It is all one big world. They learn about it by using all their senses—think of infants who put everything in their mouths. As children grow older, they explore an ever-expanding environment. Gradually they get an overall understanding of their environment.

Students constantly seek answers to questions which are generated by things they observe.

What should be emphasized in elementary school science?

This question will be answered throughout this book, but an introduction will be made here. Basically, the goal of science in the elementary school is to help students become all that they are capable of becoming and to help them learn how to live happy and satisfying lives.

A major focus of elementary school science, then, should be topics which relate to objects and phenomena in the child's real world. Studies appropriate for the primary grades include topics such as plants, animals, weather, rocks, and physical features of the land. It is important during these years to focus upon examples of those topics which are found in the child's local environment. Special emphasis should be placed upon the "processes" of science, such as observing, measuring, and classifying. Becoming skillful in using these processes will help students better understand their natural environment.

Intermediate-age children, with their knowledge of a much larger environment, should study topics further removed from their local environment, or more abstract. Topics such as historical geology, energy, physical geology, geological processes, atoms and molecules, and astronomy may be considered at this level. Emphasis should be concentrated on becoming proficient in the use of the processes of science. Processes such as formulating hypotheses, interpreting data, controlling variables, and experimenting should receive special attention.

The following three components of science (see Figure 1-2) are important in an elementary school curriculum:

1. The body of knowledge generated by and associated with science
2. The processes and procedures used to develop that body of knowledge
3. The attitudes and ideals which guide scientists in their work

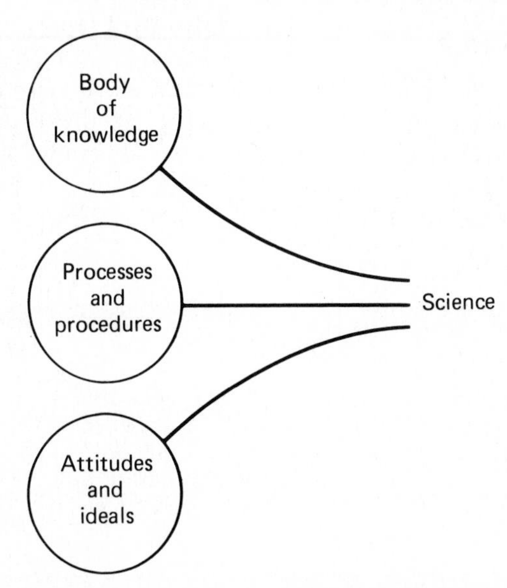

Figure 1-2. Components of science.

Do This:

Think of your concept of "plant." Is it the same as a second-grade student's concept, or that of a researcher in plant physiology? Of course not. Your concept was developed by putting together, in some way that is meaningful to you, all the information you have acquired about plants. Having a concept of "plant" gives you an economical way to "store" information on plants. When you receive new information on plants, you have a handy framework for it.

SCIENCE: A BODY OF KNOWLEDGE

Sometimes students are frightened by the tremendous amount of scientific knowledge which now exists. They would be intimidated if they were told that at some time they would be expected to know *all* of it. We believe that any attempt to have our students memorize a vast amount of scientific knowledge would be doing them—and science—a great disservice.

Knowledge in science, whether of facts or theories, becomes worthwhile and is genuinely learned only when it is presented in some way which is meaningful to the student. Scientists use basic information to produce generalizations about the natural world. Similarly, scientific knowledge becomes meaningful when students form concepts or generalizations from basic information which they have acquired.

Helping students develop meaningful concepts about phenomena in their natural environment is one of the important goals of science in the elementary school. Elementary science textbooks and curriculum guides most often group important concepts within topics such as plants, animals, and weather. In the teacher's edition of many textbooks, "major understandings," "big ideas," "important concepts," "conceptual schemes," and "generalizations" are identified. It should be pointed out that not all authors mean the same thing by these terms. One author might identify seven "conceptual schemes" for grades K–6 while another lists twenty "major concepts" at each grade level. The point is that many authors attempt to identify concepts which students are to develop.

Do This:

Examine the teacher's editions of two elementary school science textbooks at a grade level that you want to teach. How are the "units" of study identified? What "important concepts" or "major understandings" are identified? What, if anything, does the author say about *knowledge* or *content* as opposed to *process?* If you had to choose between the two books, which would you prefer? Why?

Publishers of elementary science textbooks provide many aids for the teacher, including the identification of the major concepts included in the chapters.

The elementary school science curriculum may address topics considered important for basic living: "functioning of body systems," "nutritional values of different foods," "weather phenomena and their influence on our lives." Other topics, such as dinosaurs, may be included to create interest or to show how scientists study nature. Frequently, of course, a topic is included for several reasons. The important point to keep in mind is that whatever the topic may be, there must be a good reason for including it in the curriculum.

SCIENCE: PROCESSES AND PROCEDURES

A scientist searching for new knowledge engages in several activities. Those activities can be called the "processes of science." One of the first important things that a scientist does is to become curious about something and ask an answerable question. Then data are gathered, and possible explanations are formulated and evaluated.

A scientist's job
Ask a question

Gather data
- Observe
- Measure
- Experiment

Manipulate data and form explanations
- Classify
- Communicate
- Interpret data
- Infer
- Formulate hypotheses

Evaluate explanations
- Predict
- Experiment

Scientific processes—what a scientist does—are not a rigidly defined step-by-step set of activities. There is a general pattern which describes a scientist's work, but emphasis within that pattern varies with the creativity of the scientist. Einstein once said that science starts with facts and ends with facts, no matter what happens in between. When scientists are striving to answer a question, they gather a great deal of information. Given adequate data, a theory—an explanation of the data—is finally developed. From the theory, predictions are made which can be verified. Finally, the predictions are checked against "facts" in the real world. Briefly, that is what scientists do, as Einstein suggested.

How much do elementary science students differ from this pattern? We try to discover what they want to know or to make them curious about what they are to study. (Actually, as we shall see in Chapter 2, this curiosity is already there when the child first comes to school, and the teacher's role is to kindle it and see that it is maintained.) The students then "gather data" by direct observation or reading or occasionally by being told. If they then incorporate the "data" into a concept which is meaningful to them and check their concept against additional data with which the concept deals, they have essentially developed knowledge which is new for them, much as scientists develop genuinely new knowledge.

It is clear from this description that scientific knowledge and scientific processes are related. A scientist applies processes to generate and manipulate data, in order to provide new knowledge.

An examination of what scientists do in their professional work reveals that scientific processes are important not only in science but also in many other endeavors of life. When we examined the various definitions of science, two components were found in all of them: (1) a common body of knowledge and (2) common methods by which this knowledge is established. This should suggest that a science curriculum devoted entirely or almost entirely to *either* knowledge *or* processes would distort science. Many modern elementary science units have titles that indicate a focus on knowledge (e.g., "Sound," "Weather," "Plants"), and a few have titles that indicate focus on process (e.g., "Measurement," "Observation"). But they all take an approach which reflects a mixture of knowledge and process. Since scientists use processes to generate knowledge, both are important. Knowledge without process often becomes memorization of information which will soon be forgotten; and process without knowledge becomes acquisition of skills with no apparent purpose. Using processes which generate meaningful and related knowledge—concepts meaningful to the students—should be an important goal of the elementary science curriculum.

SCIENCE: ATTITUDES AND IDEALS

Earlier we mentioned that children in science classes share some attitudes with scientists. Let's examine more carefully the attitudes and ideals which have been ascribed to scientists. These include curiosity, caution, objectivity, and skepticism.

As we mentioned earlier, *curiosity* is typical of elementary school students. As teachers, we can capitalize on our students' natural curiosity and help them expand it by introducing them to new topics. We cannot be curious about something if we do not know it exists. Introduction of a novel topic or an unusual fact is often all that is necessary to launch students into a new unit of study in science.

Caution is practiced in gathering data and doing experiments in order to eliminate error. In elementary science, caution should be a natural consequence of a real desire to obtain the best solution possible, not a result of pressure from a teacher.

Objectivity should also be promoted by a real desire to obtain the best answer or explanation possible rather than one which we would like to "prove" correct. Students will usually strive to look at the facts in an impersonal, impartial manner. The teacher's role should be to *encourage* accuracy rather than demand it.

The student, like the scientist, demonstrates *skepticism* by continually questioning data and generalizations and never accepting anything as

Children are naturally curious.

absolute. Students should learn early to avoid blind acceptance of the thoughts of "authorities." This is not to promote the extreme—not believing anything that anyone says—but rather to promote a healthy questioning attitude. Since the teacher is the only "live" example of authority in the classroom, his or her behavior becomes very important. Teachers who discourage students from questioning them indirectly discourage the questioning of any sources. Open teachers who invite students to question their statements help foster a spirit of inquiry which will generalize to life outside the classroom.

CAN ELEMENTARY SCHOOL TEACHERS REALLY TEACH SCIENCE?

Elementary school teachers are often hesitant about teaching science; they feel inadequate because they believe they know too little about science. But what we have studied in elementary, secondary, and university science courses—and what we have learned about science from newspapers, magazines, radio, and television—has familiarized us with most of the topics of elementary science curricula.

If you are dedicated to becoming a good teacher, you will also be a good teacher of science. Indeed, science can be one of the most exciting subjects which you teach. It is even permissible (and exciting) to admit that you do not "know it all" and enjoy learning along with your students. That is not to imply that we could or should approach all science teaching with little or no knowledge of the topics to be studied. The better our background on whatever we are teaching, the more confidently we approach it.

How to become a more confident teacher of science

The first step in becoming a more confident teacher of science is realizing that (as we have just noted) you already have good general knowledge of the topics taught in elementary school. But what can be done to expand this general knowledge?

One very good source of information on topics you are preparing to teach is the teacher's edition of the textbook which you use. Frequently, there are special sections of background information. Another good

Do This:

Look at the major concepts identified at the beginning of each of Parts Two, Three, and Four of this text. Look at lists of "major concepts," "important understandings," etc., in several elementary science textbooks. Are you surprised to discover that you are already familiar with most of the topics included in elementary school science?

Teachers enjoy helping their students and learning with them.

source is books which are written for children and young people. Do not feel any shame or embarrassment about using them. They can be very helpful. Specific information on many topics can be found in high school, junior high school, and middle school science textbooks. This text and similar methods books are also sources of both background information and teaching activities on selected topics.

Television can be a very important source of information. Special programs or parts of regular shows are often devoted to science topics. One example is National Geographic special programs. Useful programs are also broadcast on radio. Teachers should regularly examine the program schedules of educational television and radio stations.

Background information is available in abundance, and obtaining it is one step in gaining confidence. Another important step can be the learning activities in the textbook which you use and those in additional texts which you have discovered. Doing these activities, so that you know how something "turned out" for you, is very comforting. By the way, do not assume that the same activity will necessarily "turn out" the same way for your students.

Other very important sources of assistance are in-service workshops; lectures and presentations at local museums and universities; and special evening or summer courses provided by school districts, museums, and universities. Often special projects supported by organizations such as the National Science Foundation and the United States Office of Education are available.

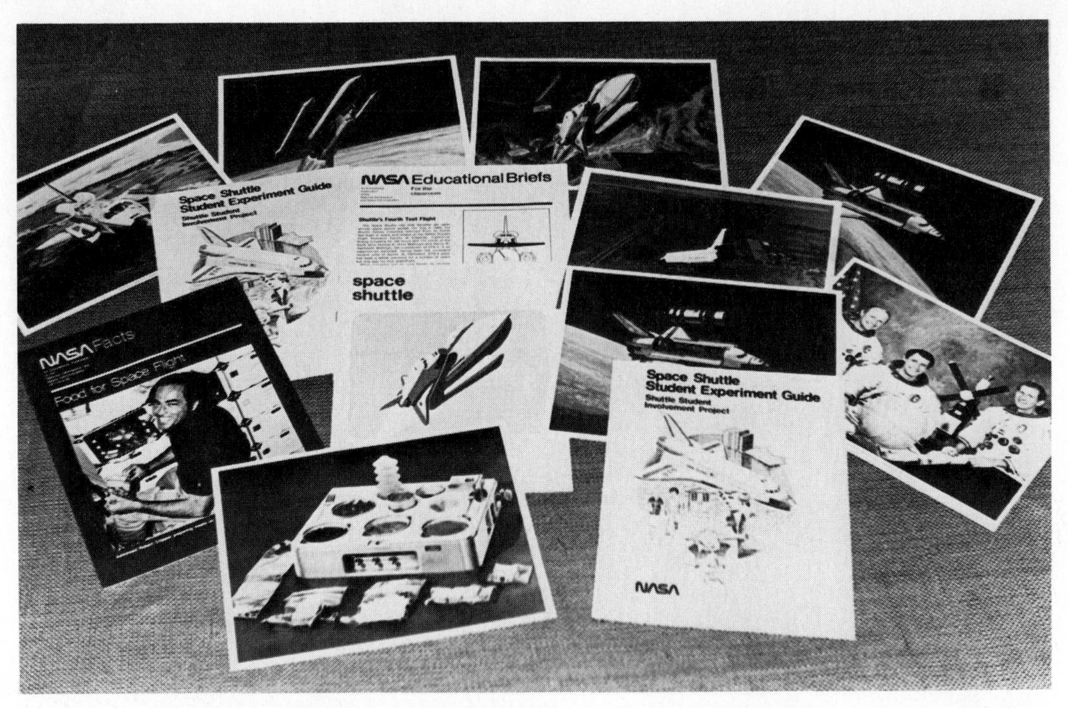

Materials such as these from NASA can help you strengthen your knowledge in areas which you will teach.

Perhaps the most rewarding and meaningful way to learn more about a topic is to experience it directly. If you want to know more about plants, examine several. If you want to know what inhabits a local pond, look! Too often education does, in fact, become a process of separating one from one's environment. Get into the real environment to obtain some of the information you need.

Do This:

Look at the major concepts identified at the beginning of each chapter in Parts Two, Three, and Four of this text, and identify some area in which you judge your background knowledge to be inadequate. Next, locate each of the following:

1. A teacher's edition of an elementary school science textbook which has background information on that topic
2. A children's book on the topic
3. A textbook on life science, physical science, or earth science (designed for grades 6 through 9) which presents that topic

Read parts or all of the identified resources. Which resource was most helpful to you? Are you more satisfied now with your level of background information on the topic? If you wanted additional information on the topic, where would you look?

Using a textbook in teaching science

Occasionally in this chapter, reference has been made to a textbook which is being used. It is not necessary to have a textbook to teach science; and some educators would even suggest that it is wrong to use a textbook. However, there is at least a 90 percent chance that you will use a textbook. You should not feel guilty about using a textbook as a basic curriculum guide, or as a reference if your school district has a curriculum guide. You will probably find that in your first 2 years of teaching you will not have time to develop your own science curriculum or even to modify or add to the basic curriculum. However, if you do not make significant developments after your first 2 years of teaching, you should begin having pangs of conscience.

This text will provide you with models for teaching materials and strategies which you can, with a reasonable amount of effort, apply in developing your own science program or modifying an existing science program. The activities, particularly, can be used to supplement or replace those in an existing science program.

Frequent questions about teaching science

The following questions are frequently asked by preservice elementary school teachers:

1. What topics are included in the elementary school science curriculum?
2. What methods of teaching science can be used to make science interesting to elementary school students?
3. Where can teaching activities be found that both teachers and students enjoy?
4. What are some good outdoor activities for teaching science?
5. What topics are taught at the various grade levels?
6. How do learners at different grade levels differ in their ability to acquire scientific concepts?
7. How can we reply to unexpected questions for which we do not know the answers?
8. Where can we find more information on science topics which is concise and reasonably easy to understand?
9. What in science motivates children?
10. How can science be related to other areas in the elementary school curriculum?
11. How can learning in science be evaluated?
12. What things—materials, and especially animals—should be found in the science classroom?

It is our goal in this text to address these and many other concerns about teaching science in the elementary school. Part One takes up general considerations; Parts Two, Three, and Four present teaching activities.

SUMMARY

Our world is amazingly complex and becoming ever more complex. Successful living requires a greater understanding of the world by each new generation than was required of the previous generation. Many of us may wish for a return to a simpler world; but this is not likely, if indeed it is possible.

Science is largely responsible for the increasing complexity of our world. But such complexity is a necessary by-product of advancements; and future generations must be equipped to handle it with confidence. This can best be achieved by introducing our students to real, everyday problems in activities which will enable them to develop the broad generalizations of science. Direct involvement with practical activities will enable them to develop in the three areas essential for success in science and in life: knowledge, attitudes, and skills.

RECAPITULATION OF MAJOR PRINCIPLES

1. Science and technology have exerted a tremendous influence upon our society and upon our personal lives.
2. Elementary school students are naturally curious about many of the topics which are studied in science, as is evident from their leisure-time activities.
3. Two common threads which run through definitions of science are that science includes (1) a body of knowledge, and (2) the processes by which that knowledge is generated.
4. Students relate best to objects and phenomena in their real environment, and they can be helped by becoming skillful in using scientific processes to study that environment.
5. Science in the elementary school should focus upon at least the body of knowledge of science, the processes and procedures used in generating that knowledge, and the attitudes and ideals of scientists.
6. The acquisition by elementary school students of meaningful concepts of science is an important goal of the science curriculum.
7. Acquiring skill in using the processes and procedures of science is an important goal in the elementary science curriculum which may have application in solving everyday problems.
8. If you are dedicated to becoming a good elementary school teacher, and if you feel confident that you will indeed be a good one, you have the qualifications for becoming a good teacher of science.
9. To obtain background information on science topics which you are preparing to teach, you can consult middle school and junior and senior high school science textbooks, teacher's editions of elementary science books, children's books, magazines, journals, and other sources.

10. A textbook is not necessary for teaching science. However, you should not feel guilty about using one—it can be a good curriculum guide and a source of information. You should feel guilty, though, if textbooks are all you ever use in your teaching career.

SUGGESTED READINGS

Althouse, Rosemary: *Science Experiences for Young Children,* Teachers College, New York, 1975.

Bainbridge, J. W., et al.: *Junior Science Source Book,* Collins, London, 1970.

Berger, Carl F., et al.: *Modular Activities Program in Science: Activity and Record Book,* Houghton Mifflin, Boston, 1974.

Caballero, Jane A.: *Aerospace Project for Young Children,* Humanics, Atlanta, 1979.

Chant, Alfred E., and Patricia Grassco: *Science Is Comparing,* rev. ed., Cambridge, New York, 1977.

Coble, Charles, R., et al.: *Mainstreaming Science and Mathematics: Special Ideas for the Whole Class,* Goodyear, Santa Monica, Calif., 1977.

Education Development Center: *A Materials Book for Elementary Science Study,* Education Development Center, Newton, Mass., 1972.

Garfield, Lee J., and Martin H. Young, *Science Is Inquiring,* Cambridge, New York, 1977.

Henson, Kenneth T., and James E. Higgins: *Personalizing Teaching in the Elementary School,* Merrill, Columbus, Ohio, 1978.

Intermediate Science Curriculum Study: *Individualized Teacher Preparation,* Silver Burdett, Morristown, N.J., 1972.

Nash, Helen O., and Gilbert A. Zinn: *Science is Looking,* Cambridge, New York, 1977.

New York University, Bureau of Research: *Conceptually Oriented Program in Elementary Science,* Center for Educational Research and Field Services, Albany, N.Y., 1972.

Platts, Mary E.: *Inquire: Suggested Activities to Motivate the Teaching of Intermediate Science,* Educational Service, Stevensville, Mich., 1977.

Stone, A. Harris, et al.: *Experiences for Teaching Children Science,* Wadsworth, Belmont, Calif., 1971.

Youngpeter, John M., et al.: *Probe: Suggested Activities to Motivate the Teaching of Primary Science,* Educational Service, Stevensville, Mich., 1976.

CHAPTER 2

THE ELEMENTARY SCHOOL CHILD

PREASSESSMENT

	AGREE	DISAGREE	UNCERTAIN
1. Rate of physical growth can affect a child's social, emotional, and moral development.	____	____	____
2. Some children do not care what others think about them.	____	____	____
3. The best way to convince children that they are important is to assign them tasks at which they can succeed.	____	____	____
4. Peer pressure is greater during preschool years than during elementary school years.	____	____	____
5. Elementary teachers should not attempt to influence the moral development of children.	____	____	____
6. Learning is a cognitive (mental) process and, as such, should be studied apart from emotional development.	____	____	____
7. Most elementary school children are unable to deal with abstractions.	____	____	____
8. Most children are capable of mastering those tasks which are commonly expected of elementary school students.	____	____	____

OBJECTIVES

Upon completion of this chapter you should be able to:

1. Describe several features of the growth of elementary school children.
2. Differentiate between human development and change.
3. Discuss several ways in which elementary students develop.
4. Explain how physical, social, psychological, and emotional development are interrelated.
5. Name several basic needs of all children, and discuss the teacher's role in meeting these needs.
6. Discuss Piaget's concrete-operations level of development and some of the implications it has for elementary science teachers.
7. Define the term "self-concept" and describe the effect of self-concept on learning.

8. Compare the influence of the family with that of the child's peers during the first 12 years of life.
9. Name and describe the three major levels of moral development as perceived by Kohlberg.
10. Justify the use of daily activities in an elementary science class.

CHILDREN AND DEVELOPMENT

Someone once said that the difference between teaching in elementary school and teaching in secondary school is that secondary teachers teach subjects whereas elementary teachers teach children. A special chapter on the elementary school child is obviously appropriate, then, in an elementary methods text.

A discussion of the characteristics of any age group must always run the risk of being misinterpreted. An even greater risk is that each characteristic described, and each adjective and adverb used in that description, is likely to be construed as a description of all the people in that age group. Like any other group of people, elementary school children are individuals. We should not stereotype them as a single group and use the same approach to teach all of them. To stereotype any child is dangerous, since it can lead to misunderstanding and eventually to faulty treatment.

However, it is practical and economical (in terms of time and energy) to discuss the characteristics of a group as opposed to those of each individual member. If we are ever to understand the nature of elementary school children, we must investigate qualities common to the group. But

All children are different.

we will do so with caution, remembering that each child is unique and that what we say about the age group may or may not apply to a particular child.

In biology courses, we learn that animals change as they grow older. The changes are not haphazard but follow definite, predictable patterns. But although development follows definite sequences, we should resist the temptation to oversimplify it. In higher animals, development is always a very complex process. Different individuals develop at different rates; and even the same individual develops in numerous ways and at different rates. Studying the general growth patterns of an age group can provide a basis for assessing the development of each individual and for understanding why people behave as they do. This is, of course, very helpful to teachers, who must guide and control the behavior of all the children in their classes.

The age span during which most American children attend elementary school (from 5 or 6 to 12 or 13) is called "late childhood." Like other ages or levels of development, it has its own characteristics; some are more important than others. Because of the complexity of development, we will limit our investigation to those changes which contribute most to the daily work of elementary teachers. Specifically, these areas of development are: physical, mental, social, emotional, and moral. A major section of this chapter will be reserved for discussing each of these areas of growth.

WHAT ARE ELEMENTARY SCHOOL CHILDREN LIKE?

Activity

Perhaps the most obvious and yet most frequently overlooked characteristic of elementary school children is their tendency to be active. A brief visit to an elementary classroom is all that is needed to remind even the most casual observer that children are highly active. Some children, of course, are much more active than others. This is to be expected. Yet teachers should be concerned about children whose behavior is extremely passive or extremely active. Either extreme may be a product of genetics, of poor diet, or of environmental factors, and is frequently caused by a combination of these and other factors.

Teachers are sometimes bothered by the high level of activity in their classrooms. A common reaction is, "Why can't I have a small portion of that energy?" A more fruitful approach is to recognize the fact that children are generally more active than adults because of their high levels of energy and mental alertness. Remembering these fundamental characteristics, the teacher will be inclined to provide activities which will utilize children's energy, channeling it in positive directions. (Parts Two, Three, and Four of this text offer activities in all areas of science that can be used to involve students to achieve this goal.)

Children are naturally active. Notice how the children here are in motion even while watching a demonstration.

Curiosity

If there is another characteristic of children that is almost as noticeable as their high level of activity, it is their tendency to be curious. You have undoubtedly noticed how alert children are to the many things in their environment and how quickly their attention shifts from one aspect to another. However, this curious spirit often backfires on children, getting

Curiosity is a common trait among children.

them into trouble. This is indeed unfortunate, since curiosity can be the most important factor in learning.

Science teachers have a unique opportunity to capitalize on children's natural curiosity about their surroundings. Science is concerned with helping children find out about their environment. As we saw in Chapter 1, science is not simply a body of knowledge but also a process for generating that knowledge, and it is guided by certain attitudes—one of which is curiosity. The role of the teacher of science is to kindle this spirit of curiosity in students; without it, the teacher cannot "teach" them anything. In fact, when children are highly curious, they will learn with or without the teacher. This is no new discovery: Socrates observed it 2500 years ago. The "Socratic method" involves asking questions to stimulate learners to think, rather than placing students in a passive role.

As you pursue your teaching of science in the elementary school, you are challenged to reflect on the different approaches your own teachers used. Which did you find most stimulating? If you were as active and curious as an elementary school student, which types of activities would you enjoy most? By the end of this term you should have built a repertoire of teaching methods which involve students in several different ways with captivating learning activities. If you begin now, you will have an opportunity to collect materials and activities using several techniques. Such a collection is a great asset to any teacher. Variety will be essential if you are going to challenge the curiosity of elementary school children.

HOW DO ELEMENTARY SCHOOL CHILDREN DEVELOP?

Physical development

For the elementary school child, physical growth occurs at a slow, steady pace (with the exception, of course, of occasional spurts). While girls tend to mature slightly more rapidly than boys, on the average elementary school children grow about 3 inches taller each year and gain about 5 to 6 pounds. This placid growth period ends at about age 9 for girls and age 11 for boys. "Most girls grow at top speed during their twelfth or thirteenth year, boys during their thirteenth or fourteenth year."*

Parents tend to overemphasize the significance of the size of their children; size and health are not necessarily correlated. Most important is the psychological, sociological, and emotional damage that may result when the child's peers become overly critical of what appear to be abnormalities, or when parents' expectations exceed a child's level of development. Usually, slight deviations from the norm (a few pounds or an inch or two) are ignored by elementary school children. But a very noticeably different child may be subjected to ridicule or even ostracism. The ways

*Mollie S. Smart and Russell C. Smart, *Children: Development and Relationships,* 2d ed., Macmillan, New York, 1972, p. 346.

in which physical growth affects other channels of development will be discussed later in this chapter.

Mental development

As children grow older, their capacity to learn changes just as surely as their physical size. But mental developments are far less obvious than physical growth. You have perhaps studied the contributions of the Swiss psychologist Jean Piaget and the American learning theorists Jerome Bruner, Robert Gagné, and David Ausubel. While these are not the only experts on learning and mental development, their contributions are recognized throughout the world. Therefore, a look at their ideas seems warranted.

JEAN PIAGET AND LEVELS OF DEVELOPMENT

According to Jean Piaget, cognitive growth, or intellectual development, occurs in four broad periods: sensorimotor intelligence (birth to 2 years), preoperational thought (2 to 7 years), concrete operations (7 to 11 years), and formal operations (11 to 15 years). The exact age at which a person may reach each period varies, and in fact some people never reach the stage of formal operations; but everyone passes through the same developmental sequence. If a teacher presents a lesson that involves a stage of thinking above the current developmental level of a child, learning for that child will be difficult or impossible. The implications for teachers, therefore, are important.

Kindergarten and first-grade children, for instance, have several limitations on their ability to learn. Two of these seem to stand out. First, they are egocentric; their thinking is limited to their own involvement—i.e., to matters which affect them. Second, their learning is hampered by their inability to focus on more than one attribute at a time. Teachers of children in these age groups should involve them with objects and processes. In other words, simple tasks which have personal meaning to them should be sought. For example, when a class is studying the human body, the teacher might group the children into pairs and have them trace each other's silhouette on butcher paper; or each child might trace his or her hand and then list some things we do with our hands.

The concrete-operations level. Since elementary teachers will most often work with students who are at the concrete-operations stage, special attention will be given to this level. A child who has reached this level can solve conservation problems; "his thought is less egocentric, he can decenter his perceptions, he follows transformations, and most important, he can reverse operations. When conflicts arise between perception and reasoning, the concrete operational child makes judgments based on reasoning."*

*Barry J. Wadsworth, *Piaget's Theory of Cognitive Development,* 2d ed., Longman, New York, 1979, pp. 97–98.

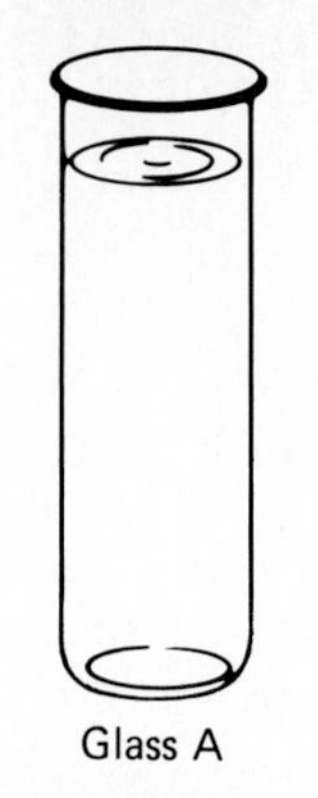

Figure 2-1. Conservation of liquids.

Before children enter school, their thinking is dominated by and limited to their perceptions. In other words, "what they see is what they know." For example, if a preoperational child is shown two glasses like those in Figure 2-1 and if all the water in glass A is poured into glass B, the child will notice that the water level is lower in glass B, and will conclude that glass B has less water than glass A had. When the concrete-operations stage is reached (usually between ages 7 and 11), thinking begins to encompass logical consequences. At this point the child begins to be able to "conserve volumes"—that is, to realize that since no water was lost, the amount in glass B must be the same as it was in glass A, despite the change in appearance. Of course this process is developed gradually and follows much concrete experience with volumes.

Care must be taken to avoid oversimplifying this process. There are other types of conservation which are more simple than conserving volume, and children learn these before they learn to conserve volume.

For example, when shown coins arranged as in row A in Figure 2-2 and asked to construct a similar row, a preoperational child will construct a row the same length as row A, but it may or may not contain the same number of coins. When given this same task, a child just entering the concrete-operations stage (at 5 or 6 years) will tend to use five coins, matching each with a coin in row A. But a 5- or 6-year-old child who is shown the two rows of coins in Figure 2-3 and asked which row has more coins will most often ignore the fact that the two rows contain an equal number of coins and will conclude that row B has more simply because it is longer. An older child who is fully "concrete-operational" will be able to "conserve number" and recognize that the two rows are numerically equal.

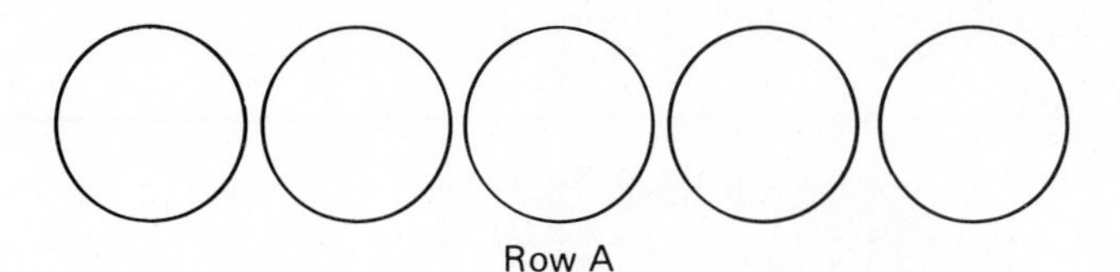

Figure 2-2. Above: Conservation of numbers.

Figure 2-3. Below: Conservation of numbers.

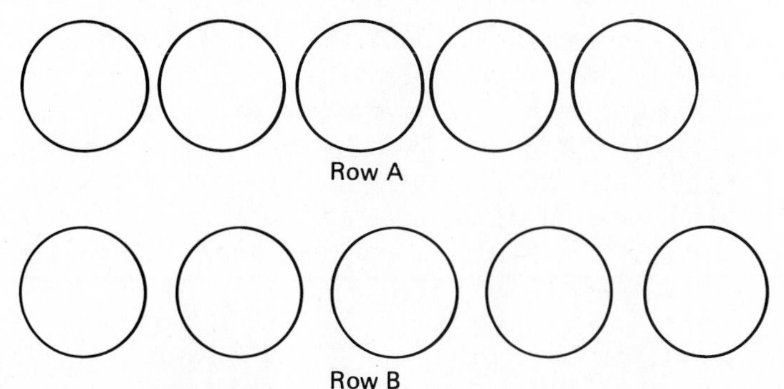

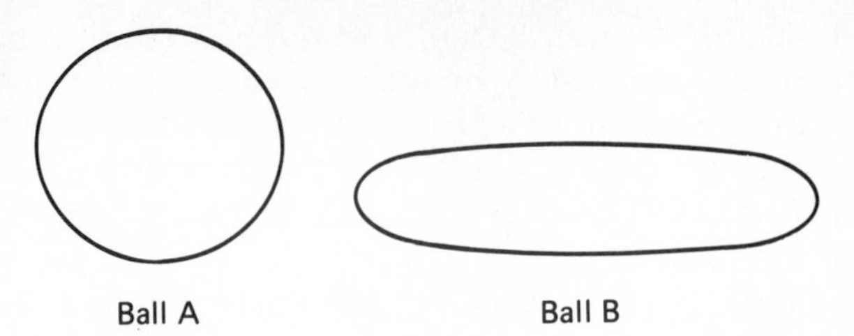

Figure 2-4.
Conservation of matter.

Piaget and his followers have conducted hundreds of such studies with children. Here is one which concerns a child's ability to "conserve substance." The child is given two balls of clay and asked to pinch off pieces from the larger ball until the two balls are equal in size. One of the balls is retained as is (ball A in Figure 2-4), while the other is reshaped to form a flat patty (like ball B in Figure 2-4) or a cylinder. Younger children do not have the ability to realize that when things are reshaped, they do not lose or gain matter. Interestingly, Piaget has found that a child develops the ability to "conserve number" before developing the ability to conserve other quantities—for example, area and volume.

The teacher's role. What implications does all this have for the teacher? Can conservation be taught? If not, then what use can be made of the concept?

Numerous attempts have been made to teach younger children conservation skills and concepts. The conclusion is clear and largely negative: teachers cannot teach students to perform mental tasks which they are not psychologically ready to handle. But teachers can use the conservation experiments and similar experiments to test their students' develop-

Students learn best, and enjoy learning, when they take active roles in the classroom—and particularly when they manipulate concrete materials.

mental levels. Knowing the level of each child can result in more reasonable expectations. Furthermore, teachers who are aware of the developmental levels are likely to use more activities which require manipulation of concrete materials. This alone should justify the effort that goes into studying the work of Piaget and other developmental psychologists.

Elementary school children need extensive opportunities to interact with concrete materials, so that they can develop the ability to think abstractly. Problems which involve transformations, reversals, and conservation can provide a means of using logic and analyzing relationships. These experiences give children opportunities to form meaningful categories. The appropriate role of the teacher may be in providing conflict rather than in resolving misconceptions. In other words, the teacher should purposefully introduce information which conflicts with students' current understanding. Figure 2-5 should help to explain this point.

In Figure 2-5, the child entered the lesson with some understanding about the content. This level remained the same from point A to point B. But something happened at point B, which initiated the learning process, and the level rose to point B_2. Points B_1, C_1, and D_1 are described by Piaget as points of "disequilibrium." Here the learner discovers inconsistencies between newly introduced or newly discovered knowledge and previous understanding. The growth areas (represented by B_1 to B_2, C_1 to C_2, and D_1 to D_2) result from the student's wrestling with these incongruities. Once the conflict is settled, the student reaches a plateau and learning levels off. Piaget would have the teacher introduce new conflicts to force the student from the state of equilibrium to a new state of disequilibrium. Conflicts can take the form of discussion, simulation

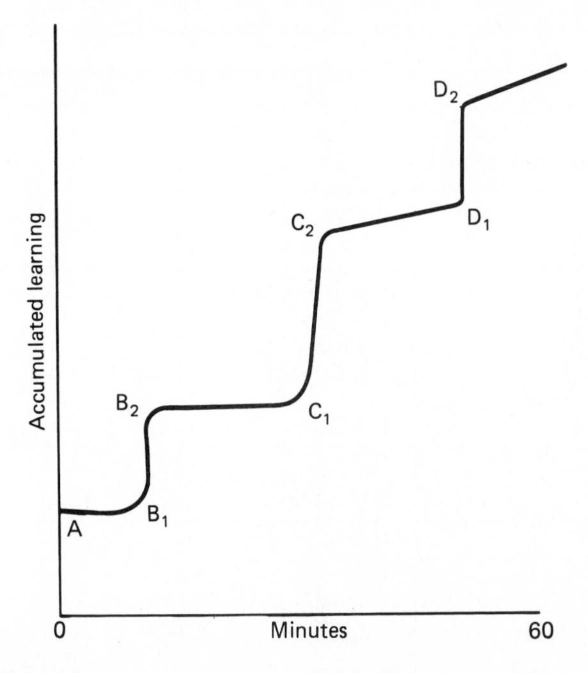

Figure 2-5. Learning curve for one child during one class period.

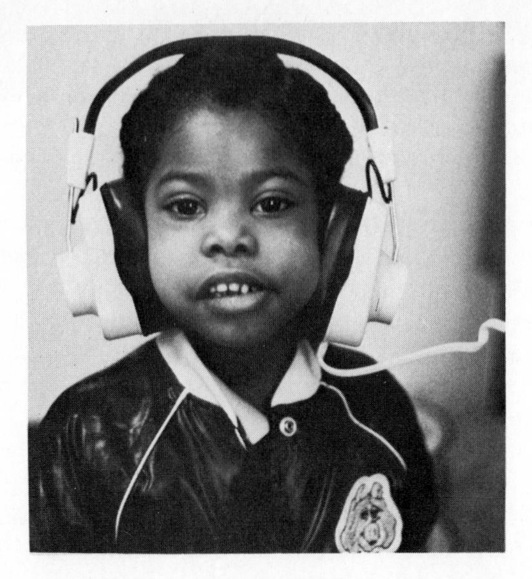

As children develop, their methods of learning increase.

games, demonstrations, laboratory problems, and so on. A variety of approaches should be used with each elementary school class.

Patricia Webb provides the following suggestions for elementary school teachers:

> Consider the stage characteristics of the student's thought processes in planning learning activities.
>
> Use a wide variety of experiences rather than drill on specific tasks to maximize cognitive development.
>
> Don't assume that reaching adolescence or adulthood guarantees the ability to perform formal operations.
>
> Remember that each person structures each learning situation in terms of his own schemas; therefore, no two persons will derive the same meaning or benefit from a given experience.
>
> Individualize learning experiences so that each student is working at a level that is high enough to be challenging and realistic enough to prevent excessive frustration.
>
> Provide experience necessary for the development of concepts prior to the use of these concepts in language.
>
> Consider learning an active restructuring of thought rather than an increase in content.
>
> Make full use of wrong answers by helping the student to analyze his thinking in order to retain the correct elements and revise the miscomprehensions.
>
> Evaluate each student in terms of improving her own performance.
>
> Avoid overuse of materials that are so highly structured that creative thought is discouraged.
>
> Use social interaction in learning experiences to promote increases in both interest and comprehension.
>
> Piaget's view on the role of a teacher can best be summed up in his own words. "What is desired is that the teacher cease being a lecturer satisfied with transmitting ready-made solutions; his role should rather be that of a mentor stimulating initiative and research."*

*Patricia K. Webb, "Piaget: Implications for Teaching," in Kenneth T. Henson, ed., *Teaching Methods,* vol. II, *Theory into Practice,* spring, 1980, pp. 69–70.

JEROME BRUNER AND WAYS OF KNOWING

At the Cognitive Center at Harvard University, Jerome Bruner has investigated learning from a psychological and experimental frame of reference. Like Piaget, Bruner has found that as children mature, their ability to use advanced cognitive processes increases; and—again like Piaget—Bruner has discovered definite stages of cognitive growth. Bruner also shares Piaget's belief that the main route to cognitive growth is through experience. But this is about as far as the similarities between the two investigators can be drawn; and fortunately so, since the contrast between their perspectives adds to our understanding of the learning process. Let us first describe Bruner's work and then contrast it with Piaget's. Bruner believes that there are three basic ways a child knows about the environment; these he has labeled "enactive," "iconic," and "symbolic."

1. Enactive stage. From birth to about age 3, a child gains knowledge about the environment only by interacting with it. The statement "We learn by doing" (attributed to John Dewey several decades earlier) reflects the enactive stage. Perhaps Bruner would modify this statement slightly to read "We *first* learn by doing," for Bruner sees a definite sequence in the cognitive development of children.

2. Iconic stage. Beginning at about age 3, children develop the ability to make mental pictures of things they have seen or done. In other words, children no longer have to be doing something in order to think about it; they can now experience it vicariously in the mind.

3. Symbolic stage. Beginning at about age 7 or 8, children develop the ability to substitute symbols for experience as they think. This is very important, since it enables them to group similar events and objects. As we shall see later, some learning psychologists believe this ability to be the basis for all learning. In essence, it eventually enables us to understand and use very complex relationships, as in mathematics and chemistry.

How does Bruner's approach differ from Piaget's? As we have seen, many of Bruner's ideas coincide with Piaget's. Both Bruner and Piaget believe that children's ability to use various mental operations depends on their level of maturity, and that while individual children develop at varying rates, most reach definite stages at about the same age. Both believe that all children progress through the same sequence of stages and that no one skips any stage.

However, partly because of their own backgrounds and partly because of their investigative procedures (Bruner used hundreds of subjects, whereas most of Piaget's studies involved only his own children), their theories contrast sharply. Piaget views children as biologically locked into a growth stage until a process of metamorphosis releases them, enabling them to move on to the next stage. Therefore, Piaget is concerned that teachers carefully choose tasks which are not beyond children's ability. Bruner, on the other hand, would encourage the teacher to present tasks that will challenge children. Although his work spans several decades, Bruner is perhaps best known for a remark he made when working with the Woods Hole Project (a project which revolutionized science education) in 1959, only two years after Americans were shocked by the Russians' launching of Sputnik I: "Any subject can be taught effectively in some intellectually honest form to any child." His implication is that a teacher should not wait for development to occur but should provide experiences which facilitate it. In other words, by having students work at skills which are slightly beyond their current reach, we can actually extend their reach.

Another significant difference between Bruner and Piaget is that Bruner views development as an internalization of technology from the culture. This leads him to stress the importance of language in learning, since he views language as the most effective technology available.

A final difference between the two will be noted here. Bruner places more emphasis on linking knowledge about learning processes to teaching. He believes that unless each discovery about how a child learns can be used to teach more effectively, the knowledge is somewhat useless. In essence, we must find ways to alter our teaching methods to utilize each particle of understanding about how children learn.

ROBERT GAGNÉ AND LEARNING HIERARCHIES

Another learning psychologist who continually ties learning to teaching methods is Robert Gagné. Consequently, most of his theories and concepts can be used by teachers to improve the conditions under which their students learn. Let's briefly examine the idea of learning hierarchies which was developed and refined by Gagné.

As children mature, they first develop the ability to make basic emotional responses, such as smiling upon hearing a parent's voice (signal learning). Eventually, they will learn to make more exact responses like saying "da-da" (stimulus-response learning). Later, they learn to connect a set of stimuli and responses in sequence (chaining); this is achieved through the use of verbal chains and motor chains (verbal association). Then they learn to identify things that are similar and to discriminate between things that are different (discrimination); and the ability to discriminate is used to form concepts. (Concept formation is discussed in detail later.) Once they have the ability to form concepts, children learn to develop or write rules to explain the relationships between and among concepts. Eventually, they learn to connect rules to solve problems. Gagné sees these steps as forming a hierarchy.

Why is this model of learning important? It is important because Gagné sees a definite sequential relationship—an order—and reminds us that as teachers we should not demand levels of thinking or performance that are beyond the abilities of the learners.

DAVID AUSUBEL AND "ADVANCE ORGANIZERS"

Ausubel's contributions to our knowledge of how people learn are numerous and varied. One of his greatest contributions to classroom teaching is the idea of using "advance organizers." This involves giving students a general introduction to material before they study it in detail. For example, a teacher might give some general information about field trips or films *before* the students participate in these activities; or a study unit on heat flow might be preceded by a general discussion about how heat makes us feel, in terms of a few broad generalizations about heat rather than numerous specific facts. Such advance organizers can help children tie previously learned concepts to others that will surface during the activity or unit.

IN CONCLUSION: MENTAL DEVELOPMENT

Admittedly, this has been a brief, cursory glance at the process of learning as viewed by a few psychologists. Our purpose is not to provide a lesson in learning theory, but to remind our readers that the effectiveness of all teachers can be enhanced by awareness of theories of learning and an ability to apply them. We believe that the best approach to learning theory is eclectic and that serious teachers must pursue this study throughout their careers.

As we noted earlier, the mental development of children affects and is affected by growth in other areas—especially social and emotional development. As we examine these other areas of development, we should look for ways in which they affect the child's ability to learn.

Social development

Normally, the process of social development is continuous from birth onward. Initially, an infant is completely egocentric. This point of view is first extended to include the family. The next extension is to neighbors such as children who live on the same street or in the same building. When school age is reached, another extension is quickly made: now the child must learn to consider and meet the expectations of a number of classmates (usually about 30). This leads to new responsibilities. The child must learn to get along and be accepted. This requires developing sensitivity to the expectations of peers. (It is at this point that the obese, skinny, or short child may begin to suffer.) The child must also learn to meet the expectations of the teacher and of the school. (Children who do not may be quickly branded as "stupid" or "troublemakers.")

Figure 2-6 shows the declining influence of the home and the increasing influence of school and other social institutions during childhood.

The comparative effect of the family and social groups on children as shown in Figure 2-6 has been determined by a series of studies at Cornell University. One of these found that by fifth grade (and on through eighth grade), children conform more to pressure from peers than to pressure from adults.*

*Smart and Smart, op. cit., p. 467.

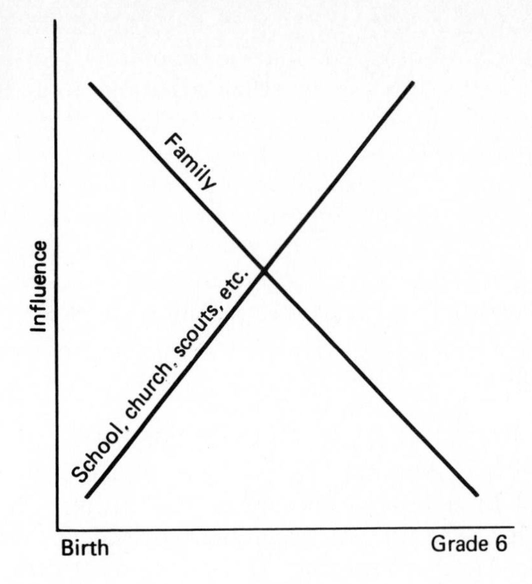

Figure 2-6. Relative influence of family as opposed to peer groups and organizations on children of different ages.

Exactly what must a child do to be socially accepted? Hurlock lists four criteria of social adjustment:

1. Overt performance
2. Adjustment to any group
3. Social attitudes
4. Personal satisfaction*

According to Hurlock, when the children's social behavior, as judged by their peers, meets the group's expectations; when children exhibit favorable attitudes toward people, toward social participation, and toward their own role in the group; and when children learn to be satisfied with their social contacts and their social roles—then and *only then* are they well adjusted. Obviously, the elementary school teacher must realize the extent of these demands on each child. Opportunities for social groups to function and for all individuals to function socially should be provided daily. The teacher who is aware of the feelings of neglected, rejected, or ostracized children will obviously be more tolerant (or at least more understanding) of their attempts to gain attention in the classroom. In some instances the aware teacher may even be able to assign responsibilities to children which will help them to earn the respect and acceptance of their peers.

Expectations of social groups vary from one group to another; traits or behaviors needed for acceptance in one group might be rejected by another group. Some factors which tend to be important in most social

*Elizabeth B. Hurlock, *Child Development,* 5th ed., McGraw-Hill, New York, 1972, p. 257.

For elementary school children, relationships with peers are strong and important.

groups are looks, intelligence, physical condition, academic achievement, sex, and social skills. By no means is this list to be considered exhaustive, but it does provide a teacher with a starting place for helping individual students gain social acceptance.

Emotional development

As children mature, the ways in which they cope with their emotions change: temper tantrums, for instance, may be expected from toddlers but are not considered acceptable for elementary school children. It has sometimes been thought that the ideal environment for children is one of total happiness, and that children should be protected from experiences which might demand concern or seriousness. But there is no evidence that children who are brought up in such environments develop into more emotionally sound adults than their counterparts who face everyday frustrations.

In the normal developmental process, children learn to experience many kinds of emotions: anger, sadness, happiness, fear, curiosity, and affection, to name only a few. A wide variety of emotions reduces boredom and can add pleasure to everyday experiences. Obviously, though, some emotions are better for us than others; for example, laughter can relieve tension. Even fear can be entertaining—otherwise, why would people like mystery and horror movies?

Perhaps our greatest concern for the emotional welfare of our students is that they receive enough love and affection to make them feel good

Each day, elementary teachers should find time to relate to small groups and to individual children. This will help children to feel good about themselves.

about themselves. This is needed not only for happiness but also for optimal learning. In other words, before children can experience academic success, they must believe themselves capable of succeeding. Educators have sometimes argued that children need to experience failure to be prepared for adult life. However, there is no evidence that merely experiencing failure equips people to cope with it; on the contrary, too much failure can result in discouragement and surrender. It has also been argued that attempts to provide successful experiences for all (or even most) students can become a cover-up for educators' own failures, and that the ability of many children to succeed in school is extremely limited. But recent studies suggest that students are far more capable than educators once believed them to be. According to Bloom, we have tended to assume that one-third of all students will be failures or near-failures, one-third will be capable of average performance, and one-third will be capable of mastering most material. Bloom challenges this assumption; he holds that, given enough motivation and enough time, 95 percent of students are capable of mastering the material with which they are confronted.*

The study of how people perceive themselves and their world is known as "perceptual psychology." According to Combs, "Emotionally healthy people feel good about themselves. Extremely adequate, self-actualizing

*Benjamin S. Bloom, J. Thomas Hastings, and George F. Madaus, *Handbook on Formative and Summative Evaluation of Student Learning,* McGraw-Hill, New York, 1971, p. 46.

persons seem to be characterized by an essentially positive view of self."* The teacher can help children develop positive self-concepts by providing tasks which are within each student's capabilities and then providing the encouragement needed for success. A positive self-concept for schoolchildren must include confidence or belief in their ability to succeed in the academic tasks expected of their age group. The science teacher is fortunate in having activities available that will continually involve students. As you read this text, you are encouraged to think about the many activities discussed and begin selecting those you will want to use in your classes.

Moral development

As we mentioned earlier in this chapter, an infant is completely self-centered: everything is viewed from a selfish, individual perspective. This egocentricity is the first of three stages of moral development described by Lawrence Kohlberg. He suggests that children pass through three major stages as their moral thinking matures:

1. *Preconventional.* Instinctive impulses are modified by rewards and punishments.
2. *Conventional.* Behavior is controlled by anticipation of praise or blame.
3. *Postconventional.* Behavior is regulated by principles embodying generality, consistency, and comprehensiveness.†

While everyone must mature in this sequence, obviously many people never reach the third level. Development from level to level requires exposure to higher and higher levels of morality. (For example, if a child were reared in a family whose members were all at the preconventional stage, that child would probably never even be aware that a higher level exists.) A teacher can, of course, serve as a model of higher levels of behavior. Ideally, each child should be introduced to the level of moral maturity which is just beyond his or her own level. To introduce principles to a child who is still thinking only in terms of rewards and punishments would probably have little or no effect on the child's behavior.

The study of science offers excellent opportunities for children to be exposed to higher levels of moral behavior. For some children, topics such as conserving natural resources and preventing pollution will be their first introduction to a way of thinking which involves concern for others. Even when teaching such general units as animals or the human body, the teacher should not assume that all children have learned to value and respect life. Rather, special efforts should be made to ensure that everyone is exposed to moral concepts.

*Arthur W. Combs, ed., *Perceiving, Behaving, Becoming,* Association for Supervision and Curriculum Development, National Education Association, Washington, 1962, p. 51.

†Lawrence Kohlberg, "Moral Education in the Schools: A Developmental View," *School Review,* vol. 72, spring 1966, pp. 1–30.

Science offers elementary school children opportunities to explore topics that call for higher levels of moral thinking. Research on topics like conservation and pollution is one example.

When possible, younger children should be given the opportunity to help care for animals. In each of the renowned British Infant Schools, there is at least one animal for which the students are wholly responsible. Younger children (5- and 6-year-olds) often are responsible for turtles, fish, birds, or other animals which require little care. (Our nursery schools and kindergartens could benefit by adopting this practice.)

Films and books which teach about caring for pets can increase children's respect for all animals. Discussion about the individual's responsibility for helping to control pollution, and participation in cleanup projects, can be designed for students of all ages. Projects aimed at conserving energy in the home can contribute to national goals while developing individual commitment to improving life for everyone. (Parts Two, Three, and Four have many activities for developing skills and attitudes of this nature.)

By the time they enter elementary school, most children have begun to move beyond the preconventional level. Since approval of peers is so important to elementary school children, they are forced to consider the opinions of others. If some children seem indifferent to others, the teacher should quickly involve them in group activities. After all, we learn to consider the opinions of others only when we are close enough to them to get their opinions. Elementary teachers should also introduce principles of behavior to get students to decide for themselves what is right and what is wrong. Most important, as teachers we should remember that our role in moral development is not to teach *our* values, but rather to lead students to consider their own behavior and their reasons for behaving as they do.

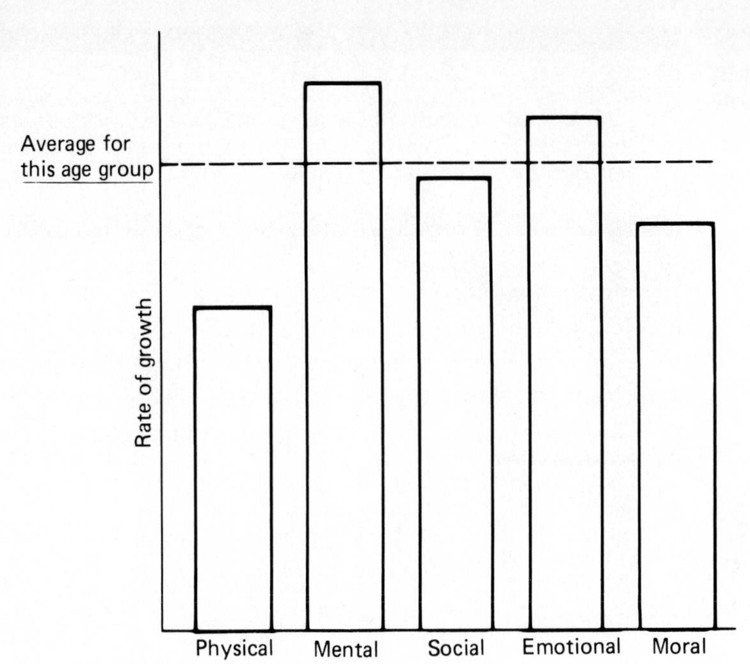

Figure 2-7. Interrelationships among growth areas.

A global look at development

EXAMPLE

Clearly, psychological, physical, social, and emotional development are interrelated. Let's take a hypothetical case and examine some of the complex effects involved in development.

The hypothetical child shown in Figure 2-7 is lagging behind most peers in physical growth. This may account for the slightly slower rate of social development, since (as we have noted) elementary school children often ridicule those who are noticeably smaller. Such ridicule could lead to emotional problems or even behavior problems, should the child decide to reject peers and other parts of the environment. Fortunately, though, this child is blessed with a good mind. Since his or her social development is only slightly below average, and emotional development is slightly above average, it appears that the child is capable of ignoring ridicule. The child's good mind allows for many interests in life.

The teacher's role involves being aware of each child's levels of development and social, emotional, moral, and intellectual needs; but it also involves more. Hoover summarizes the general role of the elementary teacher as follows: "The three most important elements that teachers should give to these children are love, understanding and stimulation."* We would add encouragement and recognition to this list.

The elementary science teacher has the opportunity to bring the world to the child and the responsibility of helping the child find and make a place in the world.

*Kenneth H. Hoover and Paul M. Hollingsworth, *Learning and Teaching in the Elementary School*, 2d ed., Allyn and Bacon, Boston, 1975, p. 7.

Do This:

Our interview below represents only one counselor's experience. Design a few questions of your own, or use the same questions, and talk to at least one or two other counselors. Compare the responses.

INTERVIEW WITH A COUNSELOR

To avoid theorizing without specific implications for classroom application, an elementary school and junior high school counselor was interviewed.

Ms. Hayes, the purpose of this interview is to gather information about how differences in growth rates affect students.

First, as a counselor, have you noticed a wide range of maturity among students the same age? Yes—especially among the upper elementary grade levels.

And do you find that the same child varies greatly in maturity in the various areas, such as psychological, emotional, physical, and social? Some children do; or perhaps several demonstrate unusual levels of development.

Is it uncommon to find that a student has developed to an advanced level in one area but is lagging in another area? No. In a class of 25 or 30, there are usually children who make good grades and cannot take kidding, children who come to me and say "I'm going to make C's so they won't pick on me." Psychologists label such a condition "mild paranoia."

Can you think of children you have known who have had problems because of their rates of growth in any areas? If so, what types of problems have you seen, and how severe were they? Those who are socially immature are still attached to the home and play a lot. They seem not to have had in-depth experiences. Life has been protected and somewhat shallow for these students. The mature ones are often from one-parent families.

As an experienced counselor, what advice could you give to teachers to help them deal with students whose level of maturity in some area or areas has been extremely fast or extremely slow? Personally, I like role playing. Teachers should be cautioned to alert the students to the fact that different reactions in similar situations are common. After role playing, think about previous classes and the level of behavior demonstrated; then have the children talk about the behaviors they saw.

I would also try (in private conversations) to get the children to verbalize what they saw as immature behavior. I'd say something like, "How did you react to a situation? Let's look back and recreate the situation. Can you tell me how you would react differently the next time you experienced this situation?"*

This interview makes it clear that growth rates vary tremendously even within the same age group, and that it is worth the teacher's time to identify students who seem to be either exceptionally advanced or unusually slow.

*Thanks to Barbara Hayes, a counselor and psychologist, who participated in this interview in Cleveland, Mississippi, in February 1983.

SUMMARY

In Chapter 1 we noted three major tasks of science in the elementary school: the development of knowledge, attitudes, and skills. We also noted that these tasks can best be accomplished by direct involvement of students in carefully selected activities. In this chapter we have seen that the ability to select and supervise appropriate activities for children requires the teacher to know and understand each group of students.

In particular, consideration must be given to children's growth—physical, mental, emotional, social, and moral. The better a teacher understands a particular age group, the better that teacher can guide students and provide stimuli, encouragement, and reinforcement. We believe that all elementary teachers and particularly science teachers should capitalize on the natural activity and curiosity of their students.

RECAPITULATION OF MAJOR PRINCIPLES

1. Children's development occurs in definite sequences.
2. While elementary school students share many characteristics, each child is an individual who possesses many unique characteristics.
3. Understanding of the needs, interests, and abilities of children is essential to planning learning experiences for any group of children.
4. A child's growth rate in some areas may exceed that of peers; in other areas it may lag behind that of peers.
5. A positive self-concept is essential to maximum progress.
6. As children mature, the influence of the family on their behavior decreases and the influence of peers increases.
7. A child's cognitive development occurs in sequential stages.
8. For learning to be optimal, experiences which parallel the child's level of development are essential.
9. Elementary school children need opportunities to interact with numerous concrete objects.
10. Since each child structures each learning experience differently, a variety of activities is needed for each group of students.

SUGGESTED READINGS

Almy, Millie, and Celia Genishi: *Ways of Studying Children,* Teachers College, New York, 1979.

Althouse, Rosemary: *The Young Child,* Teachers College, New York, 1981.

Althouse, Rosemary, and Cecil Main: *Science Experiences for Young Children,* Teachers College, New York, 1975.

Association for Supervision and Curriculum Development: *Perceiving, Behaving, Becoming,* in William Van Til, ed., *ASCD Yearbook,* Washington, 1962.

Biehler, Robert F.: *Child Development: An Introduction,* 2d ed., Houghton Mifflin, Boston, 1981.

Bloom, Benjamin S., J. Thomas Hastings, and George F. Madaus: *Evaluation to Improve Learning,* McGraw-Hill, New York, 1981.

Cohen, Dorothy, H., and Virginia Stern: *Observing and Recording the Behavior of Young Children,* Teachers College, New York, 1978.

Collier, Calhoun C., W. Robert Houston, Robert R. Schmatz, and William J. Walsh: *Modern Elementary Education: Teaching and Learning,* Macmillan, New York, 1976.

Curtis, Sandra R.: *The Joy of Movement in Early Education,* Teachers College, New York, 1982

Duck, Lloyd: *Teaching with Charisma,* Allyn and Bacon, Boston, 1981.

Evans, Richard I.: *Jean Piaget: The Man and His Ideas,* Dutton, New York, 1973; Longman, New York, 1979.

Gagné, Robert M.: *The Conditions of Learning,* 3d ed., Holt, New York, 1977.

Gagné, Robert M., and Leslie J. Briggs: *Principles of Instructional Design,* Holt, New York, 1979.

Henson, Kenneth T., and James E. Higgins: *Personalizing Teaching in the Elementary School,* Merrill, Columbus, Ohio, 1978.

Hoover, Kenneth H., and Paul M. Hollingsworth: *Learning and Teaching in the Elementary School,* 2d ed., Allyn and Bacon, Boston, 1975.

Hurlock, Elizabeth B.: *Child Development,* 5th ed., McGraw-Hill, New York, 1972.

Jarolimek, John, and Clifford D. Foster: *Teaching and Learning in the Elementary School,* Macmillan, New York, 1976.

Lanquis, M. L., T. Sanders, and S. Tips: *Brain and Learning: Directions for Early Childhood Education,* National Association for the Education of Young Children, Washington, 1980.

Miles, Matthew B.: *Learning to Work in Groups,* Teachers College, New York, 1981.

Sears, Pauline S., ed.: *Intellectual Development,* John Wiley, New York, 1971.

Smart, Mollie S., and Russell C. Smart: *Children: Development and Relationships,* 2d ed., Macmillan, New York, 1972.

Stinnett, T. M., and Kenneth T. Henson: *American Schools in Transition: Future Trends in Education,* Teachers College, New York, 1982. See chap. 16, The Human Equation.

Wadsworth, Barry J.: *Piaget's Theory of Cognitive Development,* 2d ed., Longman, New York, 1979.

Webb, Patricia K.: "Piaget: Implications for Teaching," in Kenneth T. Henson, ed., *Teaching Methods,* vol. II, *Theory into Practice,* spring, 1980.

Williams, Joyce Wolfgang: *Middle Childhood,* 2d ed., Macmillan, New York, 1981.

CHAPTER 3

ORGANIZING THE SCIENCE PROGRAM

PREASSESSMENT

	AGREE	DISAGREE	UNCERTAIN
1. Students' interests should be the most influential factor in determining the content of the elementary school science curriculum.	____	____	____
2. The textbook is most influential in determining what is actually taught in elementary science.	____	____	____
3. Processes are more important in the elementary science curriculum than concepts or attitudes.	____	____	____
4. "Real-world" phenomena are most appropriate for elementary school science.	____	____	____
5. Exploration as a step in a science teaching strategy should include gathering information.	____	____	____
6. Elementary science should not include reading as an activity.	____	____	____
7. If one chooses concepts as the central focus of a science curriculum, processes and attitudes can also be important components.	____	____	____

OBJECTIVES

Upon completion of this chapter you should be able to:

1. Name several things which influence the elementary school science curriculum.
2. Identify four important factors which should be considered in developing an elementary science curriculum.
3. Explain the place of concepts, processes, and attitudes in elementary science.
4. Identify five generally accepted ideas which guide the structure of elementary science curricula.
5. Discuss topics which should be included in the elementary science curriculum.

6. Identify and describe three possible organizational schemes for elementary science.
7. Describe the "exploration-invention-discovery" strategy for teaching science.
8. Specify the importance of both "hands-on" activities and reading in elementary science.

HOW IS THE ELEMENTARY SCHOOL SCIENCE CURRICULUM DETERMINED?

The contents of the elementary school science curriculum are determined by a number of persons and groups. Most immediately, the classroom teacher is the one who determines what is studied. But research reveals that in a great majority of cases the science textbook is the most important influence on what is taught.

All the following (arranged in no particular order) have had an influence on the science curriculum in elementary schools:

Elementary school students
Teachers
Parents
Policies of the local school district
Regulations and guidelines from state education departments
Educational researchers
Scientists
Science textbooks
Federally funded curriculum development projects

We would suggest that the most important factors to be considered in determining the elementary school science curriculum are as follows:

1. The interests and capabilities of the children
2. Reasonable expectations about elementary school teachers
3. What has been learned from educational research
4. What has been considered most important in the enterprise of science (concepts, processes, and attitudes)

Authors of elementary science textbooks, people involved in curriculum development projects, local school boards, and state departments of education to varying degrees are cognizant of these factors and consider each factor in establishing elementary science curricula.

HOW CAN CONTENT BE CHOSEN?

Elementary reading curricula have common progressions of certain skills and concepts; the same is true of mathematics curricula. But this is not the case with elementary science curricula. If a "normal" progression exists in science, it has yet to be discovered. Probably no such progression is necessary, and probably none will be established in the near future.

Some educators argue that processes and attitudes are the most appropriate components of elementary science. Others argue that concepts should be the focus of elementary science because processes and attitudes lead to concepts. In the past, acquisition of facts seemed to be the most important goal; today it no longer is.

Discussion about important goals continues; but some ideas have been generally accepted and adopted. Let's look at a few of these.

What kinds of activities are appropriate?

There is general agreement that activities should be enjoyable and active, and that they should give students successful experiences with processes, products, and problem solving.

ENJOYABLE EXPERIENCES

The primary goal of elementary science is not "fun"; but it is well accepted that given several activities which help children acquire an established objective, using those the students enjoy most is more fruitful. It is permissible for learning to be enjoyable!

ACTIVE EXPERIENCES

It is important to provide elementary school students with concrete experiences. Real rocks, animals, plants, etc., should be handled and examined. The students should *do* something. They should observe, classify, measure, and manipulate things. "Hands-on" experiences are important. Passive experiences (e.g., being talked to) are not nearly as likely to result in learning. Science at its best is not a reading program.

EXPERIENCES WITH PROCESSES AND PRODUCTS

Both the processes and the products of science are important components. "Processes" are the activities in which the scientist is involved (e.g., measuring and experimenting); "products" are the end results (e.g., ideas, generalizations, and theories). At present, the greatest emphasis in elementary science is on processes. However, processes become meaningful only as they are applied to developing concepts. Therefore, in this book we take a balanced approach, applying processes in acquiring selected concepts.

SUCCESSFUL EXPERIENCES

When we succeed at a task, we are much more likely to attempt another task. Thus, an important consideration in developing teaching activities is

Do This:

Consider: Who has failed when a student cannot successfully learn what is intended in a learning activity?

finding ones which children can succeed in doing. (That may be one of the most important jobs of a teacher! One looks forward to doing things at which one is successful; few are eager to charge forward and be a failure.)

EXPERIENCES WITH IDENTIFYING AND SOLVING PROBLEMS
In almost all modern elementary science curricula, students are encouraged to ask questions and define problems. They are then challenged to find a way of solving problems. Problem solving is a very powerful activity, since it requires students to apply all the knowledge they have and consider how they can obtain additional information which may be needed.

What topics are appropriate at the different grade levels?

In considering science topics, it is best to look separately at processes, concepts, and attitudes.

It has been suggested that the basic *processes of science*—observing, classifying, using space and time relationships, using numbers, communicating, measuring, inferring, and predicting—are to be stressed in the primary grades. The more complex integrated processes—formulating hypotheses, controlling variables, interpreting data, defining operationally, and experimenting—are to be stressed in the intermediate grades. Considering the nature of learners in elementary school, these suggestions appear to be reasonable.

The *products of science* can probably best be approached by identifying them with topics that constitute units of study in science textbooks. Examples of such topics are plants, animals, weather, and energy; these and others can be identified as areas in which scientific inquiry takes place. Most elementary students relate best to phenomena which occur in their own environment. Therefore, it would seem reasonable to deal with products that are represented by real examples in the students' everyday world: animals, plants, rocks, matter, energy, stars, weather, and so on.

Within any science topic there are *concepts* which range from the very concrete to the very abstract. For instance, matter and some of its interactions are directly observable, while mathematical models of atomic structure are very abstract. The important thing to remember is that elementary students relate best to concrete experiences. Some abstract thinking is necessary for elementary students; but in general, efforts to elicit abstract thinking will meet with limited success at most elementary grade levels.

Appropriate emphasis of scientific *attitudes* should be given at all grade levels. The application of scientific attitudes becomes especially important when integrated processes of science are being used.

Three organizational schemes

Three excellent organizational schemes were developed by prominent elementary science curriculum projects conducted during the 1960s. These are presented below as samples of organization for elementary science.

PROCESSES: SAPA

Science—A Process Approach (SAPA) is an elementary science curriculum which identifies the processes of science as its primary learning goals. Its intent is to help students become more proficient in using scientific processes. Content is taught in the SAPA program; however, processes provide the structure into which content is placed. (The actual processes used in SAPA are identified later.)

CONCEPTS: SCIS

Science Curriculum Improvement Study (SCIS) is an elementary science program which focuses upon four major concepts—matter, energy, organisms, and ecosystems—and selected subordinate concepts. Its goal is to have students actively acquire information from which they then develop meaningful concepts, either independently or with guidance. In SCIS the processes of science are employed in activities which generate concepts. In this program, then, the major concepts of science can be viewed as providing the organizational structure in which processes are applied.

SCIS includes the following units in physical and biological science at the indicated grade levels:

Grade levels	*Physical science units*	*Biological science units*
Kindergarten	Beginnings	——
First	Material Objects	Organisms
Second	Interaction and Systems	Life Cycles
Third	Subsystems and Variables	Populations
Fourth	Relative Position and Motion	Environments
Fifth	Energy Sources	Communities
Sixth	Models: Electric and Magnetic Interaction	Ecosystems

STUDENTS' INTERESTS: ESS

Elementary Science Study (ESS) is focused on neither concepts nor processes. Rather, it has units of study based on solid activities which students enjoy doing. It is a very open-ended program which allows teachers to select from a larger number of units to construct a science curriculum. Generally, each unit is appropriate for a range of grade levels rather than one specific level. The students both identify and solve problems in their program; however, what those problems will be is not necessarily known beforehand.

Among the fifth- and sixth-grade units in the ESS program are:

Mystery Powders
Clay Boats
Brine Shrimp
Ice Cubes
Rocks and Charts
Pond Water
Small Things
Kitchen Physics
Batteries and Bulbs
Behavior of Mealworms
Mosquitoes
Water Flow
Microgardening
Crayfish
Growing Seeds
Sink or Float
Balloons and Gases
Tracks

CHOOSING AN ORGANIZATIONAL SCHEME

SAPA, SCIS, and ESS—as well as a large number of other programs, including those offered by commercial publishers—are excellent science programs. There is no single best organizational scheme—or certainly none that is unanimously preferred. Perhaps the most important guideline we can offer for choosing a science program is that any program should be a part of a well-thought-out and well-coordinated K–12 science curriculum. Such a curriculum is not, of course, established by a single teacher; rather, it is a group effort within a school system.

One organizational scheme for the elementary school science curriculum focuses on science processes such as measurement.

The thing to be concerned about is whether or not your science program includes both the processes and the important concepts of science and promotes the major desirable scientific attitudes. All three components of science can be incorporated into any reasonable organizational scheme.

A strategy for teaching science

Science Curriculum Improvement Study (SCIS) has developed a teaching strategy which can be employed with a wide variety of organizational structures and which is based on current theories of how children learn. That strategy includes three phases: exploration, invention, and discovery.

EXPLORATION

In the first phase, exploration, the student is involved in activities which address the educational objective (for SCIS, the concept). If the objective were to develop the concept "organisms," the student might observe several different organisms individually or in relation to other organisms. A population of organisms might be studied. The word "organism" would not be introduced, but several examples of organisms would be encountered.

INVENTION

In the second phase of the strategy, invention, the teacher "invents" or introduces the label for the objective being addressed. The objective might be a concept, a process, or an attitude. The teacher provides a name for the concept, process, or attitude and then defines the term. It is unlikely that a student would "invent" the same name; and only occasionally can students define the term by themselves.

The name given to the concept, process, or attitude provides students with a meaningful tool for interpreting their observations. The final step in this phase is to encourage students to identify additional examples of the concept, process, or attitude.

DISCOVERY

"Discovery" is the term which SCIS uses for activities that cause students to identify new applications of a concept, process, or attitude. Discovery activities can reinforce the students' understanding of the concept, process, or attitude, or they can expand or sharpen its meaning.

To sum up: In the exploration-invention-discovery strategy the student first obtains basic information and experience, then is assisted in acquiring and defining a label to help organize the information, and finally expands and sharpens the meaning of a concept, process, or attitude by participating in additional relevant learning activities.

It should be noted that the creators of the exploration-invention-discovery strategy applied it specifically to concepts. We have suggested that it is also applicable to processes and attitudes. To be precise, applying the strategy is really a matter of developing concepts of various processes and attitudes. We should point out that in addressing a particular process (for example, measurement) the goal is both to help students develop a concept of the process and to help them acquire skill in applying the process. (We want students to understand measurement and to be able to measure skillfully.) It would be possible to have a concept of a process without being able to carry it out, or to carry out a process without really having a meaningful concept of it. Therefore both skill and understanding must be attended to. Much the same is true of scientific attitudes (for example, open-mindedness). The student should not only have certain attitudes but also understand what they mean and why they are important.

An approach to science programs

During the 1960s and 1970s, many programs emphasizing structure were developed. Most of these programs are easily recognized, since they commonly are referred to by acronyms. In addition to the three programs discussed above (SAPA, SCIS, and ESS), there are, for example, USMES (Unified Science and Mathematics for Elementary Schools), COPES (Conceptually Oriented Programs in Elementary Science), and ISCS (Intermediate Science Curriculum Study).

Many of these programs are the results of efforts by the National Science Foundation; others were developed independently by groups and individuals. The ERIC information system is an excellent source for examining these programs. An analysis of microfiche titles for one year, chosen at random, found about 10 studies of different programs.*

The excerpts below are good examples of the activities and concepts that characterize these programs.

CONCEPTS
Here is a typical concept, with subconcepts:

Content generalization:
"Earth building up"

Concepts:
1. Sediment deposition
2. Folding and faulting
3. Volcanic action†

From this example, we can see that for the most part generalizations are very broad. Each generalization is considered essential for a basic

*For example, see the 1975 ERIC catalog, Nos. ED 111 655, ED 111 656, ED 111 657, and ED 111 661. The Educational Resource Information Center (ERIC) is a cross-reference system for the retrieval of information on all types of educational materials. Because of its thoroughness and accessibility, all educators should become acquainted with this system.

†See Sandra Abernathy, *Earth Science Unit for Second Grade: A Seed Crystal Approach*, New Mexico State University, Las Cruces, 1975.

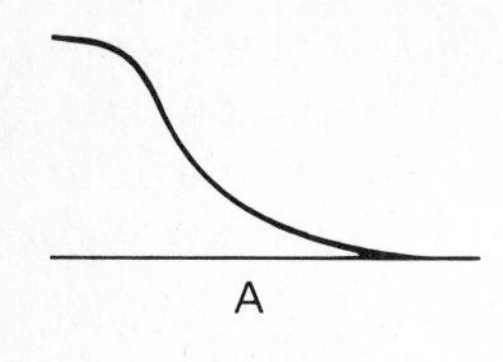

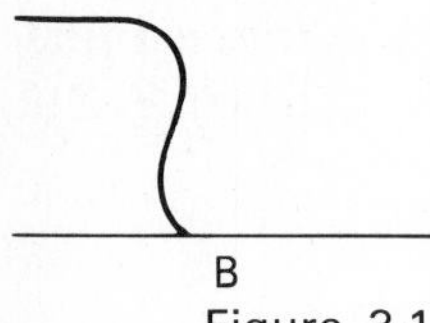

Figure 3-1.

Figure 3-2.

Figure 3-3.

Figure 3-4.

understanding of the subject. Two additional examples of generalizations taken from CIS (Concepts in Science) are: "A living thing is a product of its heredity" and "The universe is in constant change." In fact, the generalizations for CIS are so broad that it has only six such conceptual schemes.*

AWARENESS ACTIVITIES

As the following excerpt shows, activities in the science programs are many and stimulating.†

Awareness activity

1. Thought questions to bring out movement of particles
   a. Once upon a time, I was a part of a big rock on top of a huge mountain. Now I am a tiny grain of sand on a beach. What happened to me?
   b. Once upon a time, I was on the edge of a cliff. Now I am at the bottom. What forces made me fall?
   c. Once upon a time, I was a big jagged rock. Now I am a smooth pebble. Where am I and how did I get where I am?
   d. Once upon a time, I was part of the underground rock. Now I am floating around in the ocean, and the place where I was is called Carlsbad Caverns. What happened to me?
   e. Once upon a time, I was on a playground at University Hills School. Now I am near "A" Mountain. How big am I? How did I get here?
   f. Once upon a time, I was a big rock with a little tree growing next to me. Now the tree is huge and I am in pieces. What happened?
   g. Once I was a little pebble in a mountain stream. Now I am tiny and I am at the mouth of a river. What happened?
   h. Once upon a time, I was a rock on the side of the Organ Mountains. Now I make wood smooth. What happened?
2. Thought pictures to bring out that wearing down in one place leads to building up in other places—two transparencies, one superimposed over the other, or two-colored dittos
   a. [See Figure 3-1.] What happened? What might have done it and how? Put a red X on the part that was worn away. Put a blue O on the part that was built up.
   b. [See Figure 3-2.] What happened? Put a green X on the part that was worn away. Put an orange O on the part that was built up.
   c. [See Figure 3-3.] What happened? Put a brown X on the part that was eroded and a pink O on the part that was built up (anchored dune).
   d. [See Figure 3-4.] What happened? Put a yellow X on the part that was eroded and a black X on the part that was built up (free-moving dune).
3. Discuss folding. Discuss the tremendous weight (pressure) of rock and air on the rock deep underground. The temperature at that depth is very high.

*For further information about these and similar programs, see: Greg P. Stefanich, "Comparison of the Treatment of Electricity in the SAPA, SCIS, and ESS Elementary School Curricula," *American Journal of Physics*, vol. 44, no. 4, April 1976, pp 384–386.

†This extract is from Sandra Abernathy, *Earth Science Unit for Second Grade: A Seed Crystal Approach*, ERIC Reproduction Document ED 111 661, Las Cruces, New Mexico, 1976, pp 17–20. It has been adapted slightly.

The rock should be liquid at that temperature, but the pressure is so great that there is not enough space for the rock to expand, so it remains a very pliable solid. Because of internal pressures, the rock layers may be pushed out of place and folded, but they do not crack. Demonstrate with warm clay layers. We can see the folding when the rock above the folding is worn away. Show pictures of folding in rocks. Emphasize that folding takes place only deep underground.

4. Discuss faulting. When rock is closer to the surface, it is much cooler and more brittle. When it is pushed out of place (stressed), it cracks or faults. Demonstrate with cold clay layers. One side may move over the other, or one side may slide away from the other. Demonstrate by sliding two triangular pieces of wood over one another. If the fault is well lubricated, the fault blocks will wiggle and jiggle, move up, down, or sideways, and no one will be aware of it. If the fault is sticky, the pressure will build up until the fault becomes unstuck with a jolt (earthquake). We live on one of the most active faults in the world. The mountains on the east and west are moving up very fast. (Fast means a couple of inches in 40 years.) How can we tell? If the fault is moving fast, is this a good place to feel earthquakes? The San Andreas fault is in California. Californians worry about earthquakes. What is wrong with the San Andreas fault?
5. Discuss volcanic action in relation to faulting. When the fault reaches down to the superheated solid rocks deep underground, the superheated rock can expand and become molten as it fills the crack. It is forced up and out by the pressure of the superheated rock below.
6. Volcanic action. Watch the movie, LC2030, "Earthquakes and Volcanoes." Discuss faults, earthquakes, and volcanoes.
7. Volcanic action. Make a model volcano erupt. Construct a volcano model from clay or paper maché, chicken wire, and paint. Leave a small depression in the top. Line it with foil. Put a tablespoonful of ammonium dichromate crystals in the foil-lined cup. Turn the lights out, light the crystals with a match. Watch what happens. The effects can be seen more clearly if the volcano is set against a white background.
8. Discuss the life of a volcano. Have the children draw the life of a volcano in a sequence of pictures: flat ground, fault, fault block shifting, magma pushing up, cinder cone erupting, plugged, volcano.
9. Read the book *Birth of a Volcano.*
10. Watch the movie, LC 2021, "Caverns and Geysers." Review limestone caverns and discuss the action of geysers. Radium Springs is an example of a real hot spring. Fayette Hot Springs, on the way to the City of Rocks, is another.

A NOTE ON READING

In the preceding paragraphs, we have emphasized programs which stress activities; however, in no way do we intend to imply that textbooks and other printed material are unimportant to the learning of science. On the contrary, reading is essential, and we believe that teachers should take every opportunity to entice students to read. It is important that teachers make useful information available in the classroom. An activity approach to learning science in the elementary school is necessary and can be motivating and enjoyable. But it is an approach that will be enriched by extensive reading as students explore the world and their environment through literature.

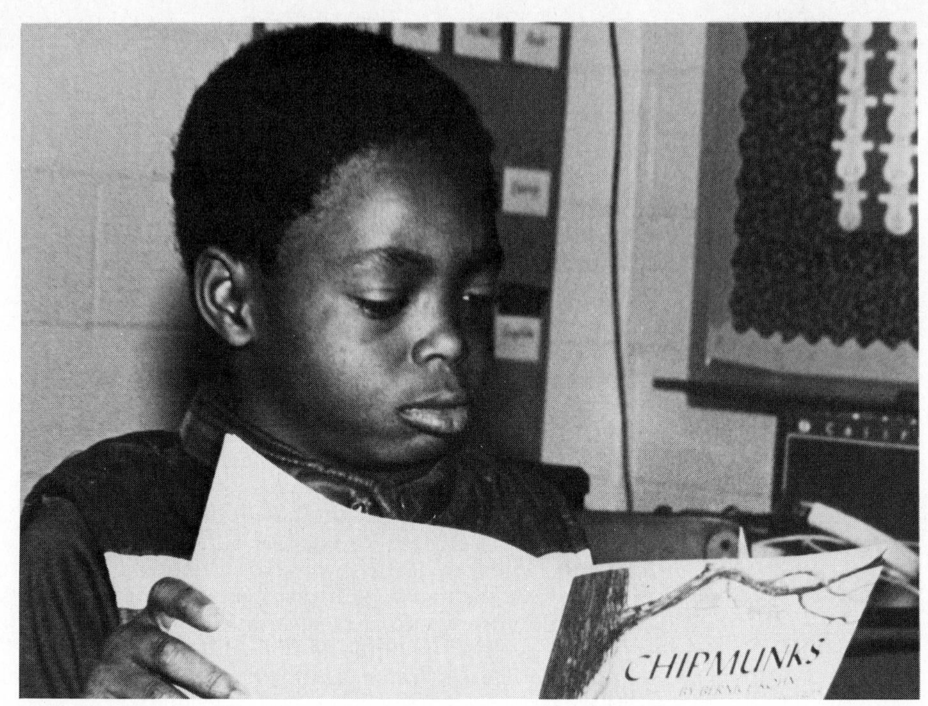

Many excellent children's books treat science topics.

LEARNING MORE ABOUT SCIENCE PROGRAMS

In this text it is neither possible nor desirable to include extensive information about the many process-type science programs. We believe that you will benefit from exploring these programs on your own. Therefore, we offer the following ideas as a few of the many approaches you can use to learn more about these programs:

1. Interview some teachers who have used one or more of the programs. Ask them about the success of these programs as compared with more traditional approaches to learning the same content. Also, ask them about other advantages, such as the acquisition of inquiry skills and attitudes.
2. Suggest to your instructor that, as a class project, you be given an opportunity to teach a generalization or concept using one of the process approaches. Have a partner use a traditional approach to teach the same content to a different but comparable class. Then compare the two approaches, using pretests and posttests to evaluate success.
3. Select as a research topic one of the process-approach programs. List its advantages and limitations. Describe any precautions the teacher must heed or special things a teacher must do to make the program succeed.

SUMMARY

Many factors influence the content of elementary science programs, including students, teachers, parents, politics, state and federal guidelines, scientists, and textbooks. We believe that the interests and abilities of the students, knowledge produced by educational research, current events, and the major generalizations in each area of science are among the most important factors that should determine the science curriculum.

The organization of the selected content and activities is crucial, since the sequence chosen will greatly influence how well the students master concepts and develop attitudes. Content and activities should be arranged so as to encourage exploration, invention, and discovery. Teachers should look for any logical sequence of the content and should consider what order will facilitate learning it—keeping in mind that attitudes are always important and that understanding major generalizations is far more important than memorizing specific facts and details.

RECAPITULATION OF MAJOR PRINCIPLES

1. The elementary science textbook is the most important single influence upon what aspects of science are taught in the elementary school. Other influences (often reflected in the content of the textbooks) include students, teachers, parents, local school policies, and scientists.
2. Four important factors which should be considered in the elementary science curriculum are the nature of the child, the nature of the teacher, what educational research has revealed, and the nature of the scientific enterprise.
3. Scientific concepts, processes, and attitudes are important components of the science curriculum.
4. Five generally accepted ideas which give direction to elementary science programs are: (1) Learning activities which students enjoy should be employed. (2) "Hands-on" activities are important. (3) Both "processes" and "products" are important components. (4) Activities in which students can succeed are to be used. (5) Identifying problems and solving problems are important experiences.
5. The elementary science curriculum can be organized around concepts, processes, or students' interests.
6. One widely employed teaching strategy includes three phases: exploration, invention, and discovery.
7. Both hands-on activities and reading are important aspects of elementary school science. Neither can justifiably be neglected in a science program.

SUGGESTED READINGS

Abruscato, Joe and Jack Hassard: *Loving and Beyond: Science Teaching for the Humanistic Classroom*, Goodyear, Santa Monica, Calif., 1976.
Blough, Glenn, and Julius Schwartz: *Elementary School Science and How to Teach It*, 6th ed., Holt, New York, 1979.
Carin, Arthur, and Robert Sund: *Teaching Modern Science*, 3d ed., Merrill, Columbus, Ohio, 1980.
DeVito, Alfred, and Gerald H. Krockover: *Creative Sciencing: A Practical Approach*, 2d ed., Little, Brown, Boston, 1980.
Esler, William K, and Mary K. Esler: *Teaching Elementary Science*, 3d ed., Wadsworth, Belmont, Calif., 1981.
Funk, H. James, et al.: *Learning Science Process Skills*, Kendall, Hunt, Dubuque, Iowa, 1979.
Jacobson, Willard, and Abby Barry Bergman: *Science for Children*. Prentice-Hall, Englewood Cliffs, N.J., 1980.
Victor, Edward: *Science for the Elementary School*, 4th ed., Macmillan, New York, 1980.

CHAPTER 4

AIMS, GOALS, AND OBJECTIVES

PREASSESSMENT

	AGREE	DISAGREE	UNCERTAIN
1. Educational aims are broader and more general than educational objectives.	____	____	____
2. Educational objectives should be written in terms of desired behavior on the part of the teacher.	____	____	____
3. Aims, goals, and objectives are needed to provide direction for learning.	____	____	____
4. Educational taxonomies have been developed which enable teachers to prepare objectives at varying levels of sophistication.	____	____	____
5. The three major domains of taxonomies of objectives are cognitive, affective, and psychomotor.			
6. All objectives should specify the conditions under which a student is expected to perform a desired behavior.	____	____	____
7. An aim is so distant and general that it can never be attained.	____	____	____
8. Objectives facilitate the attainment of goals.	____	____	____
9. All objectives should specify a minimum acceptable level of performance.	____	____	____
10. Aims, goals, and objectives are written to provide directions to both students and teachers.	____	____	____

OBJECTIVES

Upon completion of this chapter you should be able to:

1. Write an objective in each educational domain.
2. Transfer an educational aim into goals, and the goals into objectives.
3. Write a statement of philosophy of elementary science and extract from it a list of educational aims.

4. Name and give examples of each of the three essential criteria for objectives in the cognitive domain.
5. Name in sequence the various levels of the three taxonomies of educational objectives.
6. Explain the function of the taxonomies of educational objectives.

PLANNING

As we will see in Chapter 6, the teacher is responsible for both long-range and short-range planning. Both kinds of planning are essential if students are to achieve optimal understanding and develop appropriate skills and attitudes. Achieving these aspirations depends on the teacher's ability to spell out both long-term and daily expectations for each class. This chapter takes up the roles of aims, goals, and objectives in planning an elementary science program. It will differentiate each of these from the other two and explain the basis for these differences.

AIMS

Each elementary science teacher should understand his or her own attitudes toward science, children, school, and life in general. (This is a point we will take up again in Chapter 6.) Teachers must ask themselves questions such as: What are the purposes of life? What are schools for? What would we have our students become? How are children different from adults? What motivates children to learn? How do they learn? The answers to these and other questions combine to form the teacher's philosophy. Since the questions are so broad, initially they should be answered in terms of basic general expectations which the teacher has for himself or herself, for the school, and for the students. These general purposes are called "aims." An aim is an expectation which is so broad that it can never really be achieved. We tend to set aims for schools rather than for individual students or classes of students.

The "Seven Cardinal Principles" are an example of educational aims:

1. Health
2. Citizenship
3. Worthy home membership
4. Worthy use of leisure time
5. Command of the fundamental processes (i.e., mastery of the "basics")
6. Vocational efficiency
7. Ethical character*

*Formally known as the "Seven Cardinal Principles of Secondary Education," these aims were established by the National Education Association's Commission on the Reorganization of Secondary Education in 1918. Three-fourths of a century later, they still remain amazingly relevant for secondary and elementary schools in the United States.

Do This:

Each teacher should be able to think of other aims to add to the "seven cardinal principles" and our list below. Take a few minutes to list some additional aims that you would set for your elementary science curriculum.

1.
2.
3.
4.
5.
6.
7.

Notice that none of these aims can ever really be fully attained. For example, we must always work at maintaining good health or being a good citizen or a good family member; we must continue to plan to use our leisure time wisely; vocational efficiency requires continual upgrading of knowledge and skills; and we must all keep on trying to improve our behavior. Nevertheless, aims such as these are needed by every elementary teacher to provide a sense of general direction.

Science teachers also have other aims which affect the curriculum. For example, the following aims seem especially appropriate for elementary science programs:

1. Respect for life
2. Respect and concern for others
3. Conservation of natural resources
4. Safety
5. Control of pollution
6. Spirit of inquiry

While this list is not intended to be exhaustive, it does provide a starting point for developing aims for elementary science.

To summarize: Some aims should be included in all elementary science programs; others are more personal and will depend upon the philosophy of the teacher and the school.

GOALS

As important as aims are for providing general directions, by themselves they are nothing more than glittering generalities; and (as we have noted) because they are so general and broad, they cannot ever be fully achieved. Therefore, the teacher must break each aim down into statements which can be achieved during the school year. Such statements are called "goals."

For example, "health" as such cannot be achieved, but if it is made more specific and phrased in terms of a given time frame, it can become attainable. We might say, for instance, "By the end of the year the students will consistently display good personal hygiene." Similarly, the general aim "vocational efficiency," can be restated as the more specific goal, "By the end of this unit the students will be aware of several possible careers in science."

Goals, then, are more specific than aims; and, unlike aims, goals can be achieved. By definition, a goal is an aspiration which can be achieved over a fairly long period of time (from a few weeks to several months). Each goal should be derived from (and be consistent with) some more general aim. Following is an example of goals for elementary science classes:

By the end of this unit of study, each student should be able to

1. Explain the importance of conserving natural resources.
2. Display good safety habits in and out of school.
3. Show an interest in the study of science.
4. Feel competent as a student of science.
5. Work harmoniously with other students.
6. Listen when the teacher or others are talking.

You may have noticed that goals may be stated in terms of the learner's behavior or in terms of expectations the teacher has for the course. We should also point out another difference between goals and aims: aims are formulated by national committees, local schools, and even some teachers; goals may be thought of as developing at lower levels of government (state and local) and by schools and teachers. Recently, for purposes of accountability and minimum competency testing, state governments have begun writing more goals. Why? For decades, critics have said that our schools have failed to achieve their lofty aims. Many believe that aims are stated so generally that it is impossible to tell when schools are successful and when they are failing. By issuing goals, state or local governments make it easier to hold each school accountable. (For example, a state department of education, a local school district, or even a school administrator or board of trustees may formulate, mandate, or adopt a goal like this: "Before graduation from elementary school, all students will have attained a minimum reading level of the national average for grade 6.") Obviously, goals have another advantage over aims: they can be evaluated. At the end of each grading period, tests can be given to determine whether they were met, by how many students, and by which students.

Table 4-1 contrasts aims and goals. Note again that aims are intended as general guides to work toward, whereas goals are intended to be achieved. It is through the attainment of goals that aims are realized.

TABLE 4-1 CONTRASTING AIMS AND GOALS

	AIMS	GOALS
ATTAINMENT	Can never be fully attained.	Definite. Attainment is possible but not guaranteed for all.
DURATION	Infinite. No target date is set.	Finite. A target date is set for some distant time which can range from a few weeks for some goals to several years for other goals.
EVALUATION	Impossible.	Possible. May or may not be expected.
ORIGIN	Often at the federal level. Frequently, too, at state, regional, and school levels.	Usually at school and classroom levels.

Some examples of goals for elementary science classes are the following six:

By the end of this unit of study, each student will

1. Show appreciation for all forms of life by preserving the life of lower animals.
2. Understand the water cycle and be able to draw it.
3. Show the ability to work with others by completing a joint project, collecting insects.
4. Demonstrate knowledge of metrics by estimating closely the mass, length, and volume of a familiar object.
5. Develop good safety habits and follow safety rules in the classroom and outside the classroom.
6. Use the scientific method to solve problems.

Finally, we should emphasize another important characteristic of goals: the time span involved is such that a goal becomes part of students' total behavior. Even goals focused on cognitive functions often require the students to think through processes repeatedly over a period of time until the processes become natural. For example, the teacher who sets a goal requiring students to learn to measure in metrics is interested in having the students begin to think independently in metrics and will probably require them to use metrics repeatedly until they are comfortable with this system. When teaching the scientific method, a teacher would not be satisfied with a student's ability to name the steps in the scientific method but would write goals which would require its repeated use until it becomes a natural way of thinking.

Do This:

Expertise in writing goals and aims comes from experience in writing them. Now, take a few minutes to alter some of your aims so that they become goals. First, select one of your own aims or one of the aims for elementary science programs listed above. Write it below. Then write at least two goals that will facilitate the attainment of this aim.

Aim: ______________________________

____

Goals:

1. ______________________________

____

2. ______________________________

____

OBJECTIVES

As we have seen, aims are the most broadly stated purposes of schools. Since aims are so broad, they can never become fully attained. Therefore, local school systems and individual schools transform aims into more immediate goals which can be attained. Schools and school systems depend on teachers to achieve these goals. It is no exaggeration to say that without teachers, schools, school systems, and state and federal departments are powerless to realize their aspirations. It is the teacher who must see that each day the students move closer and closer to the goals. This does not imply that all important educational aims and goals come from the local, state, or federal level. On the contrary, teachers—and students—set additional goals which they consider important.

Each activity for students should have a clear objective.

To ensure that each day does carry students closer to the desired goals, the teacher must set daily expectations which more specifically describe the behavior essential for attaining goals. Such daily expectations are called "objectives." Objectives stated in terms of learners' behavior are commonly called "behavioral objectives"; those stated in terms of teachers' behavior are called "instructional objectives." But since very exact statements are needed to describe performance of designated tasks, the term "performance objectives" seems preferable to "behavioral objectives." Keeping this high degree of precision in mind, let's now see how performance objectives should be designed.

WRITING PERFORMANCE OBJECTIVES FOR THE THREE DOMAINS

Since education involves the development of mental skills, attitudes, and physical skills, objectives in each of these areas are needed. The three areas are usually referred to as the "cognitive domain," the "affective domain," and the "psychomotor domain." Since science is a combination of knowledge, attitudes, and processes, the science teacher especially will need to write objectives in all three domains.

As we begin considering how to write objectives for elementary science classes, a few preliminary comments are needed. First, the exact techniques for writing objectives may differ from one domain to another. Particularly, the degree of exactness and detail may vary. Second, the verb chosen for each objective is very important; and verbs too may differ from one domain to another.

The cognitive domain

Of the three domains, the cognitive was the first for which a hierarchy of objectives was developed. Before examining this hierarchy, let's consider the general rules for writing performance objectives in the cognitive domain.

As we mentioned earlier, there is some variation in writing techniques; but most writers agree on four criteria for cognitive performance objectives. Each cognitive objective must:

1. Identify the *a*udience.
2. Give the *b*ehavior.
3. Describe the *c*onditions.
4. State the *d*egree of behavior required.

A good mnemonic device for these criteria is "a-b-c-d." The *a*udience (a) must be the students; that is, a performance objective must describe the student's behavior, not the teacher's. The *b*ehavior (b), of course, is one that shows a student's ability to perform a certain expected way. For

Precise objectives can lead to observable and measurable performance.

objectives in the cognitive domain, a behavior must be observable and measurable. Two other criteria must be met. The objective must describe the *c*onditions (c) under which the student is expected to perform the task, and it must state the *d*egree (d) or minimal acceptable level of performance.

A performance objective is designed to transform long-term goals into short-term behaviors. Because its ability to do so depends greatly on the task involved, the verb used in a cognitive objective is extremely important. In the following list, verbs which are good choices because they describe observable and measurable behaviors are shown at the right. The verbs shown at the left are too general for cognitive objectives:

No	*Yes*
Appreciate	Build
Consider	Classify
Desire	Identify
Feel	Evaluate
Have	Interpret
Know	Label
Learn	List
Like to	Match
Remember	Measure
Think	Name
Understand	Select
Want	Write

The verbs in the "no" column do not describe types of behavior that are easily observable. For example, by observing children we cannot always tell their feelings, or whether they know, understand, or remember. The verbs in the "yes" column describe more observable behaviors. For example, merely by watching learners as they classify, list, or write we can tell whether or not they can do these tasks. Furthermore, the verbs in the "yes" column describe behaviors that are more easily measured than those described by the verbs in the "no" column. We should note, however, that measuring any of these behaviors requires breaking them down into different levels of involvement.

In 1956, Benjamin S. Bloom developed a taxonomy classifying cognitive behaviors according to levels of complexity. The levels are as follows:

Level 1: Knowledge
Level 2: Comprehension
Level 3: Application
Level 4: Analysis
Level 5: Synthesis
Level 6: Evaluation*

Research indicates that most elementary textbooks are written largely at the lowest cognitive level, knowledge. O. L. Davis and Francis P. Hunkins found that over 85 percent of the content in elementary textbooks is written at the knowledge level.† If the textbook content does not challenge students, it is up to the teacher to take them beyond the knowledge level, the simple recall of facts. Recent research indicates that teachers who challenge their students with problems at higher levels are more effective.‡

Let's now consider how to write objectives at each level of the cognitive domain.

LEVEL 1: KNOWLEDGE

Level 1 of the cognitive domain is recall of information; hence it is often called the "recall" level. At this level, nothing more than recall is measured, and the learner deals mostly with facts. Of course, there is nothing wrong with learning some things by simply memorizing them. Examples of the knowledge level are a third-grader who memorizes the multiplication table, and (say) a fifth-grader who memorizes symbols for some common elements or formulas for some common compounds. In an effort to dazzle friends and teachers, some students memorize extensive lists of information. But the length of such a list—the amount of information

*Benjamin S. Bloom et al., *Taxonomy of Educational Objectives: The Classification of Educational Goals,* Handbook I, *Cognitive Domain,* McKay, New York, 1956.

†Donald C. Orlich et al., *Teaching Strategies: A Guide to Better Instruction,* Heath, Lexington, Ma.: 1980, p. 125.

‡B. B. Rosenshine and D. C. Berliner, "Academic Engaged Time," *British Journal of Teacher Education,* vol. 4, pp. 3–16.

recalled—does not affect the level of the task, so long as the mental operation is confined to memorizing.

Here is an example of a performance objective written at the knowledge level:

When given a list of symbols of 10 common elements, the student will write the names of at least 8 of these without misspelling any of them.

First, we should notice that the student is not being asked to do anything that goes beyond simple memorization. Next, let's check to see whether this objective meets the four criteria noted earlier ("a-b-c-d"):

Audience: "the student"
Behavior: "will write"
Conditions: "when given a list of symbols"
Degree: "at least 8 out of 10" and "without mispelling any"

This objective, then, does appear to meet the criteria. Furthermore, the behavior can be observed and measured. Two final points should be made about this objective. First, it happens to have two kinds of "degrees" of expected behavior, each of which establishes a minimum level of performance. Second, the "conditions" begin with the words "when given," and these words are the first in the objective. This style is common and easy to use. It safeguards against any tendency we might have to overlook the conditions.

Some more examples of knowledge-level objectives follow:

As the teacher reads the definitions of 10 science terms, each student will write the name of at least 8 of the terms.

Given a list of the same 10 terms on the board, the students will correctly match at least 9 of them with their definitions.

Given a picture of our solar system, students will correctly label all nine planets.

Do This:

For each of the remaining objectives in this chapter, underscore the audience with a single line, underscore the behavior with a double line, enclose the conditions in a box, and enclose the degree in a circle. For example:

When given a list of symbols of 10 common elements, the student will write the names of at least 8 of these without misspelling any of them.

Then check each objective against the "a-b-c-d" criteria.

As we suggested earlier, our intent as teachers is to move students to higher levels of thinking. Let's now examine objectives at the higher levels of the cognitive domain.

LEVEL 2: COMPREHENSION

Objectives written at the comprehension level require more than simple recall of knowledge; they require learners to translate, interpret, or predict a continuation of trends.

For example:

Given a color-coded map showing the name of each state and the distribution of common minerals, the student will correctly identify and name those six states which produce large amounts of coal.

Notice that this objective requires learners to know the color symbols for various minerals (although they may not have to commit the color codes to memory). But this objective goes beyond the knowledge level. It requires the student to *use* knowledge.

Some Level 2 objectives require the learner to predict a continuation of trends. The following is an example:

Each student will complete the chart in Figure 4-1 (see opposite page) showing the continuing trend of settling.

Notice that this objective requires the learner to look at the sediment in rows A, B, and C and then predict the settling of sediment in row D.

Some examples of comprehension objectives follow:

Given five pictures showing the various stages of development of frogs, the student will correctly name the next stage in at least four pictures.

When shown three light bulbs connected in series, the student will correctly predict the relative brightness of an added fourth bulb.

When shown three plants that have had their roots in dye for 1, 2, and 3 days respectively, the student will predict the "oldest" plant's appearance after it is left in the water for an additional day.

Do This:

Our first example of a level 2 objective,

"Each student will complete the chart in Figure 4-1 showing the continuing trend of settling."

is not as clear as it could be. How could it be improved? Does it even meet the "a-b-c-d" criteria?

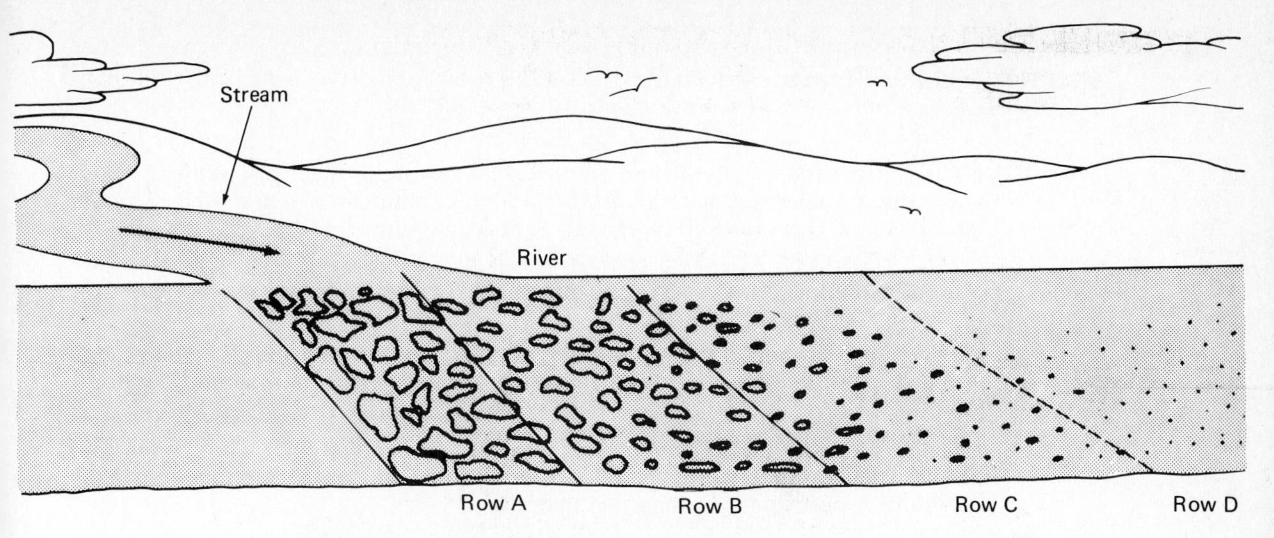

Figure 4-1. Patterns of sediment distribution.

LEVEL 3: APPLICATION

Like all other disciplines, each of the sciences has a set of principles or generalizations which must be understood. Objectives written at the application level of the cognitive domain require students to use these principles or generalizations to solve concrete problems and make predictions.

Application-level objectives often require students to make predictions. These students are learning to test the accuracy of their predictions.

For example, one principle necessary for understanding weather is that lightning produces thunder. Following is an objective based on this principle:

Given the speed of sound, the student will observe a flash of lightning, count how many seconds elapse before the sound of thunder is heard, and compute the distance from himself or herself to the lightning.

Additional examples of application-level objectives are:

Following a lesson on the effect of exercise on the heartbeat, students will predict what will happen to the rate of heartbeat when a classmate jogs for 1 minute.

Having studied Newton's third law, students will predict what will happen to a chair when a student jumps off it. (Care should be taken to choose an agile student to participate in this demonstration.)

Having learned that water expands when it freezes, students will draw a line indicating the water level in a milk carton and a second line to predict where the level will be after the water is frozen overnight.

LEVEL 4: ANALYSIS

Objectives written at the analysis level require the learners themselves to work on principles. An elementary science teacher, in a unit on electricity, might have the students analyze a complete circuit and break it down into its parts (energy source, conductor, switch, etc.) to better understand the sequence of flow. In a unit on weather, the water cycle might be treated similarly to learn the sequence involved and what happens at each step.

Here is an example of an application-level objective:

Given a chart showing the water cycle, the student will analyze each step separately and describe one physical change in the water at each step.

Other examples of analysis-level objectives are:

Given the rule "Cross the road only on the green light," the students will analyze the flow of traffic and explain the reason for this rule.

Given the rule "Ride a bicycle facing traffic," the students will defend this rule against riding with the flow of traffic.

Following an explanation of the migratory habits of certain birds, students will give the direction of the birds' flight in the spring and tell why they fly in this direction at this time.

LEVEL 5: SYNTHESIS

In a way, the synthesis level is just the opposite of the analysis level, for synthesis objectives require the student to put several parts together to form a *new* whole. This, of course, requires divergent thinking and creativity.

For example, a class studying a simulated space trip to another inhabited planet might be given the following synthesis-level objective:

Given the necessary knowledge about the planet itself and about its population, each team of astronauts will make a list of at least 10 guidelines to govern the behavior of the space team so as to protect the rights of the natives.

Additional examples of objectives at the synthesis level are:

Following a film on bicycle safety, the students, as a class, will construct their own booklet of bicycle safety rules.

Given a list of kilowatt-hours required by kitchen appliances and the maximum amount that a home kitchen can handle, each student will write at least three rules to prevent overloads on a kitchen circuit.

Following a lesson on the exclamation point, each student will write two rules governing its use.

In working with synthesis-level objectives, the teacher must be careful to give enough structure to make the assignment meaningful and yet leave enough freedom for students to use their own imagination.

LEVEL 6: EVALUATION

This is the highest level of the cognitive domain in Bloom's *Taxonomy of Educational Objectives.* It requires judgments based on known evidence.

For example, while studying a unit on safety, students might be given the following objective:

Examine the teacher's list of safety rules and formulate at least one additional necessary rule.

Here are some other examples of evaluation-level objectives:

Given a city regulation that prohibits open burning of leaves, the group will discuss this law and decide whether it is good or bad.

Following a unit on the contributions of various cultural groups to our society, the students will examine their books and rate them according to the degree of representation each gives to various cultures.

Using Moh's hardness scale and a set of minerals representing each level, the students will assess the hardness of several minerals within a level of accuracy of 1.0.

IN CONCLUSION: THE COGNITIVE DOMAIN

Perhaps by now you have concluded that writing good objectives is not always an easy task. Without planned objectives, however, students usually remain at the lowest cognitive levels. In your professional education courses and field experiences, you will want to seize every opportunity to become involved in writing objectives, since this is the best if not the only way to develop expertise in this important teaching role.

Clear objectives can often increase students' enthusiasm.

Since the upper levels of the cognitive domain are important but nevertheless are most frequently ignored, we will end this section with some additional examples of higher-level objectives. These indicate various types of activities which promote thinking at higher levels:

Draw a chart to show differences in rates of heartbeat when humans lie down, sit up, walk, jog, and climb stairs.

Record a group discussion on pollution and, using a 10-point scale, evaluate the discussion in terms of content, involvement of all members, and range of ideas.

Given the facts learned in your grade level about each planet, tell why you believe each could or could not support human life.

Examine your own daily habits, and describe three ways that you can help save energy in your home.

Examine a series circuit and determine whether the light bulb farthest from the battery is dimmest. Explain.

Given the definitions of "igneous," "sedimentary," and "metamorphic," group assorted rocks according to their types.

The affective domain

Detailed attention has been given to the cognitive domain because it is important to learning and because the writing of objectives has reached a higher level of sophistication in the cognitive domain than in the affective domain. However, recent studies show that emotions play an important part in learning; indeed, developmental psychologists recognize that psychological and emotional development are inseparable and that meaningful learning involves the emotions. Abraham Maslow has expressed the importance of the emotions in learning:

> As I go back over my own life, I find that my greatest educational experiences, the ones I value most in retrospect, were highly personal, highly subjective, very poignant combinations of the emotional and the cognitive. Some great insight was accompanied by all sorts of autonomic nervous system fireworks that felt very good at the time and which left as a residue the insight that has remained with me forever.*

Let's now briefly examine the various levels of the affective domain—the emotional domain—and consider how to write affective objectives for elementary science.

As we noted in Chapter 1, science is no longer considered only a body of knowledge. Rather, it involves attitudes and feelings. In 1964 a taxonomy of educational objectives was written for the affective domain to ensure that schools would give attention to students' affective development. The levels of the affective domain are:

Level 1: Receiving
Level 2: Responding
Level 3: Valuing
Level 4: Organization
Level 5: Characterization†

LEVEL 1: RECEIVING

The lowest level of affective behavior is simple awareness. At this level, for example, a student who is only slightly emotionally involved in a lesson will at least receive some of the information being presented. Of course, some students receive much more information than others; and selective students receive a higher quality of information than students who are less selective.

Here is an example of an objective at the receiving level:

After watching a film on pollution, the student will in a follow-up discussion ask at least one question or make at least one comment about the content of the film.

Although basic awareness may not go much beyond simply being awake and paying some attention, it is at least one step up from sleeping or daydreaming.

*Abraham Maslow, "What Is a Taoistic Teacher?" in *Facts and Feelings in the Classroom*, Louis Rubin, ed., Viking, New York, 1973, p. 159. Copyright 1973 by Viking Press; used by permission of Walker & Co.

†David Krathwohl et al., *Taxonomy of Educational Goals.* Handbook II, *Affective Domain*, Longman, New York, 1964.

LEVEL 2: RESPONDING

At the second level of the affective domain, responding, students become overtly involved. For example, a unit on plants might have an objective like this one:

When given an opportunity to select their own reading material, during the grading period each student will choose and check out at least one book on plant life.

LEVEL 3: VALUING

One of the early leaders in the area of the understanding of values was Louis Raths. According to Raths, a value must always involve someone's prizing or cherishing something.* Values are reflected in behavior. In other words, we are willing to behave according to our values and will do so repeatedly.

An example of a valuing objective would be:

At some time or times during this unit on rocks and minerals, each student will elect to do a term project on rocks and minerals such as designing a bulletin board, collecting rocks, or polishing rocks.

Notice the freedom students have in this objective. It is important—in fact, essential—that students be given alternatives.

LEVEL 4: ORGANIZATION

At the organization level of the affective domain, the student is asked to bring together different values and resolve any conflicts between them. This may result in changing or reinforcing values—in compromising between opposed values.

Here is an example of an organization-level objective:

Given an article supporting nuclear energy and an article opposing it, the student will read both articles, list at least three valid points presented in each article, and then write a concluding statement supporting or opposing nuclear energy.

LEVEL 5: CHARACTERIZATION BY A VALUE OR VALUE COMPLEX

Objectives at levels 1 through 4 do not require the student to have set values. Level 5 is different; it requires learners to act consistently with their value system, and to alter values if necessary. In fact, the learners must behave so consistently that their responses to situations can be predicted. They must also demonstrate a certain degree of individuality and self-reliance.

Example:

The student will bring to class a record of at least 10 steps taken over the past 6 weeks to conserve natural resources.

*Lewis Raths et al., *Values and Teaching*, Merrill, Columbus, Ohio, 1966.

IN CONCLUSION: THE AFFECTIVE DOMAIN
This brief look at writing objectives in the affective domain will serve to show that it is far more difficult to state exact minimum levels of performance here than in the cognitive domain. Often evidence of changes in attitude comes only over time, as students change their behavior. Some writers insist that consistent behavior over a period of time is the only way an affective objective can be adequately demonstrated.

The psychomotor domain

Elementary school teachers are always involved with the physiological development of their students. Without an understanding of the processes of physical development, teachers may form unrealistic expectations for students.

In Chapter 1 we defined science as a combination of knowledge, attitudes, and processes, or skills. Skills represent a culmination of development in science, for unless children *use* or apply their newly acquired knowledge and attitudes, the developmental process is incomplete. For example, suppose that in a unit on oceans, the students learn to appreciate the varieties of life in the ocean and to classify certain groups of sea animals. If the process stops here—if the students never actually collect and classify species—then we believe that the students have not benefited fully.

Let's now look at sample objectives for the psychomotor domain. This domain has seven levels:

Level 1: Perception
Level 2: Set
Level 3: Guided response
Level 4: Mechanism
Level 5: Complex overt response
Level 6: Adaptation
Level 7: Origination*

LEVEL 1: PERCEPTION
Purposeful motor activity begins in the brain, when our receptors pick up cues from the environment. (For example, children who are converting from centimeters to millimeters discover that rather than multiplying by 100, they should be multiplying by 10.)

Here is an example of an objective at the perception level:

Following a demonstration in which the teacher works several practice conversions on the board, students who are multiplying when they should be dividing and vice versa will notice that one must always multiply when going from larger to smaller units and divide when going from smaller to larger units.

*E. J. Simpson, *The Classification of Educational Objectives in the Psychomotor Domain*, vol. 3, Gryphon House, Washington, D.C., 1972.

LEVEL 2: SET

In the psychomotor domain, "set" does not refer to developmental readiness; rather, it means physical and emotional readiness to act. For example, a child who is learning to ride a bicycle may be seen pausing momentarily while halfway mounted before starting the forward movement. The child is thinking through the sequence of behavior required to keep the bicycle upright, and getting the courage or confidence necessary to perform the task.

Following is an example of an objective at the set level:

In a section on sound, when given the signal "ready," each student will hold his or her self-made reed instrument with the fingers in the proper position.

It would be possible to write this objective so as to specify a minimum level of performance. But would that be desirable? The answer depends on the level of the task. In our example, the task is so simple that a minimum level of performance seems unnecessary and therefore undesirable. On the other hand, suppose that the task is riding a bicycle. For this task it would be helpful, indeed essential, to include minimum levels of performance such as correct posture, correct hold on the handlebars, and correct position of feet. For example:

Upon the command "Get set," the student will prepare to mount the bicycle, holding each hand on the handlebar grip, holding the bicycle in a 60- to 90-degree vertical position, and positioning the foot and pedal at the twelve o'clock to two o'clock position.

Study of the human body offers opportunities to perform psychomotor activities at the guided-response level.

LEVEL 3: GUIDED RESPONSE
Behaviors which require a series of interrelated psychomotor movements require careful supervision or guidance. An example of an objective at this level is:

In a unit on the human body, the students will monitor the heartbeat of a classmate, testing the stethoscope, warming the cup, pressing the cup against the sternum, and counting the number of heartbeats per minute. Each student will perform each task at the teacher's command.

As students perfect their skills at this level, the teacher may wish to add minimum levels of performance. For example:

As the teacher calls for each task, the students will perform it in 5 seconds or less.

LEVEL 4: MECHANISM
At this level it is assumed that students are experienced enough to perform certain tasks without pausing to think through each step. Here is an example of a mechanism-level objective for a unit on the history of the earth:

When making a fossil mold, each student will perform the correct sequence of steps until he or she can follow through from each step to the next, without pausing to think about which step comes next.

LEVEL 5: COMPLEX OVERT RESPONSE
This level is a continuation of level 4, but it involves more complicated tasks. Here is an example of an objective at level 5:

Having heard a story about space, the students, in groups of five, will write their own ending and then act it out for the rest of the class.

LEVEL 6: ADAPTATION
At this level students are required to adjust their performance as the situation demands. Here is an example of an adaptation-level objective for a unit on sound:

As the teacher changes tempo on the piano, the students will, without being told, change their own timing accordingly.

LEVEL 7: ORIGINATION
Origination is the highest level of the psychomotor domain. Here the student creates new movement patterns to fit particular situations. An example of an origination-level objective for a unit on simple machines is:

Each student will be able to manipulate a particular machine so as to increase the amount of force produced.

As in the other levels of the psychomotor domain, the decision whether to specify minimum levels of performance should depend on the level of sophistication the students have reached with the skills involved. Generally, we feel that unless a specified minimum level is needed, it is probably best not to include one.

SUMMARY

This chapter began by discussing the use of educational aims in elementary science. Although aims are aspirations that cannot ever be fully achieved, they are essential because they tell us what our ultimate effect on our students should be. Science teachers formulate aims by asking themselves questions such as: Why should these children study science? How should the study of science help them mature and live harmoniously with the world about them? By formulating aims, teachers can begin to identify their own effects and set priorities. For example, the "spirit of inquiry" (including useful attitudes and skills) might be an aim for all elementary science classes.

Since aims are never fully attainable, educators must also set definite goals which can be attained. The length of time necessary to achieve a goal may range from a week to a year.

In recent years, administrators have realized that too often aims and goals are written only to be shelved in the principal's office. When this happens, they are no more than glittering generalities. Therefore, goals must be transformed into something that can be applied. A mechanism for doing this is the performance objective.

Taxonomies of educational objectives have been developed to indicate hierarchies of thought processes stimulated by objectives. Currently, taxonomies have been written for three domains: cognitive, affective, and psychomotor.

To be effective, a performance objective must be written in terms of the student's behavior, not the teacher's. The conditions under which the student is expected to perform the task must be specified; this clarifies communication between teacher and students. Cognitive objectives must specify a minimal level of performance, to communicate to the student what level of mastery the teacher requires. Affective and psychomotor objectives may also contain a statement of minimum acceptable level of performance; however, this is not always possible or desirable. Good judgment and practical considerations should determine whether an affective or psychomotor objective specifies level of performance.

Throughout the following chapters, we encourage you to begin identifying some goals and objectives that will contribute to your general aims for elementary science.

RECAPITULATION OF MAJOR PRINCIPLES

1. Educational aims reflect the broad philosophy of a school.
2. Federal, state, and local governments use goals to hold schools accountable.
3. Performance objectives provide for the attainment of goals.
4. To be effective, performance objectives should be carefully stated in terms of students' behavior, should describe the conditions under which students are expected to perform, and should specify minimal level of performance if that is appropriate.
5. Educational taxonomies are hierarchies of objectives which help teachers elevate the levels of thinking and behaving in their classrooms.
6. Since science involves knowledge, attitudes, and processes (skills), teachers of science should set objectives in all three of these areas.

SUGGESTED READINGS

Gronlund, Norman E.: *Measurement and Evaluation in Teaching,* 3d ed., Macmillan, New York, 1970.

———: *Preparing Criterion-Referenced Tests for Classroom Instruction,* Macmillan, New York, 1973.

Hopkins, Charles D., and Richard L. Antes: *Classroom Testing: Construction,* Peacock, Itasca, Ill., 1979.

Kibler, Robert J., et al.: *Objectives for Instruction and Evaluation,* 2d ed., Allyn and Bacon, Boston, 1981.

Lindvall, C. M., ed., *Defining Educational Objectives,* University of Pittsburgh, Pittsburgh, 1964.

Mager, Robert F.: *Preparing Instructional Objectives,* Fearon, Palo Alto, Calif., 1962.

Oliva, Peter F.: *Developing the Curriculum,* Little, Brown, Boston, 1982.

Popham, W. James, and E. L. Baker: *Establishing Instructional Goals,* Prentice-Hall, Englewood Cliffs, N.J., 1970.

Zais, Robert S.: *Curriculum Principles and Foundations,* Harper and Row, New York, 1976.

CHAPTER 5

EVALUATION IN ELEMENTARY SCHOOL SCIENCE

PREASSESSMENT

	AGREE	DISAGREE	UNCERTAIN
1. Verbal tests are inappropriate in elementary science classes.	____	____	____
2. Evaluation can and should be used to promote learning.	____	____	____
3. Evaluation of students' performance should not be determined by the teacher's value judgments.	____	____	____
4. Norm-referenced tests promote competition within the class.	____	____	____
5. Objective tests are inferior to discussion tests in letting students utilize their creative abilities.	____	____	____
6. Elementary school students lack the maturity to be meaningfully involved in evaluating their own performance.	____	____	____
7. The purpose of all types of evaluation is to help determine grades.	____	____	____
8. Teachers should use many types of evaluation.	____	____	____

OBJECTIVES

Upon completion of this chapter you should be able to:

1. Differentiate between formative and summative evaluation and give examples of each.
2. Explain why criterion-referenced evaluation is more appropriate in elementary science classes than norm-referenced evaluation.
3. List at least six criteria to be used in evaluating students' progress.
4. Explain the advantage in using a variety of criteria for evaluating students' progress.
5. Give a rule of thumb for determining the emphasis of a test score or homework assignment in evaluating students' performance.

6. Design a self-observation instrument for evaluating instruction in a science class.
7. Devise a chart to record general behavior in class.
8. Devise a chart to record the daily behavior of individual students in class.

WHAT IS EVALUATION?

Evaluation involves passing judgment on or assigning a value to something. Evaluation of students' work is not to be confused with grading, which should not be influenced by the teacher's values. Tests and other forms of measurement should also be objective and free of value judgments. In evaluation, on the other hand, value judgments are essential. Evaluation of students' work begins where testing ends. A test is a measuring device, a method of gathering information. After the testing is finished, the teacher uses the information to evaluate students' work. Figure 5-1 shows the relationship between testing and evaluation.

Evaluation has many uses in elementary science classes. It is needed to determine the strengths and weaknesses of elementary science programs, activities, content, and students' performance. Evaluation of materials used in elementary science classes is essential because new and better materials are continually being developed. Evaluation is even necessary in the teacher's long- and short-term planning. Textbooks in use should be periodically compared with newly published texts. Lesson content and activities should be evaluated against other options as they become available. And overall progress of students must be evaluated.

TYPES OF EVALUATION

Formative and summative evaluation

There are two basic types of evaluation: *formative* and *summative* (see Table 5-1). Formative evaluation never involves grading but is used to promote learning. For example, a teacher may give checkup tests to prompt students to study their assignments. A science vocabulary test may be used to give students practice and to encourage them to study. The results of formative tests should never be used to determine grades.

Figure 5-1. Relationship between testing and evaluation. Some teachers tend to think of these terms as synonymous, but they are not.

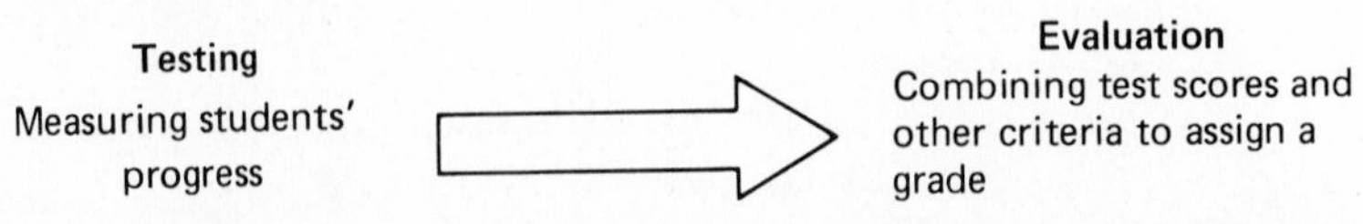

TABLE 5-1 RELATIONSHIP BETWEEN FORMATIVE AND SUMMATIVE EVALUATION

	FORMATIVE EVALUATION	SUMMATIVE EVALUATION
PURPOSE	To promote learning	To derive a grade To use in promotion and retention To use in certification
NATURE	Few questions	Many questions
TYPE OF QUESTIONS	Specific; designed to measure small bits of specific knowledge	General; designed to measure general skills
ADMINISTERED	Frequently—several times within a unit	Only once or twice during a unit—usually at the end of the unit only

In contrast, summative evaluation is often used to determine grades. Summative tests are more comprehensive and are usually given at the end of a learning unit and then again at the end of a grading period. For example, a class may study three 2-week units during a 6-week grading period. A summative test may be given at the end of each 2-week period, and a comprehensive test may be given at the end of the sixth week. Since the score on each of these tests would be used in computing the student's grade, each would be considered a summative test.

Norm-referenced and criterion-referenced evaluation

Most tests used in schools today are summative tests. Exclusive use of summative tests is not necessarily beneficial, because students need ways of evaluating their progress which are free from fear. Consistent emphasis on grades can lead students to develop a negative attitude toward all forms of evaluation.

Use of competitive grading systems is traditional in the United States. Teachers are accustomed to classroom competition. School administrators perceive grades as a way of making students and teachers accountable; test scores provide a record of achievement for the school. Furthermore, parents insist on numerical or letter grades as an indication of how their children are performing in comparison with their classmates: an A shows that without question a student is performing at the head of the class; a D or an F carries the opposite message.

Tests which compare the performances of classmates are said to be "norm-referenced." That is, the degree of success of each child depends on competition with others. Is a child performing above or below, or at the same level as, the "normal" child? Or is the child above average, below average, or average? Those who support the use of norm-referenced tests argue that the main purpose of education is to prepare children to

Through close supervision of individual students, teachers gather important feedback for evaluation—whether norm-referenced or criterion-referenced.

live in a competitive world. They insist that competition with classmates is needed to motivate students—to challenge them to achieve their full potential. (See Table 5-2.)

In recent years educators have developed serious doubts about the value of norm-referenced tests. Such tests may motivate the most capable students, but they may also discourage children who are unable to compete successfully, leaving them feeling unworthy and inept and eventually causing them to perceive themselves as misfits. Even the capable students may suffer, in that they may form superior or snobbish attitudes.

TABLE 5-2 NORM-REFERENCED TESTS VERSUS CRITERION-REFERENCED TESTS

NORM-REFERENCED	CRITERION-REFERENCED
The knowledge being tested may or may not be recognized as important by students.	The knowledge being tested is identified before testing as an important objective.
Success depends on performance of each student relative to performance of classmates or larger group.	Levels of performance needed for success are identified before the testing.
Promotes competition among peers.	Promotes self-competition.

In particular, norm-referenced tests seem to be highly inappropriate for young children. In Chapter 2 we saw that children differ in their innate abilities and in their development; and yet children are similar in that each child begins school with a curious attitude toward the world. In the elementary science class, concern for challenging each student and for kindling the flame of curiosity in all students should take precedence over the desire to have children compete with classmates of widely varying innate ability.

Clearly, there is a need in elementary classes for a different type of evaluation, one which is not based only on competition between and among students. Fortunately, teachers do have such a method at their disposal; it is called "criterion-referenced" evaluation. Criterion-referenced evaluation measures students' success not in terms of the performance of their classmates but in terms of their attainment of clearly specified objectives. Such evaluation is ideal for elementary science classes, where students can be rated on many criteria. At times elementary science students are encouraged to explore in their own ways. At other times they are asked to work in groups of two or more to perform joint tasks or to solve problems together. In such circumstances, cooperation, not competition, is emphasized. The teacher's role is to provide encouragement and reinforcement.

You may now be thinking, "How do I know how much credit should be given to each activity? How much should each count?" The final judgment will be made by you, the teacher; however, the following discussion of some of the common evaluation criteria for elementary science classes should be helpful.

CRITERIA FOR EVALUATION AND MEASUREMENT OF LEARNING

Several types of tests and projects are worthy of being included in the evaluation of a student's progress. The weight assigned to each aspect of a student's performance should be based on the time spent on the project or on the material being tested.

Written tests

Written tests may not be the best way to measure students' learning, but they are the most frequently used of all tests. Let's briefly examine the varieties of written tests, both subjective and objective, commonly used in elementary science classes.

Objective written tests may include true-false, multiple-choice, and matching items. In the elementary grades, multiple-choice and matching questions often use pictures or drawings or a combination of pictures and words. Objective tests have a major weakness in that they do not permit students to express their own knowledge and opinions. Objective tests

Written tests yield products which are easily evaluated.

tend to test for what the student doesn't know, whereas subjective tests test for what the student has learned. Another drawback of objective tests is their failure to encourage creativity or to measure a student's creative abilities.

Objective tests can be used either to promote learning (formative tests) or to derive a grade (summative tests). More often than not, they are used to determine a grade, which is legitimate if other criteria are also used in determining the grade.

Subjective tests can be used to stimulate students to think through critical issues. Such questions usually begin with a word such as "Explain," "Describe," or "Discuss." Other types of questions can be used to promote creative thinking. For example, such questions as "What would happen if . . . ?" and "What would you do if . . . ?" prompt students to use their understanding to answer a question or solve a problem.

We now give some brief guidelines for writing true-false, multiple-choice, matching, and discussion items.

MULTIPLE-CHOICE ITEMS

In a multiple-choice item, you should usually give five alternative answers. Make each alternative plausible. Each choice should help discriminate between students who understand the concept and those who do not.

TRUE-FALSE ITEMS

In true-false items, each item—whether true or false—should be plausible. After writing such questions, read them and make sure that those to

Do This:
As you read our guidelines for writing multiple-choice, true-false, matching, and discussion items, take a few minutes to design two items of each type. Be sure that each item you write meets the criteria we've suggested.

be answered "false" are completely false and those to be answered "true" are completely true. Statements which are partially true and partially false are to be avoided.

MATCHING ITEMS
In matching items, to prevent guessing, include fewer items to be matched than choices to match them with. This also makes it impossible to find correct answers strictly through the process of elimination. Check your questions carefully to make sure that only one acceptable choice is included for each item to be matched.

DISCUSSION ITEMS
Before writing a discussion item, identify specific facts or concepts that you expect to find in the answer. Next, assign credit points for each of these concepts. For example, a discussion question worth a total of 6 points may have two extremely important ideas and two less important, yet still significant, ideas; 2 points can then be given to each of the more important ideas and 1 point each to the less important ideas.

Occasionally, when answering discussion items, students will include important concepts or facts other than those you had in mind. When this occurs, credit should be given to these in proportion to their importance as compared with the ideas you have originally identified.

Oral tests

Oral tests are less common than printed tests in elementary science classes, but they do offer some advantages. If conducted properly, oral tests can reduce students' anxiety. Of course the converse is also true. Singling students out and questioning them in front of their classmates may create almost unbearable tension. One effective way to reduce tension is to conduct the oral test as a game and deemphasize the credit that may be awarded. Another tension-reducing approach is to walk around the classroom and quietly ask questions of each student individually, while the other class members are involved in their work.

Performance tests

Many objectives for science classes involve psychomotor skills, i.e., the ability to perform physical tasks. Therefore the evaluation of students should contain performance tests. For example, when studying a unit on plant life, students may be asked to classify deciduous leaves or the

These students are engaged in a performance test.

leaves of perennial plants by placing them in appropriate piles. Skills in working with weights or linear measurements are also worthy of evaluation.

In designing performance tests, work tables or work stations can be useful. Each table or station has one task to be performed, and students can rotate from table to table or station to station. For a 100-point test five such stations or tables could be used, each with a 20-point test item. An appropriate number of points can be given for work only partially completed or partially correct.

Unit projects

Some unit projects may seem too advanced for elementary school children, but many assignments can be meaningful to them. Seeds can be planted in milk cartons, for example, and their growth rates followed. Even kindergarten students can construct two-dimensional projects; and by second or third grade, students are able to construct their own three-dimensional projects out of boxes and glue.

Projects for science fairs or open houses, like term projects, offer each student an opportunity to explore a particular topic of special interest—but projects for science fairs also offer more. They allow each student to say to his or her classmates and parents, and even to the public, "Look here. This is what I am interested in. This is what I am doing. This is part of me." Teachers should create opportunities for each child to have such experiences. In science, every child should be a winner. The teacher

These two boys developed their real model rocket as a project in a class for gifted students. In science, such projects should always be a part of evaluation.

who encourages participation in projects for public display can help each child to feel like a winner. No serious project should ever be refused display space because of its untidy appearance or lack of general appeal.

Unit projects should be evaluated and graded leniently. Consideration should be given to the many limitations suffered by some students; these may include lack of materials and inadequate space at home, disruptions from siblings, and lack of encouragement by their parents. Individual interest and effort, as well as group cooperation, should therefore be rewarded.

Other criteria

In elementary science classes goals and objectives other than tests and projects are always present. For example, all science teachers should attempt to foster an attitude of inquiry, a willingness to pursue activities alone or in groups, a cooperative spirit, and the acquisition of the skills needed to search out information. These goals and objectives are worthy of evaluation. Teachers who talk about the importance of the process approach in science but who reward only correct answers are not likely to foster a belief in and mastery of process. Science teachers who talk about the importance of attitudes and yet do not reward behavior which reflects those attitudes are likely to have little positive effect on students' attitudes.

A spirit of inquiry and willingness to pursue an activity alone are important criteria for evaluation.

Teachers may avoid including attitudes in their evaluations of students' performance because they don't know how much weight to give them or exactly how to measure them. Attitudes are different from achievements that can be measured on a written test, in that they can never be measured exactly. Remember that evaluation is more than measurement; it always includes judgment. Beginning teachers may wish to count such "ambiguous" criteria very lightly until they develop a feeling for how to assess these abstractions.

OBSERVATION OF STUDENTS

Science classes should involve students in continuous purposive activities such as observing, questioning, analyzing, exploring, touching, tasting, hearing, describing, and grouping. These activities give an excellent opportunity to acquire much information about their students by observing them.

An observation chart such as the one shown in Table 5-3 will help you to remember to look for the most important factors as you observe students. Filling out this classroom chart will give a picture of the

TABLE 5-3 CLASSROOM OBSERVATION CHART

OBSERVATIONS*	Monday	Tuesday	Wednesday	Thursday	Friday
1. Children working in groups					
2. Number of students working independently					
3. Number of students seemingly bored					
4.					
5.					
6.					
7.					
8.					
9.					
10.					

ANECDOTAL RECORD:

Week of ____________________

Period ____________________

*You may wish to list your own observations in the additional spaces.

Source: Joe Abruscato and Jack Hassard, *Loving and Beyond. Science Teaching for the Humanistic Classroom,* Scott, Foresman, Glenview, Ill., 1976. Used by permission.

activities of the class as a whole. A similar chart can be made for each individual student. Such charts can be extremely useful in planning instructional activities as well as in assessing behavior for grading purposes.

TABLE 5-4 INDIVIDUAL OBSERVATIONS CHART: WEEK OF ——

STUDENT	Monday	Tuesday	Wednesday	Thursday	Friday
SHARON					
ALFRED					
GENE					
DAWN					
BOBBY					
COLLEEN					

Table 5-4 shows an example of an individual observation chart. Recording observations of individual students on a daily basis makes it possible to compare current behavior with behavior on previous days. Such a comparison can give clues to problems at home, medical problems, and other persistent abnormal conditions in students' lives. Here are examples of the kinds of observations you may wish to record:

Seems bored
Looks interested
Cooperates with others
Fusses a lot
Listens well
Talks a lot
Follows instructions
Dominates (bosses) others
Appears withdrawn
Seems sleepy or tired

PERSONAL CONFERENCES

The personal conference is probably the least commonly used method of obtaining information about students and their progress, but it can provide an excellent opportunity to get to know individual students. The conference should, at some point, focus on science content and on the studies currently being pursued by the class; but the conference should not be limited to science. Instead, the teacher should use conferences as opportunities to find out how their students feel and to learn what may be bothering them.

This college student is interviewing a counselor to learn skills she will later need to use in conferences with her own students.

From the student's viewpoint, the personal conference can have a positive psychological effect. Unfortunately, many students go through an entire school year without being asked by their teachers how they, as individuals, feel about school, science, or themselves. A student who finds out that someone as important as the teacher cares enough to talk with him or her in private may develop a better attitude toward classes, studies, and classmates. Students who have a positive outlook toward the world and a positive, self-confident feeling about their own abilities are able to make maximum progress in school.

Do This:

We have introduced you to several criteria that are important in evaluating elementary science students, but we realize that the degree of emphasis we place on each criterion is somewhat personal and subjective. So that you may achieve a more balanced view, we encourage you to interview at least two or three elementary science teachers. Find out how much emphasis each places on the different criteria we have discussed. Also, find out about additional important criteria used by elementary science teachers to evaluate their students.

EVALUATION OF INSTRUCTION

In education we may tend to think of evaluation primarily in terms of students' progress, but the instructional process and the curriculum can also be evaluated. Periodically, each teacher should assess the curriculum in the science classes, as well as his or her own effectiveness in helping students to attain the desired objectives. In this section, we will show you how to analyze your own instructional approaches. Since you may have an opportunity in a general methods course to examine some of the widely distributed self-observation systems (such as interaction analysis), we will concentrate on skills needed to develop your own instruments. Science teachers use many dialectic approaches in their classes, such as discovery learning, inquiry learning, laboratory method, and simulation games. Since the teacher's role in such methods is often very different from that in more didactic methods, it is important to learn to develop your own observation strategies so that you can look at any aspect of your own or your students' behavior.

A systematic approach to self-observation should include the following:

1. Statement of the problem
2. Observation instrument
3. Presentation of findings (objective)
4. Discussion of findings
5. Visual presentation of findings
6. Discussion of implications

The observer should begin by identifying an area to be analyzed. The area you choose may be a problem, or it may be simply a question about students' behavior or even about your own behavior which you need to answer in order to evaluate your classes.

For example, if you are wondering whether the students waste time, you may state the problem as a question: "How much time does each student spend on the lesson topic?" The instrument you use to observe this might be a drawing of the classroom seating arrangement, providing a space for analyzing each student and using symbols to simplify and expedite the recording of various behaviors. (See Figure 5-2.)

In Figure 5-2 each square represents a student. Crossed-out squares indicate students who were absent. The three squares with circles drawn around them indicate three students who worked on the assigned lesson throughout the period of observation. They are to be eliminated from the analysis, since their inclusion would not contribute to an understanding of the fluctuation of variables—an understanding that is essential to the analysis of classroom behavior. Had there been other students whose behavior remained unchanged throughout the hour (e.g., a student who slept throughout the hour and a student who had been given permission to work on a different assignment), then those students would also have been eliminated from the analysis, for the same reason.

Teacher

1 A 5 A
2 A 6 A
3 A 7 A
4 A

1 A 5 F
2 A 6 F
3 A
4 A 7 A

1 E 5 A
2 E 6 A
3 A 7 C
4 A

1 A 5 F
2 A 6 A
3 A 7 A
4 A

1 A 5 A
2 A 6 A
3 A
4 A 7 A

1 A 5 F
2 A 6 F
3 A 7 A
4 F

1 A 5 A
2 F 6 D
3 A
4 F 7 A

1 A 5 F
2 A 6 A
3 A 7 A
4 A

1 D 5 A
2 C 6 A
3 A 7 A
4 F

1 D 5 A
2 D 6 A
3 D
4 A 7 A

1 C 5 F
2 A 6 F
3 A 7 C
4 F

1 A 5 F
2 A 6 A
3 A 7 F
4 C

1 A 5 F
2 A 6 A
3 A 7 A
4 A

1 5
2 6
3 7
4

1 D 5 D
2 D 6 D
3 D
4 D 7 D

1 A 5 C
2 A 6 F
3 C 7 A
4 A

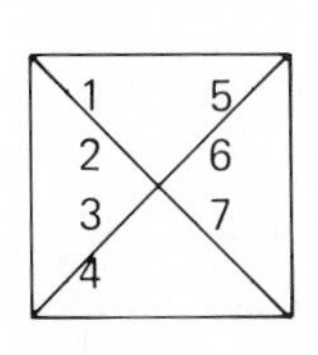

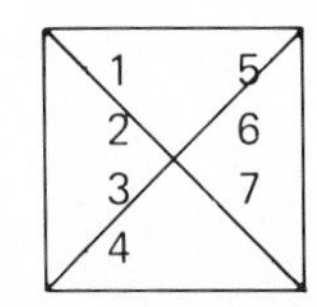

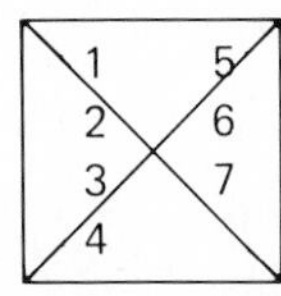

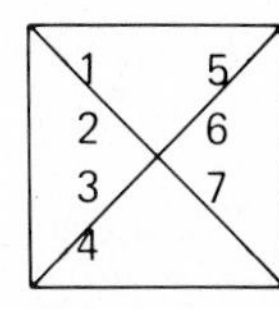

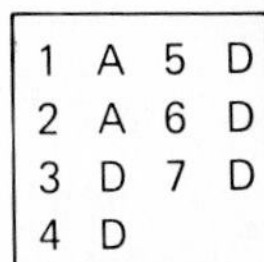

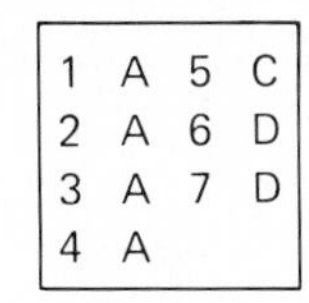

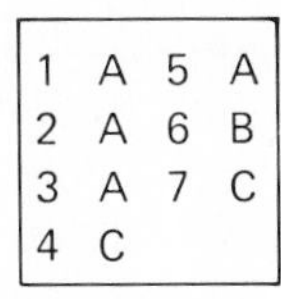

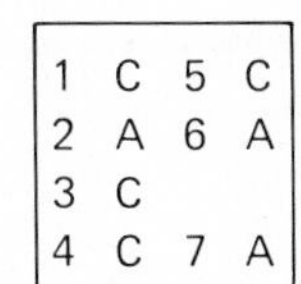

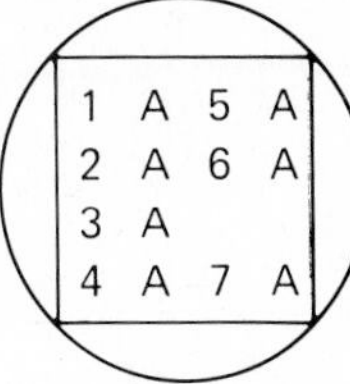

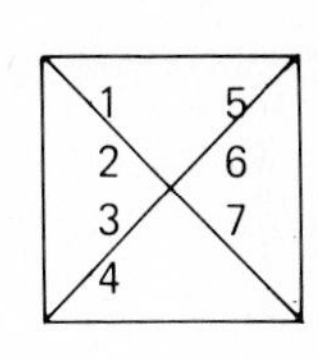

1 C 5 A
2 C 6 F
3 A
4 A 7 C

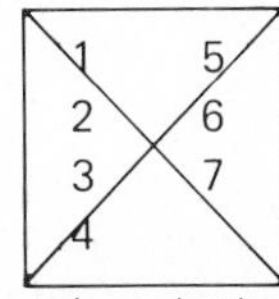

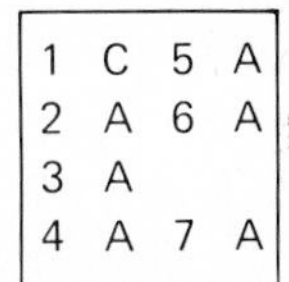

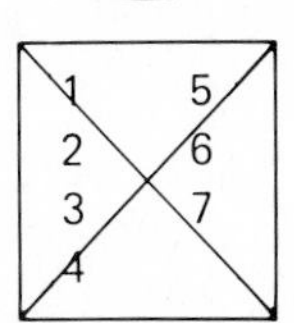

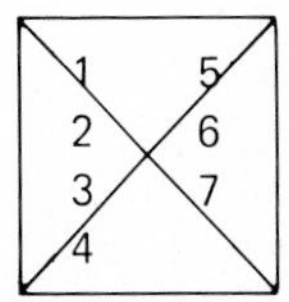

Purpose: To determine whether individual learners require as much time as provided to complete in-class assignments and to determine how individuals spend their class time.

Code:

☒ Student absent
A. Working on task
B. Working on another subject
C. Daydreaming
D. Out of seat
E. Sleeping
F. Talking to neighbors

1. 9:35
2. 9:40
3. 9:45
4. 9:50
5. 9:55
6. 10:00
7. 10:05

Figure 5-2. Time-on-task recording chart.

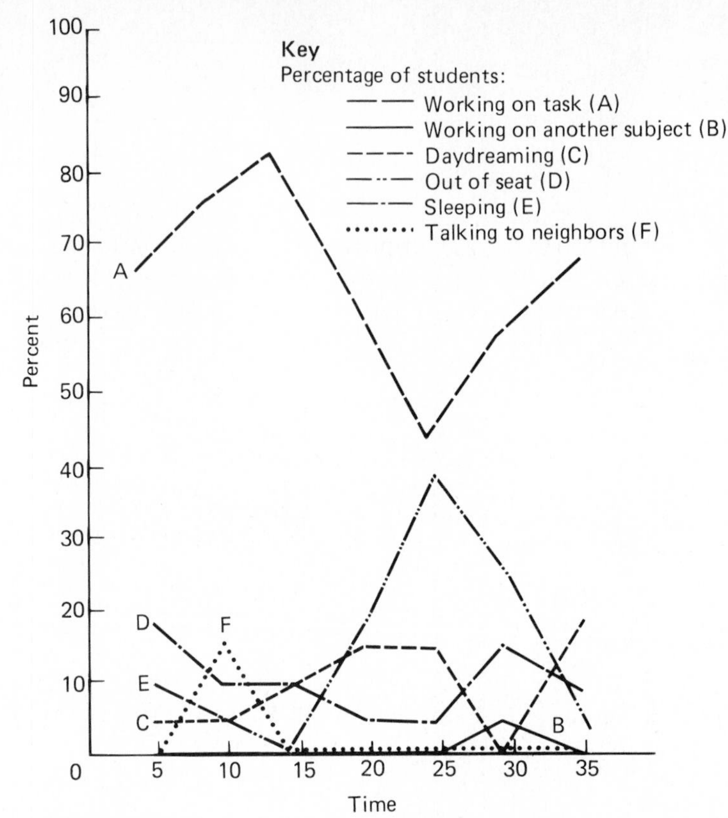

Figure 5-3. Time-on-task graph.

Since a sample of behavior of each student is taken every 5 minutes during the period, this information can be transferred to a graph which will show the average percentage of time students spend on a given task at 5-minute intervals throughout the hour. (See Figure 5-3.)

Transferring the collected data (in Figure 5-2) to a pictorial graph (as in Figure 5-3) makes several phenomena clear. First, line A shows that the average percentage of students working on the assignment peaks at 15 minutes into the period and then rapidly diminishes—indicating that the teacher should vary the task, introduce a new task, change to a different teaching method, or make another alteration to retain the students' attention.

Second, line E shows that more students are asleep at almost 25 minutes into the period than at any other time. A comparison of line E with line A quickly shows that the time when most students are asleep coincides with the time when fewest students are working on the assignment. This information confirms the earlier suspicion that something needs to be done to regain students' attention. Can a similar conclusion be drawn from comparing lines C and D? What does line B tell us? Why doesn't it fluctuate like the other lines? To answer these questions, carefully examine Figure 5-2.

The findings should be presented objectively, i.e., without judgmental comments such as "Too much (or too little) time is being spent on the lesson." Such opinions, if given prematurely, can hamper the analysis; however, once the objective data have been gathered, the teacher should begin to make evaluative statments and draw implications. For example, the teacher may infer from the graph in Figure 5-3 that a new class task is needed by the end of the first 15 minutes to sustain a high level of involvement.

The example we have given is but one of many types of studies which could lead to the improvement of instruction in the classroom. Some other potential topics for analysis include: the teacher's position, the students' positions, time spent on various factors (introducing the lesson, summarizing the lesson, asking or answering questions, performing psychomotor tasks as compared with mental tasks, studying as a group versus studying independently, rate of reinforcements during the period, and disciplinary actions). The number and variety of possible studies is endless.

Look at your goals and objectives for the course, then decide which major categories of your own and your students' behavior you need to analyze to help you attain these goals. The observation instruments you devise should be sophisticated enough to gather comprehensive data, yet simple enough so that you can implement and interpret them easily.

INVOLVING STUDENTS IN EVALUATION

Recently there is much discussion of involving students in their own evaluation, but, actually, as early as 1919 school systems were beginning to involve students in both planning and evaluation. Even so, talking about self-evaluation by students is much more acceptable to teachers than actually putting it into effect in the classroom. Teachers have a major problem with involving students in the evaluation of their own achievement: reluctance to relinquish even part of this responsibility to students.

How willing would *you* be to share responsibility for evaluation? Table 5-5 (on pages 98–99) provides a test to help you find out.

Do This:

Take the "open learning environment" test in Table 5-5 to measure your willingness to involve your students in their own evaluation.

While taking the test you may wonder, "What is the right answer? What should I do?" An abundant number of responses in the "often" and "always" column is viewed as healthy; but probably none of us could fairly check all statements in these columns. Most of us would like to be more open than we currently are, but it may take a long time to make the necessary changes. We must first learn to become confident teachers and shed our defensive attitudes.

TABLE 5-5 TEST FOR OPEN LEARNING ENVIRONMENT

TAKE THIS TEST

The following statements are about some ways of operating a classroom. Please respond to each statement under two categories: (A) What you are *now* doing, (B) What you would *like* to do. For each statement circle the *two* most appropriate dots.

	AM NOW DOING					WOULD LIKE TO DO				
	ALWAYS	OFTEN	SOMETIMES	SELDOM	NEVER	ALWAYS	OFTEN	SOMETIMES	SELDOM	NEVER
1. Each student can decide whether or not to take part in a particular assignment.	•	•	•	•	•	•	•	•	•	•
2. Each student is allowed to decide how to study each topic.	•	•	•	•	•	•	•	•	•	•
3. Each student determines how much time to spend on a topic.	•	•	•	•	•	•	•	•	•	•
4. Students are free to group as they want.	•	•	•	•	•	•	•	•	•	•
5. Every student has free access to all the materials in the classroom.	•	•	•	•	•	•	•	•	•	•
6. Students determine what is removed from or added to the classroom.	•	•	•	•	•	•	•	•	•	•
7. Every student is completely free to move about the classroom.	•	•	•	•	•	•	•	•	•	•
8. Students freely ask for help whenever they need it.	•	•	•	•	•	•	•	•	•	•
9. Students are graded on a curve.	•	•	•	•	•	•	•	•	•	•
10. Students have the responsibility to evaluate themselves.	•	•	•	•	•	•	•	•	•	•
11. Students initiate their own activities whenever they choose.	•	•	•	•	•	•	•	•	•	•
12. Each student has his or her own personal space in the classroom (a drawer or cupboard).	•	•	•	•	•	•	•	•	•	•
13. Peer-group teaching is a primary activity in the classroom.	•	•	•	•	•	•	•	•	•	•
14. Many diverse activities simultaneously go on in the classroom.	•	•	•	•	•	•	•	•	•	•
15. Students are encouraged to report on topics in any way they want.	•	•	•	•	•	•	•	•	•	•

SUMMARY

Although for most teachers evaluation is not one of the more enjoyable teaching tasks, it is an essential part of every teacher's work. Unlike measurement, which requires only an objective assessment of students' performance, evaluation requires the teacher to use judgment. One of the basic problems of evaluation is deciding what is important enough to be included in the evaluation and how much emphasis to place on each activity. In other words, the teacher must weigh the importance of each test and assignment as compared with all other grading criteria.

Generally, the teacher should include in the evaluation those assignments and test scores which reflect important objectives of the class—not

TABLE 5-5 CONTINUED

	AM NOW DOING					WOULD LIKE TO DO				
	ALWAYS	OFTEN	SOMETIMES	SELDOM	NEVER	ALWAYS	OFTEN	SOMETIMES	SELDOM	NEVER
16. I have contacts with the students about their learning.	•	•	•	•	•	•	•	•	•	•
17. I am confident my students will learn if left to themselves.	•	•	•	•	•	•	•	•	•	•
18. I leave my students alone in the classroom.	•	•	•	•	•	•	•	•	•	•
19. I lower students' grades when they make mistakes.	•	•	•	•	•	•	•	•	•	•
20. I follow a school outline, a manual, or a text in planning course content.	•	•	•	•	•	•	•	•	•	•
21. Students have fixed places to sit.	•	•	•	•	•	•	•	•	•	•
22. My class is child-centered rather than subject-centered.	•	•	•	•	•	•	•	•	•	•
23. Students can leave the classroom whenever they want to.	•	•	•	•	•	•	•	•	•	•
24. Students are free to bring anything they want into the classroom.	•	•	•	•	•	•	•	•	•	•
25. Students are encouraged to consult other teachers as resource people.	•	•	•	•	•	•	•	•	•	•
26. I encourage teachers, parents, and administrators to enter the classroom as resource people or observers.	•	•	•	•	•	•	•	•	•	•
27. The operational rules of the classroom are made by the students.	•	•	•	•	•	•	•	•	•	•
28. Students are free to do nothing.	•	•	•	•	•	•	•	•	•	•
29. Individual, novel solutions are rewarded more than concensus solutions.	•	•	•	•	•	•	•	•	•	•
30. Students are encouraged to develop personal goals.	•	•	•	•	•	•	•	•	•	•
31. I talk individually with students about their personal goals and then follow up their development.	•	•	•	•	•	•	•	•	•	•
32. Students are free to talk to each other at any time.	•	•	•	•	•	•	•	•	•	•
33. Students are encouraged to pursue their own interests.	•	•	•	•	•	•	•	•	•	•
34. I like to go to school in the morning.	•	•	•	•	•	•	•	•	•	•
35. Students consider their total community as a primary resource.	•	•	•	•	•	•	•	•	•	•

Source: Joe Abruscato and Jack Hassard, *Loving and Beyond: Science Teaching for the Humanistic Classroom,* Scott, Foresman, Glenview, Ill., 1976. Used by permission.

only content objectives but also the attitudes and skills espoused in science education. The evaluation of attitudes and general behavior such as social skills is necessarily somewhat subjective. A good approach in evaluation, and indeed an essential one, is to include a variety of criteria such as classroom assignments, homework, group and individual projects, and a mixture of written, oral, and skills tests. Criterion-referenced evaluations are best, since they emphasize and reward individual achievement.

The credit awarded to each activity should be proportional to the emphasis given to the topic in the classroom and to the time invested in the assignment or project. Voluntary projects for science fairs and open-school exhibits should be rewarded liberally, since the attitudes and enthusiasm which engender such projects are a major goal of science education.

Evaluation should also be used by teachers to improve curriculum and instruction. Observation instruments made by the teacher are usually preferred to popular instruments found in the literature, which often require expertly trained operators. The design of these instruments should be kept simple, so that they can be easily applied and interpreted.

RECAPITULATION OF MAJOR PRINCIPLES

1. In addition to being used to derive grades, evaluation can also be used to promote learning and to improve instruction.
2. Various types of evaluation are needed to determine students' performance, including performance tasks, oral questioning, private conferences, participation in daily class activities, open-school and science-fair projects, and attitudes toward science and toward working with others to learn more about the environment.
3. Voluntary participation in science fairs or open-school programs should be evaluated generously.
4. Criterion-referenced tests belong in science classes, since norm-referenced tests often discourage cooperation and produce feelings of defeatism.
5. Personal conferences can be used to build self-confidence and positive attitudes toward science, school, and peers.
6. Elementary students should be involved in the evaluation of their own work.

SUGGESTED READINGS

Bloom Benjamin S., J. Thomas Hastings, and George F. Madaus: *Evaluation to Improve Learning,* Macmillan, New York, 1981.

Brown, Frederick G.: *Measurement and Evaluation,* Peacock, Itasca, Ill., 1971.

Dizney, Henry: *Classroom Evaluation for Teachers,* Brown, Dubuque, Iowa, 1971.

Hopkins, Charles D., and Richard L. Antes: *Classroom Testing: Administration, Scoring and Score Integration,* Peacock, Itasca, Ill., 1979.

Mager, Robert F.: *Measuring Instructional Intent,* Fearon, San Francisco, 1973.

Marshall, Jon C., and Loyde W. Hales: *Classroom Test Construction,* Addison-Wesley, Reading, Mass., 1971.

Nelson, Clarence H.: *Measurement and Evaluation in the Classroom,* Macmillan, New York, 1970.

Smith, Fred M., and Sam Adams: *Educational Measurement for the Classroom Teachers,* 2d ed., Harper and Row, New York, 1972.

Ten Brink, Terry D.: *Evaluation,* McGraw-Hill, New York, 1974.

Terwilliger, James S.: *Assigning Grades to Students,* Scott, Foresman, Glenview, Ill., 1971.

Tuckman, Bruce Wayne: *Evaluating Instructional Programs,* Allyn and Bacon, Boston, 1981.

CHAPTER 6

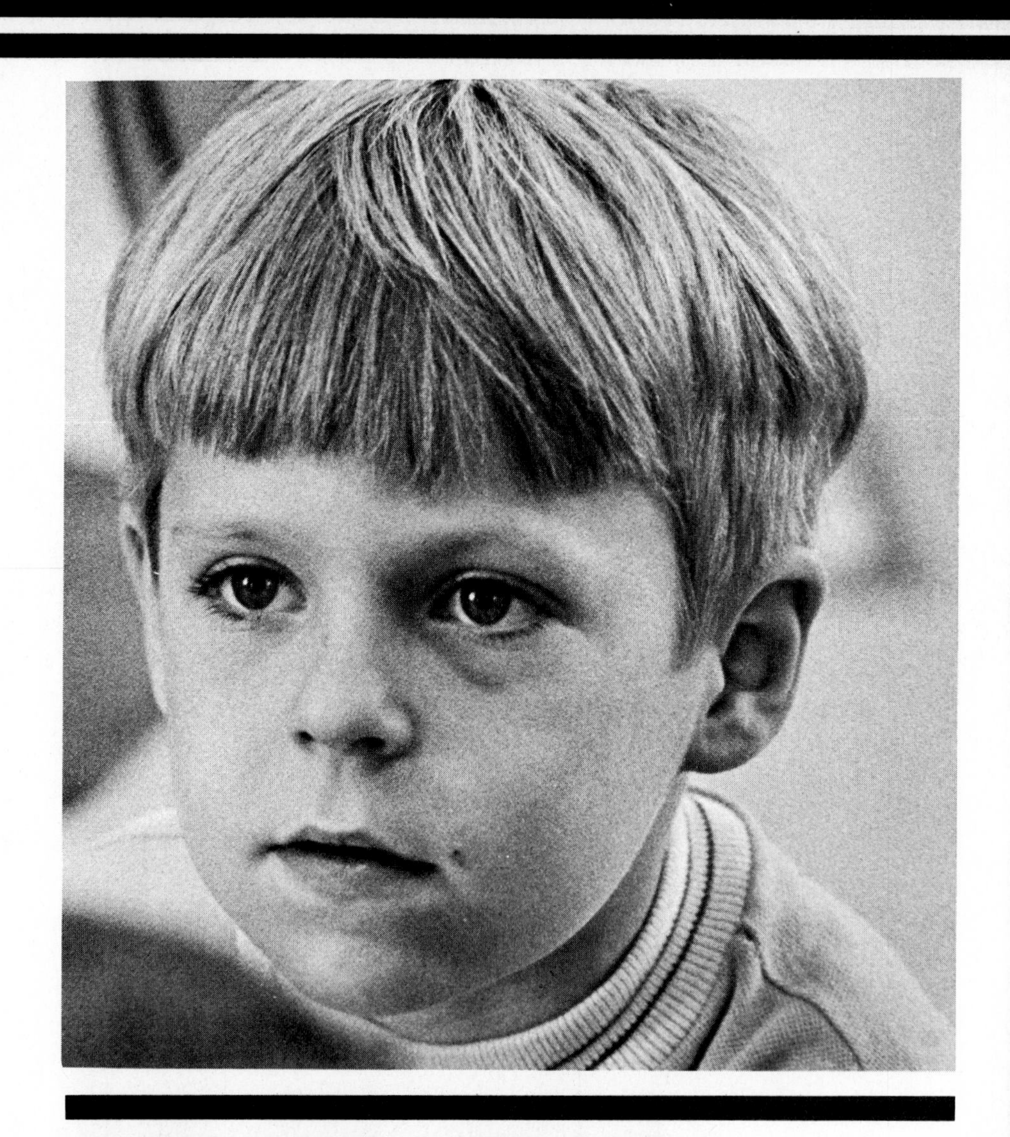

UNIT AND LESSON PLANNING

PREASSESSMENT

	AGREE	DISAGREE	UNCERTAIN
1. Some of the best lessons occur spontaneously, without being planned.	____	____	____
2. Teachers cannot and should not attempt to divorce their own values and beliefs from their teaching.	____	____	____
3. Science classes should be inquiry-oriented and therefore cannot be planned.	____	____	____
4. All learning units should reflect the philosophy and values of the teacher.	____	____	____
5. An attempt should be made to reflect all levels of the learning hierarchies in the development of all learning units.	____	____	____
6. Evaluation should occur throughout a learning unit rather than only at the end of the unit.	____	____	____
7. Each part of the content of a lesson plan should have corresponding activities to help students learn it.	____	____	____
8. Lesson plans frequently obstruct the spontaneous, insightful type of learning that is desired in science classes.	____	____	____

OBJECTIVES

Upon completion of this unit, you should be able to:

1. Develop a convincing rationale for long-range planning in education.
2. Develop a learning unit for an elementary science class.
3. Write your own philosophy of education, including some particular goals for science.
4. Discuss the role of evaluation in planning.
5. Write or select an affective objective and develop learning activities appropriate for it.

WHY PLANNING IS IMPORTANT

Why plan? This question may seem unnecessary. Everybody knows that planning is needed for most of the things people do, from developing our careers to more immediate activities such as going on vacations or spending paychecks. But is planning needed for teaching elementary school children? Many beginning teachers don't think so. Let's see.

The lives of today's teachers are, to say the least, hectic. Some demands on teachers' time come from their principals, who must insist that they fill in a large number of state, local, and federal forms. Many of these forms are required by law, and others are essential for smooth operation of the school. Most teachers consider these tasks necessary evils of teaching. Often teachers remark, "If only we had time to teach!" This situation may seem sad or even ridiculous, since the central purpose of schools is learning. Nevertheless, a teacher can quickly get into trouble for failing to fill in required forms or keep an accurate daily record of students' attendance.

With all these pressures, teachers may find that their schedules do not permit adequate time for planning. Ironically, there seems to be time for everything except the one thing teachers are hired to do—teaching! The need for planning classroom lessons and activities becomes clear when we consider why nonteaching activities must be planned. For example, if we did not plan our curricula in college, we would never graduate. If we never planned ahead how to spend our money, many of us would run out of money before the next payday and probably before we could buy some of the most essential items.

Putting teaching on a cost-effective basis would show that many teachers waste valuable time on relatively insignificant matters—time which they could save by identifying and focusing on the topics of highest priority. This is the first reason for planning.

A second reason for planning concerns the effect it can have on the clarity of lessons and subsequently on the level of understanding in the classroom. The teacher who relies on flying-by-the-seat-of-the-pants teaching is sure to have to make a crash landing.

Basically, there are two types of educational planning: long-range planning and short-range, or daily, planning. The major differences between the two lie in their respective degrees of detail and specificity. We will now examine both kinds of planning.

LONG-RANGE PLANNING

Once convinced that planning is necessary, beginning teachers often do not know where to begin. Many questions quickly arise: "Where do I start? Do I begin by making out several daily lesson plans and later assemble them to form a long-range plan? Or should I make my long-range plans first and use them to derive daily lesson plans?" The sequence of planning is important.

At first, these children were reluctant to participate in this activity. Thoughtful planning (and patience) will help the teacher to involve all children.

To grasp the meaning of long-range planning and the teacher's role, consider the K–12 curriculum. So that each student's experiences may lead to optimal learning, the knowledge and experiences at each grade level must build on those of the previous year. In other words, the program must have appropriate sequence and continuity from grades K through 12. "Sequence" is the chronological order in which the content is arranged, and "continuity" is an even flow—or, to put it negatively, the absence of disruptions or gaps. Continuity causes the curriculum to proceed smoothly. If a teacher fails to cover the topics, concepts, principles, and generalizations expected of the grade level, continuity is broken and a disruption occurs. Every student in such a class suffers by being poorly prepared for the next grade level.

This does not imply that every teacher must master the content covered at *all* grade levels. To know all the content at all grade levels would be impossible. Yet, it is essential that all teachers have a general understanding of what happens in the grades immediately below and above their own. The easiest and most accurate way to acquire this understanding is to read a curriculum guide which outlines the content for each grade level. The guide for the grade immediately below his or her own can give the teacher a good "feel" for what the children have already learned (or at least for the content and activities they should have been exposed to).

Ideally, each school develops its own statement of philosophy to show the general levels the students are expected to attain. This statement should be used by every teacher in planning the curriculum. All goals for each grade should be consistent with the philosophy of the school. The goals may be

This is a well-planned learning center for a unit on safety.

included in a course outline or may simply be a list of items. The outline format enables teachers to show many subgoals which are part of the larger goals, whereas the list format only identifies the goals and the sequence in which they are to be approached.

Units

Units are often developed by teachers, since teachers know better than anyone else the needs and interests of the students, the goals of the school, the resources offered by the community, and the expectations of the community. To develop a unit, a teacher must plan a cohesive set of interrelated concepts and activities which will make it possible for students to reach the goals set for each grading period. This combination of philosophy, aims, goals, objectives, content, activities, and methods of evaluation is called a "curriculum unit" or "learning unit." (See Figure 6-1.)

A good unit begins with a brief statement of philosophy and goals. Each goal should be one that can reasonably be achieved during the unit. For example, "All students will learn to classify rocks into these groups: igneous, sedimentary, and metamorphic" is more reasonable than "All students will become experts in classifying rocks and minerals."

In deciding the sequence of goals for a grade, attention should be given to the school's calendar. Most school systems arrange their programs so that grade reports are sent to parents every 6 weeks. Many informal indications of their children's studies and progress should be made available to parents during this period, but the formal 6-week report card with numerical or letter grades is held in high esteem by parents.

Figure 6-1. Anatomy of a unit.

Philosophy ⟶ Aims ⟶ Goals ⟶ Objectives ⟶ Content ⟶ Activities ⟶ Evaluation

SAMPLE UNITS

The best units are often those developed by student teachers. Following are two good examples of units made by teachers.*

Unit: Animals
Level: Primary

Philosophy
I believe that children are naturally active. I believe that they learn more when they are actively involved, with much opportunity for manipulation. In other words, I think that children learn from experience. But not all children learn the same way; each child is unique. Hence, I believe that children should be encouraged to relate science subjects to themselves.

Aims
1. Appreciation of animals.
2. Enjoyment of studying animals.

Goals
1.1 Appreciate the varieties of animals.
1.2 Appreciate the habits of animals.
1.3 Appreciate the uses humans make of animals.
2.1 Enjoy learning about how animals get their foods.
2.2 Enjoy learning how to group animals.

Objectives
By the end of this unit, each student will be able to:
1. List at least 10 types of animals without the aid of a book or poster.
2. Explain two differences in plants and animals.
3. Give two major reasons why all animals need food.
4. Name two major ways that we depend on animals.
5. List two responsibilities we have toward animals.

Activities
1. *Characteristics of animals.* The teacher should bring to class a plant and an animal (such as a kitten, gerbil, or fish). Ask the students probing questions leading them to realize the basic differences between plants and animals. For example: How are this plant and this animal different? Would you want a plant for a pet? Why? Emphasize that both plants and animals grow and are alive. Answers to the questions might be written on the board as in a language-experience activity.
2. *Animals grow.* Talk about how baby animals grow and change. Ask the children to bring their own baby pictures. Discuss the changes that occur in the children themselves. Make a "Guess Who?" baby chart. (Children guess who is shown in each baby picture.) Also, construct a baby-animal chart; children match baby animals with adult animals.
3. *Animal food.* Let children look at and examine different types of animal food. For example: hay, oats, rabbit pellets, seed. Explain that these are healthful foods for animals. Have pictures of food for people. Let students compare their own balanced meals with animal food. Make a picture with different kinds of animal food.
4. *Animal stories.* Have six or seven large pictures of animals. Let the children discuss each one for a very short period of time. Let them choose their favorites. Let students write a short story about one animal. Each student should be encouraged to contribute one sentence. Copy the story on a chart and let the children try to read it. Display it in the room.

*We have adapted the units slightly for use in this text.

5. *Group drama.* Students will go through several warm-up pantomime activities. With the teacher's guidance, students, all together, will act out various animals. The teacher will tell several facts of interest about each animal as it is being portrayed.
6. *Animal art.* Each child will be given clay to make a model of an animal. If a child chooses an animal which he or she does not know much about, reference books should be available for the child to look at, with help from the instructor.
7. *Animal pantomime.* Each student will have a turn to act out an animal in front of the class. Other students will guess what animal is being acted out. The child who portrayed the animal will then have an opportunity to tell something about the animal, and other students in the classroom may add to the information.
8. *What's wrong here?* Show pictures of animals in inappropriate habitats (fish in air, bird under water, etc.). Have students point out why the animals are in the wrong places. Students should then create and draw their own "What's Wrong Here?" pictures. Let them exchange their drawings and see if they can find the mistakes.
9. *Habitat mural.* On a long sheet of butcher paper, have the children draw scenes symbolizing land, sea, and air. Let the children look through magazines and cut out pictures of animals. They should paste the animals in the correct habitat.
10. *Animal game.* Each child will have on his or her back a picture of an animal. (The child does not see this picture.) Students ask questions of other students until they can guess what is shown on their backs. The following kinds of questions should be encouraged: Does it live on land? In water? In the air? Does it have a house? What does it eat?
11. *Animals for fun.* Discuss how animals provide recreation or fun for people (going fishing, having pets, riding horses, etc.). Have the students draw pictures showing how they have used animals for fun.
12. *Animals for food.* Conduct a class discussion which will lead the students to realize that humans use animals for food. Start by asking students whether they have ever eaten eggs and whether they know where eggs come from. Show pictures of chicks. Talk about fried chicken and chicken soup. Discuss how we raise some animals for food. Show pictures of cows and tell about how we get milk from them. Next, discuss butter—and make some.
13. *Animals for clothing.* In small groups, talk about how people make clothing from animal skins and furs. Display a kit which describes the process of making wool. Point out woollen items in the classroom. Discuss how leather is made from cowhides. Point out several leather items. Have the students tell what they are wearing that comes from animals.
14. *Farm animals.* After showing several pictures and perhaps a filmstrip, have students discuss farm animals. Have the students draw a picture of a farm, showing each animal in its proper place.

Evaluation

Each of the objectives is to be evaluated in the following manner.

A. *Activity 1*: The teacher should watch to see if students participate in the discussion; and the answers should be evaluated for understanding.

B. *Activities 2 and 3*: The teacher should evaluate the discussion. In Activity 2, children should be observed as they match the baby and grown animals.

C. *Activities 4, 5, 6, and 7*: Students will be given and asked to use information on a variety of animals. During all these activities, students should be observed to see if they use the information adequately. Activity 7 is an excellent one to check for understanding as students pantomime various animals they have learned about.

D. *Activities 8 and 9*: The teacher should check the students' drawings and where they placed the animals on the mural.

E. *Activity 10*: This is a good opportunity for review and evaluation. Students will be using knowledge to discover what secret animals are on their backs.

F. *Activities 11, 12, 13, and 14*: The students' drawings should be evaluated for understanding in Activity 11. Discussion should be evaluated in Activity 12. For Activity 13, the students should be observed as they point out the objects they are wearing which come from animals. In Activity 14, the drawings can be evaluated to see if students understand farms.

Unit: Prehistoric life
Level: Intermediate

Philosophy
I believe that children learn best through inquiry and that learning occurs best when children are having fun. Competition is a great motivator of children, especially when they are allowed to compete against their classmates.

Aims
1. Excitement about learning.
2. Understanding of change.
3. Appreciation of the age of the earth.

Goals
1.1 Develop inquiry skills
1.2 Develop discussion skills.
2.1 Understand that the rate of change of animals is very slow.
2.2 Know that change in plants and animals occurs in an orderly fashion.
3.1 Be aware that the earth is very old.
3.2 Realize that humans' time on earth has been relatively short, when compared with that of some other animals.

Objectives
Upon completing this unit, each student will be able to:
1. Name and draw at least three types of dinosaurs without misnaming any of them.
2. Given the dates of existence of 10 plants and animals, draw a time line and enter all dates without missing more than two places.
3. Without making an error, when given the date of the first existence of humans, enter this on the time line.
4. When viewing a film or filmstrip, ask at least one pertinent question.
5. Participate in a discussion by adding at least one fact (or opinion) and asking at least one question.

Activities
1. *Paleontology slide show.* This activity will be used with the entire class as an introduction to the unit. The slides are from a field trip to the University of Texas paleontology laboratory.
2. *Fossils.* The teacher and the students will discuss what fossils are and how they are formed. The idea that fossils are a link to the past will be emphasized. Directions will be given for the students to make their own mold-type fossils.
3. *Time line.* Students will make a simple time line showing dates when different animals lived, with emphasis on the adaptation of animals to different environments.
4. *Puzzler activity.* Why did the dinosaurs become extinct? The children will be asked to form hypotheses and present them. They will then be given the hypotheses that scientists have come up with and will compare these hypotheses with their own ideas.
5. *Fossils again.* Students will display on a table the fossils they made in Activity 2. Students will try to guess what was used to make the mold for each fossil.
6. *Guided fantasy.* This fantasy will be a trip back into some prehistoric time to retrieve information for people today.

Do This:

Now that you have seen two sample units, take a few moments to do a critique of them. Each has its own limitations. In each of these units, do the aims relate to the philosophy? Are the goals attainable? Do the objectives reflect the "a-b-c-d" criteria discussed on pages 64–65 in Chapter 4?

IN CONCLUSION: UNITS

Planning is essential to ensure that the goals of the science program are consistent with the philosophy of the school. Since the teacher's philosophy is also important, it too will be reflected in good planning.

The learning unit is the main vehicle through which expectations for learners are achieved. It should include objectives at various levels of all three educational domains. Each objective should have activities to help the students attain the objective. The various parts of a unit should be coherent. Good units should facilitate rather than hinder both teachers' and learners' creativity. Although we recognize that a good lesson may result from spontaneous inspiration, we believe that teachers should not rely on spontaneous teaching but rather should carefully plan each lesson throughout the year, leaving room for flexibility and spontaneity.

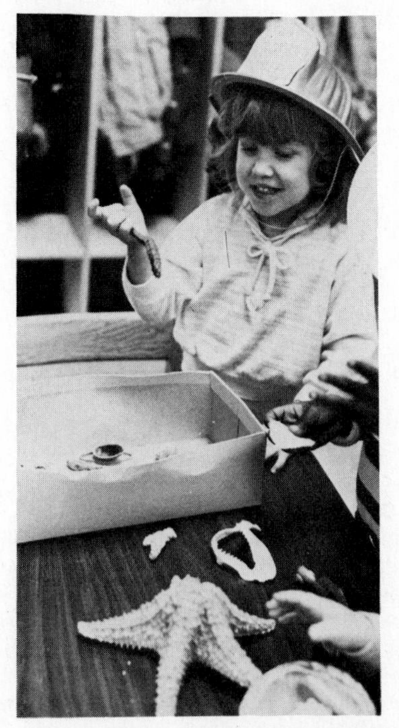

Curriculum guides planned by local schools often reflect the nature of a community. Children who live on or near a seacoast, for example, may have better opportunities to study ocean life—and may be more interested in it—than children who live far inland.

Curriculum guides

You are probably aware that most school districts provide curriculum guides for their teachers, which are intended to be used in selecting generalizations, objectives, and often activities. Here is an abstract of a guide which is currently being used in a public school system, the Abilene Independent School District.

Figure 6-2 (below) is a letter which introduces this guide to the elementary science teachers in the district.

Figure 6-2. Covering letter for the curriculum guide.

ABILENE INDEPENDENT SCHOOL DISTRICT
ALLISON KOONCE, SUPERINTENDENT

York H. Clamann, Ph.D.
CONSULTANTS OFFICE

To the Reader of This Guide:

As you will see, this Guide is purely a framework for the curriculum being established in the Abilene Independent School District. The individual teacher is being supplied with the major target areas of his/her teaching field as well as the concepts previously established as those necessary to adequately impart the cognitive material required; measureable for student attainment in the designated area.

The purpose of this Guide is to allow the teacher to be aware of the prerequisites of his/her course of instruction, as well as what is required of him/her as prerequisite material for following grades.

The format of this Guide is such that the teacher-user will easily be able to add activities, ideas, questions, methods and plans in each of the designated areas. Materials will be added on an approved, district-wide basis, as well as on an individual basis. For reasons of economy, this particular Guide has been combined into one publication for all six grades. The actual Guide is bound into a 3-ring loose-leaf notebook with all 13 dividers for each notebook. The loose-leaf additions for each teacher include only the pertinent information for each grade level.

York H. Clamann, Ph.D.
Science Consultant
Abilene Independent School District
Abilene, Texas

P. O. BOX 981. ABILENE. TEXAS 79604. 915/677-1444

ABILENE INDEPENDENT SCHOOL DISTRICT
ALLISON KOONCE, SUPERINTENDENT

York H. Clamann, Ph.D.
CONSULTANTS OFFICE

THE CURRICULUM GUIDE FOR TEACHING
SCIENCE IN GRADES 1-6

Introduction:

This guide is the product of two years of hard work including research, curriculum design and testing, and pilot testing in the Abilene Independent School District. The current arrangement was designed and polished by a committee of Abilene teachers representing the broad geographical areas of the AISD, as well as each of the grade levels. Special thanks go to this excellent committee:

Ms. Jean Brown	First	Taylor
Ms. Berta Fields	First	College Heights
Ms. Pauline Wall	Second	Jane Long
Ms. Mildred Woody	Second	Travis
Ms. Geraldine McEntire	Third	Valley View
Ms. Marilyn Threlkeld	Third	Jackson
Ms. Linda Chuk	Fourth	Jones
Ms. Carolyn Freeman	Fourth	Taylor
Ms. Johnny Moutray	Fifth	Johnston
Ms. Melba Holt	Fifth	Austin
Mr. Alex Koperberg	Sixth	Jane Long

Purpose:

The purpose of this guide is to give the individual science teahcer a framework upon which to base planned classroom instruction. The science program for grades 1-6 has been divided into 13 subject areas in which are found a number of essential concepts to be understood by students before they enter the seventh grade.

Concepts:

The concepts listed in each subject area are the ones essential to the area, and must not necessarily be taught in each grade level. If a particular concept can be taught in your particular grade level, then a number will be found in the chart to the right of the listed concepts. This is the page number of the grade level text-book on which will be found information regarding that concept. If the number has an asterisk (*), the concept is essential to your grade level and should be taught if at all possible (A = material not available).

The concepts given in each subject area are key concepts and must be taught in grades 1-6. Naturally, there are many other concepts not listed which may be taught either as prerequisite concepts or as supplementary material depending upon the students' ability.

P. O. BOX 981, ABILENE, TEXAS 79604, 915/677-1444

Figure 6-3. Introduction to the curriculum guide.

Figure 6-3 (above and opposite) more explicitly explains to the reader how the guide is to be used. Notice that it specifically states that the purpose of the guide is to "give teachers a framework" upon which to plan. As you examine sections from this guide (which includes 13 different sections), note that each statement is content-oriented. Is this approach useful? Not helpful? Why? Notice, too, that each statement is a science concept (a particular type of generalization based upon a recurring pattern).

Page 2

Materials and Methods:

This guide is designed to be a framework (really, a skeleton) of elementary science teaching. As you teach science this year, you will be collecting activities, materials, methods and other things which you will use to teach the various concepts. Put this material in the appropriate place in your notebook for future use (now you are putting "meat" on the "skeleton"!). If you wish to share some of your outstanding material with other teachers of your grade level, please send me a copy of the material and a short explanatory note. I'll copy the material and send it to your grade level colleagues.

Evaluation:

The concepts are so designed that they can easily be adapted to a diagnostic testing program presently being developed. This will in no way replace your individual testing procedures. When you have completed the concepts of your grade level in each unit, please take a minute and fill out the CURRICULUM GUIDE CRITIQUE SHEET, the pink sheet found after each unit divider. Please send the completed pink sheet to me and soon as you have finished that unit. The information will be compiled and used in the second phase of expanding this guide.

The first unit contains a sample of what we wish to accomplish in the future for all 13 units. Please use this format (Concepts, Pearls of Wisdom, Activities, Media) for material which you develop for the remaining 12 units and hopefully send to me.

If I can be of any further help to you in the coming year, please do not hesitate to come, call, or write. I am looking forward to working with you in continuing the development of a useful, and valuable guide for our program of excellence in science in Abilene.

York H. Clamann

York H. Clamann, Ph.D.
Science Consultant
Abilene ISD
Abilene, Texas

I

LIVING PLANTS

SUGGESTED GRADE PLACEMENT
(Pages in the Silver Burdett Series)

Concepts	1	2	3	4	5	6
1. Some plants are big and others are small.	80	A	59			
2. Green plants have leaves, stems, roots, and sometimes flowers.	86	A	A			
3. Plants need many things from the soil.		A	49			
4. Each part of a plant has a specific job to do.	86	A	66			
5. Seeds need water in order to grow.	85	A	A			
6. In the fall, many plants lose their leaves.	A	A				
7. New leaves and flowers grow in the spring.	A	A				
8. Fruits, berries, and nuts are seeds which are seasonal.	A	A	67	A	115	
9. Plants are for food, clothing, shelter, and enjoyment.	A	38 142	A		103	
10. Plants grow in different places.	82,94	A	74	A	105	A
11. Seeds are scattered in many ways.	92	A			115	
12. Green plants make their own food.			60*	A	111*	236
13. To make food, plants use carbon dioxide and water with the help of light.			62*		111	239*
14. Flowers produce seeds, often in the form of fruit.			67		115*	
15. A seed contains a new plant.			67		123 118*	
16. Roots grow downward in soil and leaves grow upward toward the sun.				A	108*	
17. Liquids travel to all parts of a plant.	87		63	A	110	
18. Plants respond to light, gravity, temperature and water.	96	145	60	A	130*	A
19. Photosynthesis changes carbon dioxide and water into sugar and oxygen in the presence of light.					112*	A*
20. Photosynthesis is a food-making process in green (chlorophyll-bearing) plants.					112*	A
21. Green plants exchange oxygen and carbon dioxide twenty-four hours a day.			64		112	A
22. Plants compete with each other for the energy available in the environment.		144*	70	17*	137	A

Figure 6-4.
"Living Plants."

A subject section is shown for each of the major divisions of science (biological, physical, and earth science): Figure 6-4 (Living Plants), Figure 6-5 (Heat and Its Behavior), and Figure 6-6 (Solar System; page 116).

Finally, we have included a sample of a critique sheet (Figure 6-7; page 117).* In this unit, such a sheet followed each of the 13 concept lists. Notice that this sheet asks you for *your* reaction to this study unit and, more specifically, that it asks you to identify concepts which you would

*Our thanks to the Abilene (Texas) Independent School District for providing all this material.

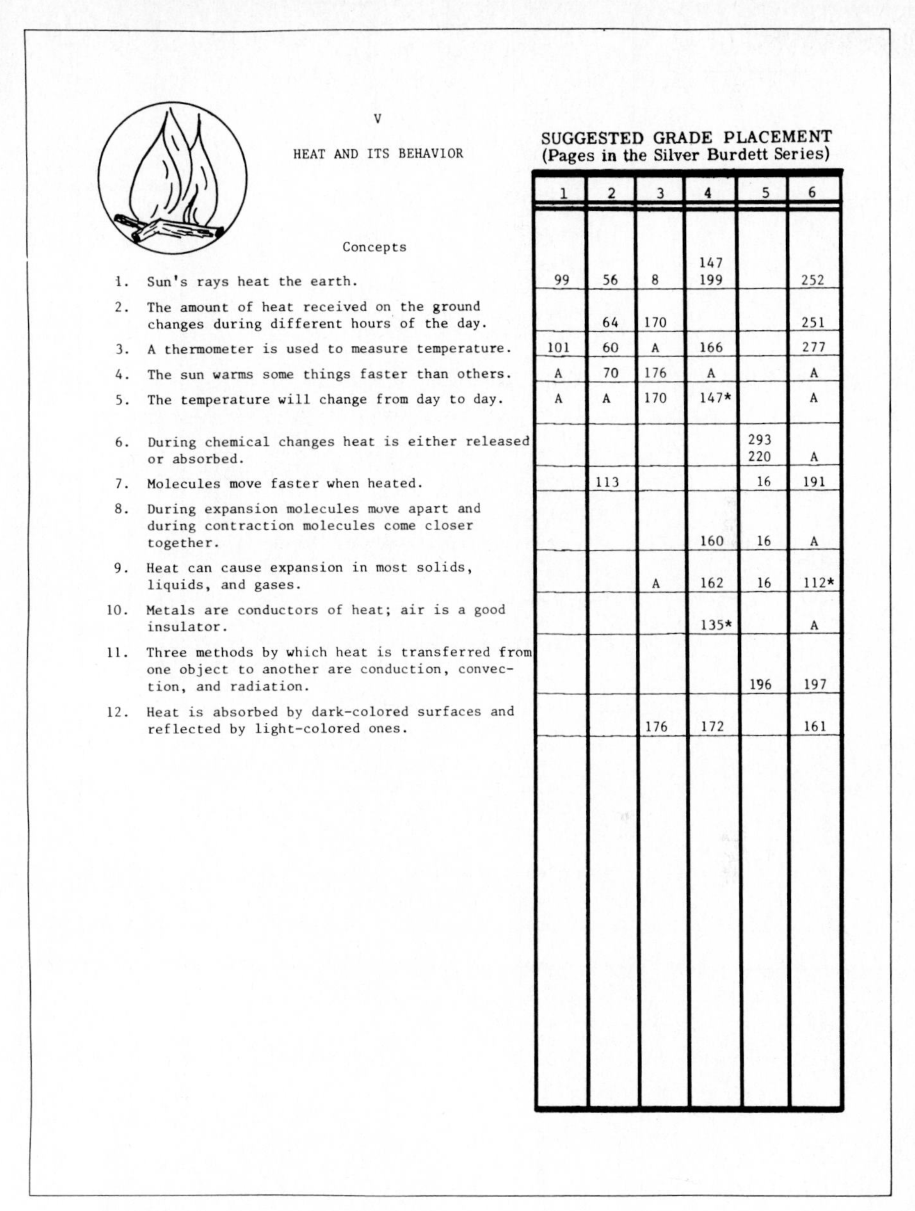

V

HEAT AND ITS BEHAVIOR

SUGGESTED GRADE PLACEMENT
(Pages in the Silver Burdett Series)

Concepts	1	2	3	4	5	6
1. Sun's rays heat the earth.	99	56	8	147 199		252
2. The amount of heat received on the ground changes during different hours of the day.		64	170			251
3. A thermometer is used to measure temperature.	101	60	A	166		277
4. The sun warms some things faster than others.	A	70	176	A		A
5. The temperature will change from day to day.	A	A	170	147*		A
6. During chemical changes heat is either released or absorbed.					293 220	A
7. Molecules move faster when heated.		113			16	191
8. During expansion molecules move apart and during contraction molecules come closer together.				160	16	A
9. Heat can cause expansion in most solids, liquids, and gases.			A	162	16	112*
10. Metals are conductors of heat; air is a good insulator.				135*		A
11. Three methods by which heat is transferred from one object to another are conduction, convection, and radiation.					196	197
12. Heat is absorbed by dark-colored surfaces and reflected by light-colored ones.			176	172		161

Figure 6-5. "Heat and Its Behavior."

like to see eliminated from the unit. In addition, an opportunity is provided for getting feedback from students. Since all units are designed to serve students, the value of knowing their reactions to the units should be obvious. Such feedback can be used to modify and improve the guides, which will eventually affect the curriculum.

When you begin teaching, it would be a good idea to ask whether your school provides such a guide. If it does not, you can ask to see a copy of the statewide guide for the state in which you will be working. If possible, study these guides *before* you begin teaching, so that you can use them in developing your own curriculum.

VII

THE SOLAR SYSTEM

SUGGESTED GRADE PLACEMENT
(Pages in the Silver Burdett Series)

Concepts	1	2	3	4	5	6
1. The sun is far away from the earth.	103	58				A
2. The sun gives light and heat.	99	58	7		A	
3. The earth spins like a top, making day and night.	102		162		A	253
4. Stars are far away and look small.	108	A	26		A	
5. Gravity keeps us from floating into space.	A		13		156	369 103
6. Planets move around the sun, and their "circular" paths (ellipses) are called orbits.	104		13		178	309
7. The sun is a star that gives off direct light.	99	68	7		173	
8. The moon and planets give off reflected light.	107	69	A		181	
9. The earth rotates every twenty-four hours and only part of it receives light from the sun at any given time.	102		162		178	
10. As the earth revolves around the sun each year, seasons change.			168		178	253
11. The tilt of the earth's axis makes for differences in the heat absorbed by the earth from the sun.			162 170		178	253
12. The Northern and Southern Hemispheres experience opposite seasons.			172		178	253
13. Each planet is different in size and distance from the sun.	104		12		174	
14. The sun, planets, satellites, asteroids, meteors, and comets make up the solar system.			7*		174*	A
15. A group of stars that appears to outline a figure is known as a constellation.					167	
16. Time is determined by the rotation of the earth on its axis.			162		178	253

Figure 6-6. "The Solar System."

NAME:__________________________ U N I T __________

SCHOOL:__________________________

GRADE LEVEL:________

CURRICULUM GUIDE
CRITIQUE SHEET
1980-81

Please fill this form out as completely as possible when you have finished work with your classes in this unit.

Month in which you taught most of this unit: (please circle)

Sep. Oct. Nov. Dec. Jan. Feb. Mar. Apr. May

Concepts completed by you in this unit: (please circle the concepts' number)

1	2	3	4	5	6
7	8	9	10	11	12
13	14	15	16	17	18
19	20	21	22	23	24

Concepts which you feel are inappropriate in this unit. Please give the concept number, and the CHANGE which you desire: (If none, leave this blank).

Concept No. :____________________________

Concept No. :____________________________

Concept No. :____________________________

Concept No. :____________________________

If there are additional concepts needing comment, please mark here ______ and continue on the back of this sheet.

What was your classes' overall reaction to this area of science?

______ Enthusiasm: want more of this area to study.

______ Enthusiasm: had enough and are ready to move on.

______ Attention: would like to move on sooner!

______ Boredom: what more can you say?

______ Confusion-frustration. Do you know why?____________________

RETURN THIS COMPLETED SHEET TO THE SCIENCE CONSULTANT VIA THE SCHOOL MAIL

THANK YOU!

Figure 6-7. Critique sheet for the curriculum guide.

DAILY LESSON PLANNING

What is daily planning?

We have examined learning units and curriculum guides. Both of these tools can help the teacher ensure complete coverage of objectives, goals, aims, philosophies, content, activities, and evaluations. But if each lesson is to help carry out these long-term plans, it will be the result of daily planning. Planning can be compared to taking a cross-country drive: the driver needs a specific road map (daily lesson plan) in order to reach the major cities along the way (goals) and eventually the final destination (aims).

Usually the daily lesson plan does not contain a statement of philosophy. It may or may not contain a statement of a general goal or aim. Most often it begins with objectives stated in behavioral terms and contains statements about content, descriptions of activities, and ways to evaluate students' performance.

A sample daily lesson plan

The sample plan shown in Table 6-1 is the format that one teacher uses. This plan was chosen because it is rather typical. Look at the plan to see whether there are any missing parts. You will notice that there is no statement of philosophy. But remember that this is not a 2- or 3-week unit plan. Usually we expect to find a statement of philosophy at the beginning of such a unit. To repeat this statement, or even part of it, on every daily plan would be unnecessary. The same is true for a rationale.

Daily lesson plans are essential for meaningful activities.

TABLE 6-1 SAMPLE DAILY LESSON PLAN

GRADE 6
Topic: How Rocks Are Formed
General goal: Students will learn how to distinguish among the three major types of rocks.

OBJECTIVES	CONCEPTS	ACTIVITIES	EVALUATION
By the end of this lesson each student will be able to:			
1. Without any aids, describe the three major processes of rock formation.	Sedimentary rocks are formed by the settling and packing of sediment in bodies of water. Igneous rocks are formed by intense heat resulting from the weight of the rocks above them. Metamorphic rocks are formed from existing igneous and sedimentary rocks that are subjected to intense pressure and heat.	Students will watch a 16mm film on rock formation. The teacher will draw each type of rock on the board and label its identifying characteristics.	Each student will be given one rock and asked to tell how he or she knows that it is a particular type and not the other two types.
2. When given samples of different types of rocks, identify the three different types.	Sedimentary rocks contain small, identifiable particles or granules of sand and minerals and often identifiable layers. Metamorphic rocks have no identifiable granules, and may have streaks of different colors, and are usually harder than sedimentary rocks.	Students will form three groups, each according to a station having several rocks of a particular type. They will examine these rocks and label them. Next, they will turn over a label which identifies the type.	A discussion will be held, and students will be asked to design a chart to be used to characterize rocks.

Do this:

Reread the sample daily lesson plan (Table 6-1), and then reread our discussion of it. Now, consider what additional comments you would make about the sample daily plan. Do you think that some of your comments make more important points than some of ours? Remember: Criticizing and evaluating other teachers' lesson plans is an excellent way to improve your own plans.

Now examine the objectives. Ask yourself whether two objectives are enough for one lesson. The answer depends on how detailed the objectives are. These particular objectives demand the mastery of several concepts, and therefore they are probably all a teacher could manage in one period. Then ask yourself whether the objectives are stated completely. In this instance, perhaps they could be improved by adding a minimum acceptable level of performance.

Notice that each of these objectives has more than one corresponding concept. This is all right. The converse is also true; a concept could have more than one objective.

Examination of the activities quickly shows that there are only two activities for objective 1, and yet there are three concepts for this objective. This is acceptable so long as these two activities provide opportunities for students to master all three concepts.

Consider the method of evaluation. Lesson plans often include only written tests; but it is good practice to include a variety in each unit evaluation, if not within each daily plan. The main question is whether the evaluations chosen actually measure achievement of the objectives. In this instance, they do seem to be an acceptable means of evaluation.

INITIATING YOUR OWN PLANS

Now that you have read a rationale for planning, as well as several suggestions (accompanied by sample plans) which can serve as a general guide for planning, you are ready to begin initiating your own plans. We should warn you that not all planning goes quite as smoothly as we wish it would.

In planning a unit, there are many obstacles that must be removed, and there are other obstacles that cannot be removed and therefore must be dealt with. The school program and regulations often inhibit planning. For example, a policy that forbids all field trips can erase any hope of taking students to an off-grounds site. Other school programs constantly interrupt daily plans by getting students out of class for various reasons. A blasting public address system or a large number of nonteaching assignments that come your way may leave you thinking, "If only I had time to

Planning has provided these students with a very timely activity. But remember that planning does not always go smoothly.

plan and teach!" Even your own lack of preparation in some areas will prevent you from developing plans of desirable depth and scope. You can gradually overcome such personal weaknesses by getting involved in in-service programs, by keeping up with your professional reading, and by being active in professional associations. Breaking down institutional barriers may be far more difficult and may require years of work.

Even your colleagues may impose restrictions on your planning. It is not uncommon for aggressive and eager novice teachers to be perceived by established teachers as agitators whose innovative ideas threaten the security and comfort of other teachers. Especially when you are new to a school system, it is good to spend more time planning than talking about your plans. You may also find that making friends with established teachers who put a lot of effort into planning will help you to become comfortable in your new job.

A final word of advice about your planning and teaching: Remember that even the best-planned lesson doesn't always work well. When this happens, don't be insulted or personally offended. Try to step back and objectively analyze what happened, so that you can pinpoint the causes of failure. In other words, plan thoroughly and then try to relax, roll with the flow, and make adjustments as they are needed. For example, suppose a day's lesson plan takes 2 or 3 days of class time—or even worse, takes only half a day. Neither of these common situations is cause for panic, but both warrant adjustments. That plan which ended up being three lessons long can be extended for another day or two; and the teacher who runs out of material can use the extra time to give further examples or can have the students give examples, discuss what they have learned, or role-play the lesson.

Do This:

To prepare for the possibility that you may someday finish an hour's plan in 30 minutes, explain how you would handle each of these situations:

1. The lesson is about reptiles, and you have told all you know about reptiles. Having 20 to 30 minutes left in the period, you would ______________________________

2. The lesson is on fission and fusion. In 10 minutes you have covered all the information on these topics to be found in the textbook. (There was a total of ½ page devoted to the two topics.) You would ______________________________

3. While preparing a unit on rocks and minerals, you planned to take your class on a field trip to collect rocks for later classification. Now, you learn that the school policy forbids off-grounds trips. You would ______________________________

SUMMARY

Planning is essential to ensure that the goals of the science program are consistent with the philosophy of the school. The teacher's philosophy is also important and will be reflected in good planning.

The learning unit is the main vehicle through which expectations for learners are achieved. It should include objectives at various levels of all three educational domains. For each objective, there should be activities to help the students attain the objective. The various parts of a unit should be coherent. Good units should not hinder learners' creativity but rather should encourage it.

Most school systems provide curriculum guides which show how any subject fits into the total curriculum (K–12) across all subjects. Curriculum guides by nature are broad; they do not usually have the depth, detail, and general comprehensiveness that characterize curriculum units. Therefore, guides should not be used to replace units; instead they should be used to show how each unit fits into the total curriculum, and thus prevent unintentional omissions and overlaps.

Curriculum guides provide general direction, and some units provide much more detail, but daily lesson plans are needed to ensure that the objectives are met. Although good teaching may result from spontaneous inspiration, teachers should not rely on inspiration but should carefully plan the year's lessons. This does not imply that teachers should be enslaved by their plans. Used correctly (i.e., flexibly), good planning actually frees the teacher and students to be creative. The needs of the students should never take second place to any plan; rather, plans should be continually adjusted to fit the needs of the students.

RECAPITULATION OF MAJOR PRINCIPLES

1. Effective instruction in the elementary science classes is unlikely to occur without adequate planning.
2. Long-range planning in science should begin with an examination of the total science program. Special attention should be given to the content and objectives of the grades immediately preceding and following the course being planned.
3. Learning units of approxmately 2 weeks' duration are popular and effective in elementary schools.
4. Each part of a learning unit should relate to the other parts.
5. Teachers should preface their lesson plans with a statement of philosophy spelling out their own values and beliefs about the purpose of education and the nature of children and learning.
6. Science teachers should strive to include, in each unit, objectives at various levels of the cognitive and affective domains.
7. Thorough planning should not preclude flexibility, but should enhance it by including alternative learning activities.
8. Elementary science planning should include objectives and activities designed specifically to use students' creative talents.
9. All planning for elementary science classes should help shape students' attitudes toward science and toward their total environment.
10. Elementary science plans should involve students in a variety of activities.

SUGGESTED READINGS

Association for Supervision and Curriculum Development: *Considered Action for Curriculum Improvement*, Arthur W. Foshay, ed., ASCD, Washington, 1980.

Gage, N. L.: *The Scientific Basis of the Art of Teaching*, Teachers College, New York, 1978.

Gagné, Robert, and Leslie J. Briggs: *Principles of Instructional Design*, 2d ed., Holt, New York, 1979.

Harnischfeger, A., and D. E. Wiley: "Teaching-Learning Process in Elementary School: A Synoptic View," *Curriculum Inquiry*, vol. 6, 1976, pp. 5–43.

Henson, Kenneth T., and James E. Higgins: *Personalizing Teaching in the Elementary School*, Merrill, Columbus, Ohio, 1978.

Henson, Kenneth T.: "Teaching Methods: History and Status," *Theory into Practice*, vol. 19, winter 1980, pp. 2–5.

Hyman, Ronald T.: *Strategic Questioning*, Prentice-Hall, Englewood Cliffs, N.J., 1979.

Oliva, Peter F.: *Developing the Curriculum*, Little, Brown, Boston, 1982.

Posner, George J., and Kenneth A. Strike: "A Categorization Scheme for Principles of Sequencing Content," *Review of Educational Research*, vol. 46, 1976, pp. 665–690.

Posner, George J., and Alan N. Rudnitsky: *Course Design: A Guide to Curriculum Development for Teachers*, Longman, New York, 1978.

Taba, Hilda: *Curriculum Development: Theory and Practice*, Harcourt Brace Jovanovich, New York, 1962.

Zais, Robert S.: *Curriculum Development: Principles and Foundations*, Thomas Y. Crowell, New York, 1976.

CHAPTER 7

SKILLS, ACTIVITIES, AND RESOURCES TO HELP CHILDREN LEARN SCIENCE

PREASSESSMENT

	AGREE	DISAGREE	UNCERTAIN
1. Effective teachers are excited about their work.	____	____	____
2. Activity cards can be used with individual children as well as with groups.	____	____	____
3. An electronic game board is useful only in teaching science.	____	____	____
4. A guided fantasy can be used to introduce a topic but would have little application as a culminating activity.	____	____	____
5. A shoebox science kit should contain only one activity.	____	____	____
6. Educational outdoor activities should be a normal part of every teaching week.	____	____	____
7. Activity bulletin boards require students to actually do something.	____	____	____
8. Learning centers usually cover a very broad topic.	____	____	____
9. Educational games usually involve an element of chance.	____	____	____
10. Value-clarifying techniques are especially useful in teaching science.	____	____	____
11. Teachers' editions of science textbooks contain few real aids for the teacher.	____	____	____
12. Resource persons who are invited to the classroom should not be given topics to address: rather, they should be allowed to present whatever they choose.	____	____	____
13. Students can assist teachers in acquiring needed teaching materials.	____	____	____
14. Teaching materials are often available from industries, professional associations, and government agencies.	____	____	____
15. Teacher-made information files are valuable teaching resources and can be easily put together and maintained.	____	____	____

OBJECTIVES

Upon completion of this chapter you should be able to:

1. Identify the types of help which a teacher can find in teachers' editions of elementary science textbooks.
2. Identify five sources of potential resource persons for science classes.
3. Outline a procedure for effectively using a resource person.
4. Name several sources of free and inexpensive teaching materials for science.
5. Specify a plan for involving students in collecting teaching materials.
6. Name one caution about using free and inexpensive printed reference materials in the classroom.
7. Describe a procedure for filing information valuable for teaching (such as resource persons and teaching ideas).
8. Name at least five kinds of information files which would be helpful for a teacher.

WHAT MAKES A SUCCESSFUL TEACHER?

A little excitement can do great things in our lives. It may carry us into new places and ventures, and it certainly helps sustain us when we meet an obstacle. In teaching, new experiences are frequently encountered—and yes, obstacles are present too.

If we are genuinely excited about helping children become all that they can become, many things which might otherwise be problems often become opportunities.

To be excited about teaching does not mean that we know all the answers and that we have total confidence that everything we attempt will work perfectly. Rather it means that we have the dedication and the confidence that will enable us to succeed in doing great things with our students in spite of an occasional obstacle and an occasional failure. Dynamic, successful teachers are excited about what they are doing.

The most important characteristic of a successful teacher is the ability to have relationships with students that make them feel important. Each student must have confidence that the teacher values him or her as a person. All students must feel sure that the teacher is sincerely doing everything possible to help them become the very most that they can become. If you think back to the teachers from whom you learned the most, you will probably realize that most or all of them were people whom you considered friends—friends who would support you in any way possible. To be a great teacher, you must have that commitment which says to each student, "I am going to do the very best I can to help you become all that you are capable of becoming"—even if the child disobeys, smells bad, says nasty words, or dis-

Do This:

Find out by the "grapevine" the names of some of the most successful and dynamic teachers in a local elementary school. Arrange to visit those teachers in their classrooms on an afternoon after school. Ask them what they attribute their success to. Try to get them to share a few of their "secrets" with you.

agrees with you. That kind of commitment is never phony, and your students will detect, appreciate, and respond to it. It is not easy to make such a commitment, and other people sometimes do not appreciate it; but over the long term, it can give you the satisfaction and rewards that you need to keep going.

Another important characteristic of a successful teacher is knowing the learning goals for students and, furthermore, believing that these are important goals. Knowing what the goals are is an obvious necessity, but could you imagine doing a great job of teaching about plants if you did not believe it to be important for your students to have knowledge of plants? If you question the importance of a topic you are asked to teach, talk it over with your supervisor. If you can still find no compelling reason for having that topic in the curriculum, begin to work toward having it removed.

Successful teachers are organized. You have a feeling of confidence when you know what is next on the schedule for your class. This does not mean that you should have a rigidly structured and inflexible schedule, but it does mean that you should have a plan for the next activity. Plans, however, need to be frequently modified or temporarily suspended to take advantage of an unexpected event. If your class were studying hurricanes and a tornado struck your town, would you wait until next week to talk about tornadoes just in order to follow your plan? Or if you found out that one of your students, at the tender age of 10, was an expert on propagation of African violets, would you avoid tapping his or her expertise because you had not built houseplants into your plants unit?

Successful teachers must understand and use a recognized theory (or theories) of how children learn. Whether the teacher adopts the explanations of Piaget or some other learning theory, the activities used must be evaluated with regard to their appropriateness for the particular group of students.

Successful teachers use a variety of teaching methods and activities. Variety is important in teaching. Think of some activity in which you were involved, which you enjoyed, and from which you learned a great deal. How would you like to do that type of activity and only that type of activity for 6 weeks? The major portion of this chapter describes activities to increase the variety of techniques and materials which you can use in teaching. We believe that you will find these activities widely applicable and useful.

Finally, successful teachers recognize that they are human. They can be happy; they can be sad. They can succeed; they can fail. They behave like other humans. Many present-day students suspect that teachers are human, but some teachers feel that it is inappropriate for them to be merely human. We want to let you in on what we hope will become less and less of a secret—it is all right for a teacher to be human. Teaching, like every other profession, requires that we put forth the very best effort that we are capable of; but we must realize that we will occasionally fall short of our goals. What we must do then is move on to the next challenge with renewed resolve to do a great job.

The remainder of this chapter presents skills, activities, and resources which we think will help you become a more exciting teacher. We have used these activities and resources with students, and we are confident that they can help students learn. Many of the ideas which you will encounter, including the examples in this chapter and those in Parts Two, Three, and Four of this book, have a great deal of appeal. You must, however, resist the temptation to use every exciting new activity or resource immediately with your students. You should use only those ideas which will help your students achieve the learning outcomes which you have established for them. To do otherwise will waste your students' time. Unless each activity or resource can be meaningfully tied to others and to building toward established outcomes, the student will not have a framework into which to fit the knowledge gained, and the bit of knowledge will very soon be forgotten.

SKILLS AND ACTIVITIES

Activity Cards

Designing and developing activity cards is one way of making a collection of activities for various topics. The activity cards described here were designed for use by individual students or small groups (two to four children). However, almost all the activities can also be used with larger groups. An example of an activity card is given in Figure 7-1.

The front side of an activity card usually contains both words and an illustration. Sometimes it may contain only one or the other, but an activity card that includes both is usually more meaningful to students. The word or words should suggest the activity or the topic. Some examples from this text are "Dried-Flower Arrangements," "Colors of the Rainbow," "Stringed Instruments," and "Shadows." The illustrations may be your own sketches or drawings, or appropriate pictures taken from magazines, travel folders, old books, or any other source you choose.

The reverse side of the activity card gives directions for one or more activities related to the topic. It is better to list only one or two activities, for a card with a dozen different things to do may overwhelm a child. If you want to list more than one or two activities, make additional activity cards. The main activity or activities for a card should stand out, but related activities or questions may be printed along the sides.

Front side

WATCH ONE PARTICULAR WILD BIRD OR KIND OF WILD BIRD. WHERE DOES THIS BIRD SPEND MOST OF ITS TIME? WHAT DOES IT EAT?

PRETEND YOU ARE ANY BIRD YOU WANT TO BE. WRITE A DIARY FOR ONE DAY OF YOUR LIFE AS A BIRD.

FIND SOME UNUSUAL BIT OF INFORMATION ABOUT YOUR BIRD.

Reverse side

Figure 7-1. Activity cards can challenge students with interesting things to do.

Most of the activities should be things the students can do in a short period of time in the classroom, but a few more challenging and time-consuming ones may be included. Examples of brief classroom activities are:

How many products made from wood can be found in your classroom?

Measure the top of your desk.

Examples of activities which may have to be completed at home are:

Determine where the moon is (or whether it is even visible) at 8 P.M. for 5 nights in a row.

Collect three different kinds of plants in the yard (or in sidewalk cracks) around your home.

Some activities may be longer-term, such as:

Paint a dot on the south-facing window in your classroom. Record the location of its shadow at 11 A.M. each school day for a month.

Record the types of clouds that are present and what the weather is like for several weeks. Can you detect any patterns connecting cloud types and the weather which follows?

Do This:

Construct two activity cards for each of these topics: insects, astronomy, matter.

Ideas for activities can be found in students' and teachers' editions of elementary science textbooks, childrens' science magazines, books of science activities, and books such as this one. One very fruitful source of ideas for activity cards is elementary students themselves. Conduct a brainstorming session: Ask them what activities they think of when you mention a topic for which you wish to develop activity cards. Some of your students will have been exposed to related activities in 4-H Club, Boy Scouts, or Girl Scouts; on television; or in books and magazines.

Activity cards can be used in interest centers to supplement a unit of study. As part of a unit on weather, for example, you might require your students to:

Do 3 of the 30 activities on the weather activity cards.

The shining light: An electronic game board

"Shining light" is a game which students will find fascinating and which is a good educational tool for learning and reviewing. The game board (Figure 7-2) is a homemade device upon which students can place game cards containing questions, problems, identification activities, and other learning activities. Examples of game cards are shown in Figure 7-3.

You can build a game board if you want to, but a game board is not essential in using "shining light" game cards. The student can put a finger or a pencil through the hole of the answer selected and then flip the card over to see if the hole is circled or checked (Figure 7-4).

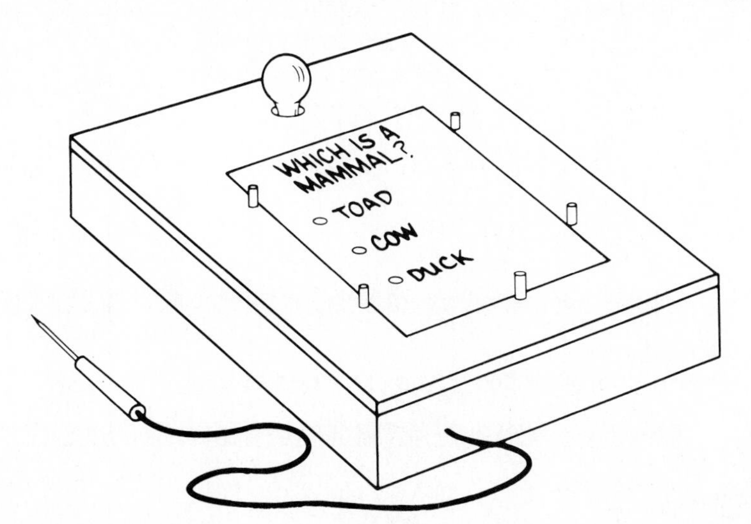

Figure 7-2. "Shining light"—a battery-operated game which students enjoy.

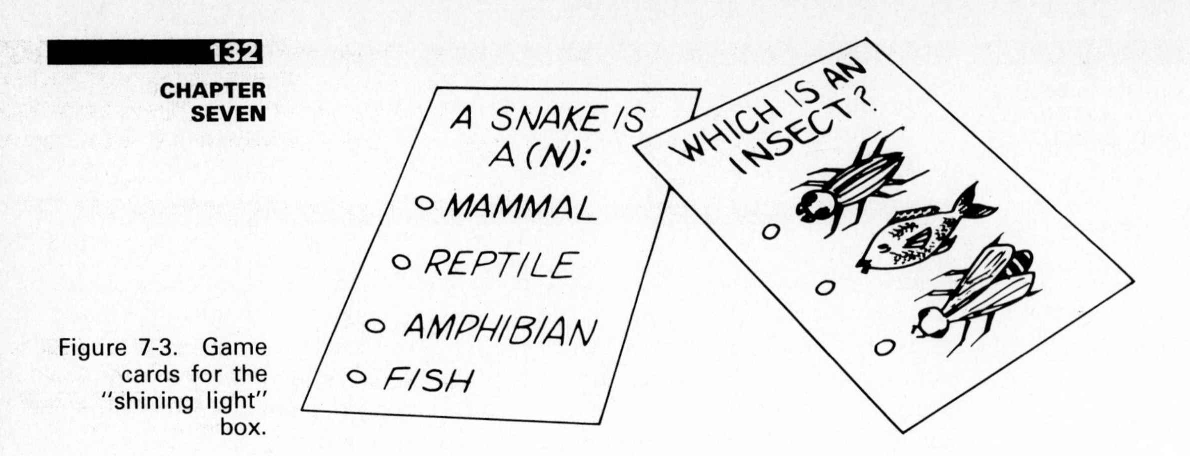

Figure 7-3. Game cards for the "shining light" box.

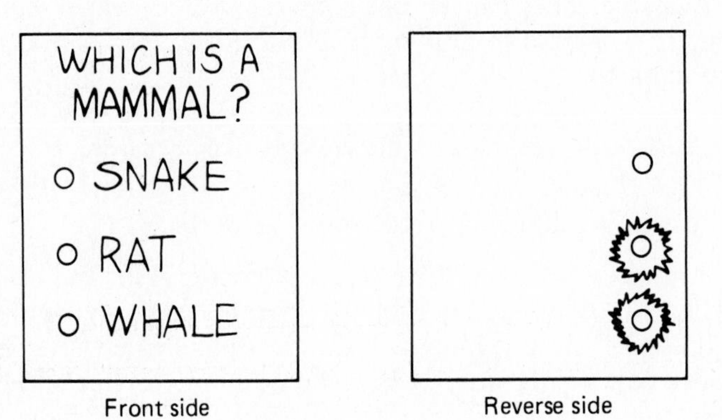

Figure 7-4. The correct response or responses can be indicated on the reverse side of a game card.

To facilitate handling and storage, the game cards should all be the same size. Inexpensive, standard-size game cards can be made from 4- by 6-inch index cards. These can be purchased at any office supply store and are a convenient size for students to handle.

Students are, as you know, a rich source of ideas for many class activities, and game cards are no exception. Just ask them! Or you may tell your students to construct game cards as one of their assignments.

Do This:
Construct five game cards on the topic "Plants."

Guided imagery

Not only are guided-imagery techniques fun for both students and teacher, but they are especially valuable for encouraging the use of the right cerebral hemisphere. Recent research on the differences between the right and left cerebral hemispheres of the human brain has shown that, in general, the right hemisphere is concerned with aesthetic and emotional experiences (music, visual arts, imaginative writing) whereas the left hemisphere usually processes logical, orderly thoughts and information (mathematics, language, scientific experimentation). Much scientific work, from elementary school through the most advanced research, is conducted primarily through left-hemisphere activity; yet imagination and intuition do play a large role in science. Part of the job of the elementary science teacher is to help students understand the many dimensions of science; thus, undue emphasis should not be placed on left-hemisphere activities. Guided imagery can enhance childrens' abilities to pull information together in meaningful ways and to look at phenomena from new perspectives.

Through guided imagery your students can become fish, soaring eagles, or ants. They can visit places in imagination that they have not yet been able to go to in reality: the desert, mountains, or seashore; the Grand Canyon; Washington, D.C. They can also learn new things about places to which they do have easy access: the vacant lot across the street, a local city park, the downtown business district.

Below is an example of guided imagery—a guided fantasy. To prepare the students for an activity like this, the teacher should use a relaxation technique with them. Then, after asking the children to close their eyes, the teacher slowly reads the guided fantasy to them in a low, clear voice, pausing for a few seconds at significant points. (In our example, the teacher should pause for 3 to 5 seconds whenever a slash mark appears.)

Desert fantasy trip

Today we are going to take a trip into a well-known desert near Phoenix, Arizona: the Sonoran Desert. Picture yourself in a van traveling to the desert. / You are dressed for a 4-hour hike into the desert. / In your backpack you have placed the supplies that you will need. / We park the van and get out. It is a bright sunny day, and at 9 o'clock in the morning it is just beginning to get quite warm. / We are walking along through the desert in single file. An animal scurries out from under a bush and moves quickly away. / You look around and see the various kinds of plants present. / As we continue to walk along, it is getting hotter and hotter. / Suddenly a shadow passes over you, and you look up to see what caused it. / Off to the side of the trail you see many flowers. They are so beautiful. / It is getting hotter, and because we are getting a little tired, we sit down to rest for a few minutes. / As we again begin to hike, it is now approaching noon and becoming very hot. You take a drink to satisfy your thirst. / You notice that there are even fewer animals out now than there were earlier. / As we continue the hike, you stop to take a picture of a beautiful scene. / Finally we make it back to the van and a shaded picnic table. It feels so good to be in the shade. / Now, remain quiet and slowly open your eyes. /

Do This:
Write a script for an imaginary walk as an ant through the vacant lot across the street.

After reading a fantasy the teacher asks some appropriate questions. For our example, appropriate questions might include the following: How were you dressed for your hike into the desert? (Was the clothing light enough to be comfortable? Did it cover you well enough to protect you from sunburn?) What did you have in your backpack? (Enough water? Did you or your leader have first-aid supplies? How about food? Something to make shade? A whistle?) What kind of animal ran out of the bush? What kinds of plants were present? What caused the shadow? What were the beautiful flowers like? What kind of plant were they on? Why did you see fewer animals later in the hike? What did you take a picture of? Why did you take a picture of it?

The children's answers to questions like these can reveal to the teacher what they already know or what they have learned in their study of a topic.

A guided fantasy can introduce a topic or can be a culminating activity at the end of a unit. For example, the "Desert Fantasy Trip" could introduce the study of deserts. It introduces ideas of desert plant and animal life, heat and water characteristics of deserts, beauty in the desert, and planning and supplies necessary for a desert trip. If you use it at the end of a unit on deserts, it can help the students to pull together the information they have learned about desert life and about the physical and aesthetic characteristics of the desert. The feedback that you get from students' answers to your questions will help you assess the degree to which they have learned what you wanted them to learn.

In another type of guided imagery, you can have your students become something—perhaps a fish, an eagle, a rock, or a river. Again, the fantasy can help them pull together what they know. You can later ask them to write or talk about their experiences.

As you develop imagery experiences, keep in mind that your students must be able to relate to the situations you describe. They will have some degree of familiarity with the vacant lot across the street, but having them climb Devil's Tower might be meaningless. Or they might be able to imagine being a mouse, but asking them to become a marmoset could be frustrating.

Each chapter in Parts Two, Three, and Four of this text includes one or two activities based on guided imagery. Other types of guided imagery are presented in books identified in the "Suggested Readings" for this chapter (see page 152).

Shoebox science kits

A shoebox science kit can be designed to serve any of several purposes. Each such kit is a complete package of activities related to a specific science topic (usually of limited scope). It may be an integral part of a science unit being studied in class, or it may contain supplementary activities. Then again, it may contain activities on topics which are not being formally studied in class but which may be appropriate for a science interest center.

The shoebox science kit should fit into a shoebox or other container of similar size, and it should include complete instructions as well as the materials required to do the activities. It is designed to be used by a single student, but it could be used by small groups; or the activities in a shoebox kit could be carried out by a whole class. Normally shoebox kits are not required assignments, and therefore you should choose activities for them that will be especially interesting to students.

The first step in constructing a shoebox kit is to decide on the topic and the learning outcomes. Next identify or create approximately six activities for inclusion in the kit. If there are fewer than six activities, a given child may not find an activity that he or she would like to do, but many more than six activities can overwhelm children and give them the impression that there are simply too many things to do. If you have more than six or seven activities in mind, make two kits.

The next step is to collect or develop the materials needed to do the activities. The materials for a given kit might include a hand lens, some feathers, a tape measure, crossword puzzles, word-search games, or special types of paper, for example. Never place potentially dangerous items (such as matches or iodine) in a kit. Materials which are bulky and easily obtainable (pencils, soil, water, milk cartons) should also be omitted.

The box which contains the kit should be labeled on a side that will be visible when the box is on the shelf. It can be decorated with appropriate sketches, drawings, or cutouts from magazines or books. Information on the activities should be immediately visible when the box is opened. The information may be written or typed on index cards bound into a small booklet or attached to the inside top of the box, or it may be given on the outside of envelopes which contain the materials for the activities. A description of a shoebox science kit is given below.

Spider shoebox science kit

Activities

1. Collect a web. (*How do you collect a spiderweb in such a way that it can be permanently saved?*)
2. Read a book on spiders. (*The book* Be Nice to Spiders* *is one good suggestion. Include in the kit a few questions about the reading on a "checkup" sheet.*)

*Margaret Graham, *Be Nice to Spiders*, Harper and Row, New York, 1967.

Shoebox science kits can include directions for individual activities along with the materials needed to do the activities.

3. Observe a spider. (*Tell the students to locate a spider in its web and check it every 10 minutes on one or two evenings to try to learn what the spider eats. Challenge them to learn the name of their spider and read about it.*)
4. Search for words about spiders. (*Challenge students to find 12 spider words.*)
5. Match pictures with spiders' names. (*Glue small pictures of spiders to cards. Instruct the students to match each picture with another card you make up, containing the spider's name and some interesting information about it.*)
6. Make a spider. (*Give students clay and ask them to make a model of a spider. Also offer them the option of writing a story about a spider or drawing a picture of it.*)

Materials
Include these materials in the kit: spray cans of white enamel and clear lacquer, an assortment of colored construction paper, information on a question sheet about the book *Be Nice to Spiders*, an information sheet on observing spiders, a word-search puzzle (laminated), pictures of spiders on index cards, information cards on spiders, and modeling clay.

An interest center which contains several shoebox science kits is usually popular with students. The use of uniform-size boxes facilitates stacking and storing the kits.

Do This:
Make a shoebox science kit on flowers. Include at least one activity which deals with vocabulary and one which requires observation of living flowers.

For students who do not yet read, you can include an audio-cassette tape recording (if a cassette-tape player is available in the classroom) or you can make kits with activities which do not require written instructions. To make a kit on plants, for example, you can cut up a picture of a plant and place the pieces in an envelope. Upon opening the envelope, the student will recognize the contents as pieces of a puzzle, and no further instructions will be needed. You can get more ideas for shoebox kits for nonreaders by observing the activities in their classes.

Outdoor activities

Studying science offers many opportunities for meaningful outdoor experiences. While field trips off the school grounds are potentially very beneficial, staying on the grounds has nearly as many possibilities and is much easier to arrange: no special administrative approval, parental permission slips, or arrangements for transportation, lunch, and toilet facilities are required. Furthermore, the students can easily revisit the field-trip site if it is on the school grounds. Grounds with trees and perhaps even a planned outdoor education site are very useful and pleasant, and should be a goal of every elementary school; however, not having such grounds does not make outdoor activities impossible.

Both students and teachers enjoy being outdoors. A teacher who plans at least one outdoor educational activity each week, whether for science or another curriculum area, is much appreciated by students. But how can you make an outdoor learning activity really productive for the students? Probably the single most important factor, after selecting the learning objective and an appropriate activity, is to tell students exactly what they are to do while outside. If you do not clearly specify what they are to do, they will certainly find something else to do.

If you want your students to collect tree leaves to be classified later, your instructions might be "Today we will spend 10 minutes outside. In that time I want you to collect at least five different kinds of leaves from trees on our grounds." Outdoor activities do not have to take a long time, as this example indicates. Many meaningful outdoor learning activities can be done in 20 minutes or less. Of course, longer outdoor experiences are appropriate if more time is necessary for the learning objective.

Each chapter in Parts Two, Three, and Four of this text includes several examples of outdoor learning activities. The checklist on page 138 will help you plan outdoor activities. The list is designed to include off-grounds trips. For on-grounds outdoor activities, eliminate the unneeded steps.

Do This:

Plan an on-grounds outdoor activity for your students (or for your own classmates, if you are not yet teaching). Include item 1 and items 9 to 14 from the list on page 138. Conduct the outdoor learning experience and evaluate it, following the guidelines in item 16.

Checklist: Outdoor activities

*____ **1.** Establish educational objectives. (These can often be taken from the unit plan.)
____ **2.** Identify several sites at which the established objectives could be attained.
____ **3.** Select one site.
____ **4.** Obtain approval for the trip from your administration.
____ **5.** Obtain permission from the organization that controls the site to take your class there.
____ **6.** Arrange transportation.
____ **7.** Secure permission for the trip from your students' parents.
____ **8.** If necessary, arrange for extra supervisory help.
*____ **9.** Inform your students of the details of the activity.
*____**10.** Explain the objectives of the activity to your students.
*____**11.** Suggest to your students things they can watch for during the trip.
*____**12.** Establish rules for behavior during the trip, and explain them to the students.
*____**13.** If appropriate, assign special tasks to the students.
*____**14.** Explain follow-up activities, if any, to the students.
____**15.** Write any thank-you letters that are appropriate.
*____**16.** Evaluate the experience:
    ______**a.** Did the students learn what you wanted them to learn?
    ______**b.** Was the amount of time invested reasonable?
    ______**c.** Should this activity be done another year?

*The asterisk signifies on-grounds activities.

Activity bulletin boards

Activity bulletin boards are bulletin boards with which the students are involved in some active manner. For example, if a class is studying light and shadows, the bulletin board could be similar to the one illustrated in Figure 7-5. The sun and a few objects are represented on this bulletin board. In the pocket at the bottom are shadow images for each of the objects. Students are to place the shadows where they should go. Observ-

---

Do This:

Make both an informational bulletin board and an activity bulletin board, each relating to a different topic, and leave them on display in your classroom for 2 weeks. On the day after you remove them, administer a short quiz covering the lessons taught by each bulletin board. From which bulletin board did the students learn the most? Ask the students how you could have made the other bulletin board more effective.

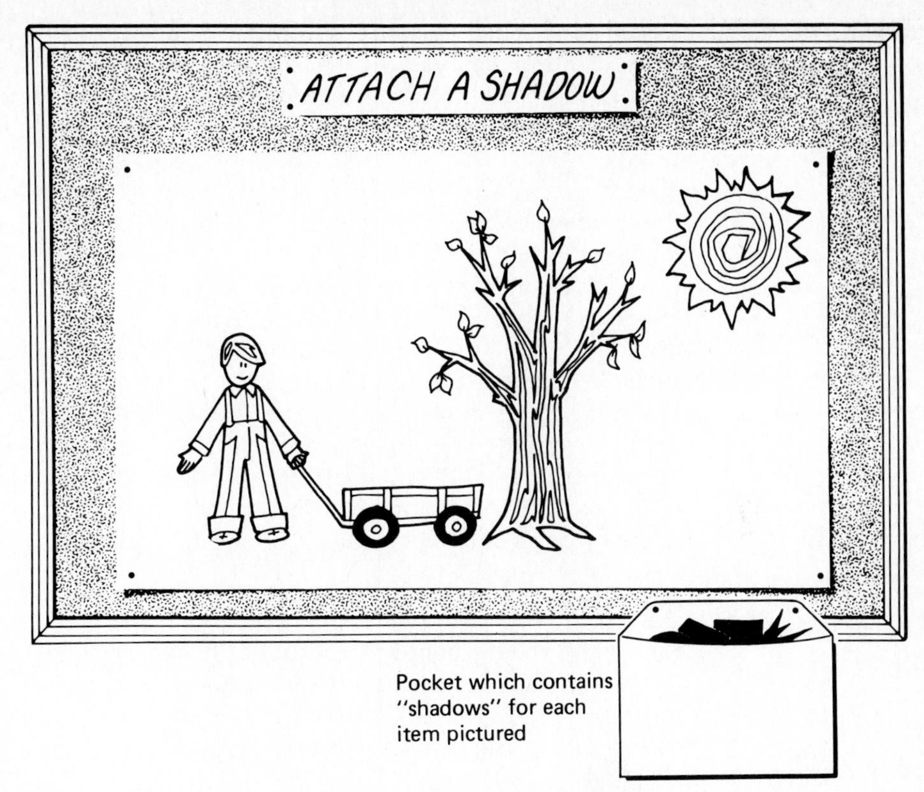

Figure 7-5. Example of an activity bulletin board.

ing each student helps the teacher know whether or not the student has developed the right concept of the relationship between light source, object, and shadow.

Information bulletin boards can also be appropriate. However, unless they are dramatic or are called to students' attention by the teacher, students may not learn as much from informational bulletin boards as from activity bulletin boards.

Examples of and suggestions for bulletin boards are found in each of the chapters in Parts Two, Three, and Four of this text.

Games and simulations

Educational games and simulations have become popular activities for use in achieving educational objectives. Educational supply houses offer many good educational games and simulations. Once a teacher has become familiar with a few educational games and simulations, he or she can produce many more for use in teaching various subjects. "Games" and "simulations" are labels which are often used interchangeably. However, they are generally differentiated by two characteristics: Games usually incorporate some element of chance, while simulations usually include role playing.

The element of chance in a game may be throwing dice, spinning a spinner, or drawing a card. One such game is presented in Chapter 16,

Oceans and Their Resources. Similar games could be made up for organisms of other ecosystems. In addition, card games can be made up for almost any topic. Students can be asked to match similar named pictures or to match a named picture with an information card. Board games usually involve moving a marker along a path, as on a Monopoly board, thus encountering rewards or penalties which give information about the topic being taught. Such board games are often published in teachers' journals.

As mentioned above, simulations generally involve role playing (see below). Often the format of a simulation is that a governing body—for example, a city council—has to make a decision on an important topic, such as land-use zoning. In this simulation, students take roles such as city council member, Sierra Club member, land developer, unemployed worker, retired worker, and the air and wildlife of the area. Students research the issue from the point of view of their role and present the results to the city council. Simulations can be invented for many other science topics related to issues of social importance.

Role playing

Role playing is a popular technique. It can be used for clarifying values (see below) but is also useful in many other ways. To introduce students to role playing, the teacher can design a role-playing situation and indicate some of the initial attitudes pertaining to the roles; for example:

"Mary, you play the role of a company owner. Bill, you are a company employee, and you have just learned that you have a serious illness which is probably caused by the dusty and smoky working conditions at the company. Betty, you are presently unemployed and want to get a job at the company to support your family. Joe, you are the leader of a local environmental quality group. Each of you is to react to the proposal that the company must either spend a very large amount of money to reduce air pollution or close permanently."

But after the students have become more familiar with the role-playing techniques, the teacher can simply present a situation or an issue and have them take it from there. Students will often express a need for more information. Great! We can learn more and develop more meaningful values if we are better informed. Ask the students how they might be able to get more information (especially, encourage suggestions other than "Look it up in a book" and "Ask someone"), and then help them carry out their suggestions.

Do This:

Choose any science topic and briefly describe one game or role-playing situation that could be used for teaching it.

Do This:
Working with three classmates or fellow teachers, develop a learning center (see below) for a selected science topic. Include at least one activity from each of the types described in this chapter.

Learning centers

Learning centers can vary greatly in scope. They can focus on a very narrow concept and include a few activities which address that objective, or they can focus on a broad topic and include activities which address several concepts or processes. A learning center can be concerned with a topic as broad as deserts or as narrow as how to care for a pet hamster.

To develop a learning center, follow these procedures:

1. Identify the purpose of the center.
2. List the objectives (conceptual, process, and attitudinal) of the center in terms of what the students are to learn.
3. Identify or create activities to help students achieve the objectives.
4. Design evaluation techniques which allow you and the students to know when they have achieved the objectives.
5. Plan for modifications of the center based on students' achievements and reactions.

A learning center can incorporate whole units of study. To design an entire unit in a learning-center format, so that individual students can select activities which appeal to them, will be a very big job, but it will also offer many rewards.

In general, it is much easier to modify activities designed for individuals so that they can be used by a whole class than it is to modify activities designed for whole classes so that they can be used by individuals. Thus, you should initially plan activities for individual use, so that you will have flexibility in teaching. Students who prefer working individually most of the time will be free to do so, while students who prefer or need more direct help from a teacher can get assistance.

Any of the materials or activities described in this chapter can be incorporated into activities for learning centers. In developing a learning center, set yourself a goal to provide several activities keyed to each of the class objectives, so that students can select the activities they prefer. You may not be able to achieve this goal the first time you put a given center into use; but as new activities become available, you can add them.

Value-clarification activities

Values cannot be taught; they can only be acquired. However, some teaching activities can offer opportunities for learners to clarify their own values. It is *not* the role of the teacher of science to identify values to teach. It is, however, the teacher's responsibility to offer students chances to consider possible choices. Then, if students want to, they can form their own values.

Teachers only deceive themselves if they try to avoid the issue of value clarification. It is much more realistic to consider what is known about values and value clarification and then to use appropriate teaching strategies.

Raths et al. include the following seven requirements for the process of developing values:

1. *Choosing freely.* What is chosen must be a result of free choice.
2. *Choosing from among alternatives.* No choice can be made if there are no alternatives available.
3. *Choosing after thoughtful consideration.* The consequences of each alternative must be weighed.
4. *Prizing and cherishing.* We must be happy with our values.
5. *Affirmation.* We must be willing to affirm our values publicly.
6. *Acting upon choices.* Our values must be reflected in our lives.
7. *Repetition.* Actions as a result of value choices tend to be repeated and to make a pattern in life.*

In this section we will consider a few value-clarifying techniques. For more techniques as well as for a much more thorough treatment of the topic, consult the book by Kohlberg and Turiel listed in "Suggested Readings."

VALUE CONTINUUM

In this technique the teacher identifies for students two polar positions on a selected issue. Here are examples:

Very bad for health _______ SMOKING CIGARETTES _______ No harmful effects
Avoid at all costs ________ NUCLEAR-GENERATED POWER ________ No dangers
Necessary ______________ WILDERNESS AREAS ______________ None needed

The teacher then asks students to discuss alternatives that are available between the two extreme positions. This is a very useful value-clarifying technique in that it makes students aware of many possible value choices on the selected issue. They may develop value choices of their own as a result of the discussion. After an appropriate level of trust is achieved in the class, you may wish to offer students the option to identify their positions by placing their initials on a continuum line for that issue on a bulletin board.

*Louis E. Raths, Merrill Harmin, and Sidney B. Simon, *Values and Teaching*, Merrill, Columbus, Ohio, 1966.

ON THE SOAPBOX

Getting up on a "soapbox" gives students an opportunity to express their positions on science topics which are being studied. The questions below can serve as a catalyst for soapbox discussions. Ask a question and allow each student who wishes to speak as much time as necessary to develop his or her points.

1. When we travel to other planets, do you think we will find life there?
2. Should the amount of energy each household uses be limited?
3. Who should be responsible for cleaning up water pollution?
4. Should the government support research on nuclear energy? Solar energy?
5. Should hunting wildlife be permitted?
6. Should cigarette smoking be outlawed?
7. How could you identify a scientist?
8. Would you like to be a scientist? A weather forecaster? A geologist? A wildlife biologist?
9. If you could visit any environment that you chose to study, which one would it be? Why?
10. Are some kinds of plants more important than others?
11. Will science provide solutions for energy problems?

GROUP DISCUSSION

Discussion in small groups can help students to identify their feelings about controversial areas of science. The teacher selects a group of students who are interested in a controversial issue and asks the group members to identify their value positions between the extremes on the issue. The group may then, if they wish, share their positions with the whole class.

Here are some examples of controversial issues:

Use of animals for research
World population
Use of nuclear energy
Use of drugs
Colonizing space
Land-use zoning
Pollution control
Food from the sea
Weather modification

VOTING

It is sometimes helpful to allow students to vote on a number of issues. First, secret ballots can be used for each issue, and then students can inverview other children (outside the class) or selected adults (e.g., parents, teachers, factory workers) to find out their views on the issue.

If your students are very interested in a given topic, select several students to study the issue and "campaign" on each side. Then have them vote on the issue again. Ask them to suggest reasons for the change, or lack of change, between the two votes.

Do this:

Choose a science topic, and describe one way in which each of our four value-clarification techniques could be used in teaching it. Remember, the techniques are: (1) the value continuum, (2) the "soapbox," (3) group discussion, and (4) voting.

RESOURCES FOR THE TEACHING OF SCIENCE

In teaching science in the elementary school, a number of resources can assist the teacher. One of the most readily available resources is the class textbook. Other books, resource persons, free and inexpensive materials, and teacher-made files can also be very helpful. The teacher's own imagination is another very important resource; don't forget to use yours.

The elementary science textbook

Many elementary science textbooks are published, and the following aids usually appear in the teachers' editions: background information on the unit topic, activities, questions which help to identify important concepts or processes related to the topic, a bibliography of related children's books, a bibliography of references for teachers, a list of related films and filmstrips, a list of learning objectives, and suggestions for evaluating students' achievement. Thus, the elementary science textbook can help you in several ways. Whether or not you choose to use a given textbook as the primary source for your science curriculum, referring to it can be useful.

Other books

In addition to the science textbook which you may use in your classroom, you will find that other elementary science textbooks can be a rich resource of ideas and activities. A collection of teachers' editions of textbooks is also invaluable. Begin as soon as possible to form your own personal collection. When you begin teaching, also ask your principal or supervisor whether a collection of teachers' editions is available in the school system.

Secondary school science books can also be valuable references for the elementary school teacher. They can give you additional information to strengthen your background in a selected topic, or they can serve as a reference when you need to answer a student's question. Trade books, often found in an elementary school library, can also be used as sources of information.

Resource persons

If you visit the classroom of the most dynamic teacher you know while a resource person is interacting with the class, you may be surprised at how much "extra" attention the classroom visitor will get from students. Resource persons are useful for creating a change of pace, which students enjoy, but they can be most useful as "experts" on their topics.

Resource persons can be parents, students, other teachers, or other persons such as those identified in "Teachers' Files" later in this chapter. Resource persons, like other resources, should be used only to help students achieve the educational objectives for a given unit of study.

The following checklist will help you to plan your use of resource persons:

Checklist: Resource persons

____ **1.** Establish educational objectives (often taken from a unit plan).
____ **2.** Identify and evaluate potential resource persons.
____ **3.** Tentatively select a resource person.
____ **4.** Obtain approval from your administration.
____ **5.** Invite the resource person to come to your class.
____ **6.** Discuss details with the resource person:
    ______**a.** Describe your students and the classroom environment.
    ______**b.** Agree upon topics to be covered in the visit.
    ______**c.** Arrange the time and place, and give the person directions on how to reach the school, if needed.
    ______**d.** Discuss any special equipment that will be needed and make definite arrangements about who will provide it.
____ **7.** Tell your students the visitor is coming.
____ **8.** Discuss with the students questions which they may want to ask the resource person.
____ **9.** If appropriate, assign special tasks to students.
____**10.** Consider having the students do special reports or projects after the visit.
____**11.** Write a thank-you letter to the resource person.
____**12.** Evaluate the experience:
    ______**a.** Did the students learn what you wanted them to learn?
    ______**b.** Was the time invested reasonable?
    ______**c.** Should this person be invited another year?

Do This:

Select an appropriate educational objective for a science class. Use the checklist to plan for a resource person.

The use of a resource person can be enlightening to both you and your students. Frequently resource persons give students ideas about career opportunities; and seeing your students' responses to the visitors may give you new ideas about effective ways of helping them.

Free and inexpensive materials

To create a science program that will get students actively involved in learning, you will need materials. Aware of the limited funds normally available to the elementary school teacher, textbook publishers tend to include in elementary science textbooks activities with easily obtained and inexpensive materials. The materials needed are frequently available from the publisher or a science supply house for a relatively low cost. However, teachers can usually save money by collecting the materials themselves.

Begin by identifying inexpensive, easily obtained materials which will substitute for materials or equipment called for in the textbook. Ask students to suggest substitutions. Allow the students to gather as many materials as possible—they really do want to help you. They want and need recognition and a sense of accomplishment. Obviously, allowing students to help also conserves your time and makes it possible to collect materials for many more learning activities than you could collect by yourself.

In gathering materials for a science unit, you should consider the following checklist.

Checklist: Free and inexpensive materials

____**1.** Make a list of the required materials.
____**2.** Identify free or inexpensive sources, as well as free or inexpensive substitutes for the materials on your list.
____**3.** Have students gather as many materials as possible.
____**4.** Gather or order the materials which you must personally obtain.
____**5.** Arrange the materials in a suitable spot in the classroom.
____**6.** After completing the unit, store reusable materials in a suitable box. Write the unit title and a list of contents on the side of the box that will be visible while it is in storage.

Many government agencies, businesses, organizations, and special-interest groups offer teaching materials which are either free or very inexpensive. Printed material can be obtained with relatively little effort. All you have to do is write a few postcards or letters. Books which identify free and inexpensive teaching materials are listed in "Suggested Readings," and other free and inexpensive materials are identified in Parts Two, Three, and Four of this book. Such materials are also identified monthly in several teachers' journals.

The materials you and your students gather can be used by the students for reference, as background information, on bulletin boards, and in some cases as learning packages or activities.

Before you use free and inexpensive materials in the classroom, you

Many governmental agencies, private industries, and professional organizations supply free or inxpensive reference materials such as those pictured.

must carefully evaluate each piece of material: Is it merely a promotional tool for some narrow point of view (such as a heavily pro-industrial or anti-industrial viewpoint on environmental pollution)? You might use materials with two opposed viewpoints if you judge that your students are capable of evaluating them.

Do This:

Make a collection of at least 20 free and inexpensive materials obtained from government agencies, businesses, organizations, and special-interest groups, as well as from similar local sources.

Design a filing system for the materials. It is often convenient to file materials under the unit titles which commonly appear in elementary science textbooks.

File the materials in a sturdy cardboard box. (It will then be ready to move when you do.)

Indicate what educational objectives at least six of your materials would be useful for, and evaluate their suitability for use in the elementary classroom.

Sites for local field trips

Communities offer many opportunities for field trips which can be useful in teaching science. Plant and animal life can be observed in parks. Cemeteries often offer the same possibilities as parks and are usually accessible if the persons in charge of them are confident that students will behave appropriately. Water-treatment plants, power-generating stations, sewage-treatment plants, and solid waste disposal facilities usually allow visits by elementary classes; such visits offer illustrations of lessons which may have been studied in the science class.

A visit to a nearby farm will not only be fascinating but also demonstrate how people manage plants and animals for our benefit. Visiting a pet shop will help make students aware of the diversity of life and also can help them learn how to care for classroom or home pets. Once you begin to think of ways local businesses and industries can be used to illustrate science concepts, you will probably identify more field-trip sites than you can ever visit. If the whole class is unable to visit certain sites, you may choose to assign individuals or small groups of students to make the visits and report their experiences to the class.

An exciting possibility which should not be overlooked is local museums. Large, well-known museums are excellent, but we should not overlook the smaller museums which are too often unknown to teachers but which eagerly welcome visitors and often provide delightful hands-on experiences. For example, many small communities have museums of natural history which focus on the local environment.

In preparing for field trips, refer to the checklists for outdoor activities and resource persons.

Teachers' files

It is a rare person who never says, "If only I could remember the name of the person who can . . ." There is a way to solve this problem. The investment is small and the dividends can be great. All you need are index cards, a file box, and the commitment to record important information. Whenever anyone offers to do anything for you related to your teaching responsibilities, immediately record the person's name and address and a description of what was volunteered on an index card. But do not just wait for people to volunteer; seek out helpers. Most people are more than happy—they are even proud—to help schools and teachers educate children, if the requests are reasonable.

Many kinds of resource files can be made by the teacher. Two kinds, resource persons and teaching ideas, will be described.

FILES ON RESOURCE PEOPLE

A possible format for recording information on resource people is shown in Figure 7-6. Such a filing system can be organized in categories reflecting the unit titles in elementary science textbooks—"Animals," "Plants," "Ecology," "Weather and Climate," and "Astronomy," for example.

ANIMALS

Jones, David
125 South Brazos Drive
Ourtown, Texas 77843

Raises animals for pet shops. Volunteered to donate some animals and loan others for classroom use.

Can talk with students about caring for pets and about unusual animals.

Figure 7-6. File card for a resource person.

Here are some resource people whom you may want to include in your file:

1. *Parents of your students.* Send out a survey form. Tell parents you would like to know about their occupations and special interests or hobbies because, if possible and appropriate, you may call upon them to share these things with your class. Each year, after the forms are returned to you, enter the information on an index card of a distinctive color, so that you can readily identify each person as a parent of either a present or a previous student. Don't forget to record the child's name on each parent's card.
2. *Your students.* Your students have special interests and hobbies. One of them may just be the expert you need for some science topic (e.g., almost every classroom has at least one expert on dinosaurs). Have all your students fill out survey forms and place cards (of the appropriate color) for them in your files.
3. *Other students.* Perhaps youngsters who were previously your students can help with a special topic. Or ask teachers—they may be able to suggest one of their students. High school and college students are other possibilities.
4. *Other teachers.* Most people—including your fellow teachers—have special areas of expertise. As these become known to you, make file cards. Do not overlook one very rich resource—retired teachers.
5. *Professional organizations.* Many professional organizations have speakers' bureaus. Also, consider government agencies and civic organizations.
6. *The yellow pages.* The yellow pages of your telephone directory can help you find businesses or organizations which may supply or identify resource persons for science topics.

Do This:

While you are in school: Select two science topics which are commonly taught in the elementary school, and list five possible resource persons who could be used for each topic. How were you able to identify each resource person? Do you think similar sources will be available to you when you are teaching?

After you start teaching: Meet with the other teachers at your grade level for 2 hours and construct lists of potential resource persons for four science topics. Each teacher can contact one or more people on the list to see if they are willing to serve as resources. Make a permanent file of those willing to serve.

IDEA FILES

Teachers, both while they are studying and after they begin teaching, often have "really great" ideas for activities, but then they forget about them. An idea file can come to the rescue. An idea file, like a resource file, will probably be most useful if the filing categories reflect units or topics to be studied.

Whenever you are exposed to or think of a good teaching idea, write a brief description of it on an index card and file it. The most creative time for designing new activities is usually while you are teaching a given subject or just after you finish it. Write those ideas down—or else you will probably not remember them. Then, the next year, when you begin to design the unit on the same topic, get out the file and surprise yourself with your own good teaching ideas.

Many other kinds of files can be useful—ideas for bulletin boards and field trips, activities for a rainy day, activities for the first day of school, children's books on science topics. If you think we are suggesting that teachers carry a few index cards with them at all times, you are correct.

An offer: If you decide to give up a file that you have started, send the cards with all your good ideas recorded on them to us (the authors of this book), and we will send you an equal number of blank cards (limit 500 per customer). Most teachers would rather give up an arm or a leg than the valuable ideas they have recorded.

Do This:

Develop a classification system for filing teaching ideas, and begin filing ideas which really appeal to you.

SUMMARY

A proper attitude toward teaching science is essential for maximum effectiveness. When teachers are enthusiastic, their students become excited. When teachers are curious, their students become inquisitive. And when teachers want their students to experience deep enjoyment from science, their students are likely to reap positive rewards from their studies.

To do good teaching, you must put your good attitudes into action. You must transform some of your emotional energy into kinetic energy. You must learn how to select stimulating activities and organize them to achieve the goals of each teaching unit. You must design your classroom so that it will be a stimulating environment.

As you prepare to teach science to elementary students, you will want to start building a good resource library and a repertoire of activities. A good filing system will help you build and maintain superior resources: books, ideas for bulletin boards, activities, and people you can invite to visit your class.

RECAPITULATION OF MAJOR PRINCIPLES

1. Excitement about teaching can help sustain teachers through occasional difficulties.
2. As teachers, we must let our students know that we are dedicated to helping all of them become the very most that they can become.
3. Successful teachers identify important learning goals, get organized, apply a reputable theory of how children learn, use a variety of teaching methods and activities, and allow themselves to be human.
4. Activity cards are designed for use by individual students. Each has a title and an illustration on one side and suggested activities on the other side.
5. Game cards, for use with or without an electronic game board, can help students to learn or review information.
6. Guided imagery uses enjoyable and powerful techniques which engage the right cerebral hemisphere.
7. Shoebox science kits contain descriptions of educational activities as well as the materials necessary to perform the activities. They are often used in interest centers.
8. Outdoor activities do not have to be exotic experiences which take a class far from school. Many can be conducted on the school grounds in 10 to 20 minutes. A few simple guidelines facilitate the planning of effective outdoor activities.
9. Activity bulletin boards give students something to do and thus increase the probability that they will learn.

10. Many educational games and simulations are available. Simulations generally involve role playing. Games frequently have game boards and include an element of chance.
11. Role playing is a popular technique with many kinds of applications.
12. Learning centers can be very narrow or very broad in scope. To develop a learning center: identify the purpose, list learning objectives, identify or create activities which address the objectives, include evaluation techniques, and plan to incorporate future modifications. Learning centers can include any of the activities and techniques described in this chapter, as well as others.
13. The value continuum, the soapbox, group discussions, and voting are four value-clarifying techniques described in this chapter. Role playing can also be used to clarify values. Value-clarifying activities enable students to consider a variety of possible viewpoints on selected topics and develop their own values if they choose to do so.
14. Teachers' editions of elementary science textbooks often contain helpful background information for the teacher on the unit topic; a description of activities; bibliographies of related children's books, references for teachers, and films and filmstrips; a list of learning objectives; and suggestions for evaluating student achievement.
15. The use of resource persons can be made more effective by following the "Resource Persons" checklist.
16. Free and inexpensive materials may be used in many activities.
17. Students want to help the teacher to collect materials, and teachers should encourage them to help.
18. There are many easily accessible sources of free and inexpensive teaching materials.
19. Resource people for science classes include parents of students, your current students themselves, your former students, other teachers, members of professional organizations, and representatives from local businesses and industries.
20. Teacher-made files are valuable tools to aid the teacher's memory in identifying resource persons, teaching ideas, ideas for bulletin boards, activities for a rainy day, activities for the first day of school, and children's books on science topics.

SUGGESTED READINGS

Abruscato, Joe, and Jack Hassard: *Loving and Beyond: Science Teaching for the Humanistic Classroom*, Scott, Foresman, Glenview, Ill., 1976.

Cornell, Joseph: *Sharing Nature with Children*, Ananda, 1979.

DeVito, Alfred, and Gerald H. Krockover: *Creative Sciencing: A Practical Approach*, 2d ed., Little, Brown, Boston, 1980.

Grady, T., ed.: *Free Stuff for Kids*, 3d ed., Meadowbrook, 1979.

Gross, Phyllis, and Esther Railton: *Teaching Science in an Outdoor Environment*, University of California Press, Berkeley, 1972.

Harmin, Merrill, Howard Kirschenbaum, and Sidney B. Simon: *Clarifying Values through Subject Matter*, Winston, Philadelphia, 1973.

Harty, Sheila: *Hucksters in the Classroom: A Review of Industry Propaganda in Schools*, Center for the Study of Responsive Law, 1980.

Hendricks, C. G., and Russell Wills: *Centering Book: Awareness Activities for Children, Parents and Teachers*, Prentice-Hall, Englewood Cliffs, N.J., 1975.

Houndshell, Paul B., and Ira R. Trollinger: *Games for the Science Classroom*, National Science Teachers Association, 1977.

Kohlberg, L., and E. Turiel, eds.: *Recent Research in Moral Development*, Holt, New York, 1973.

Raths, Louis E., Merrill Harmin, and Sidney B. Simon: *Values and Teaching*, 2d ed., Merrill, Columbus, Ohio, 1978.

Russell, Helen Ross: *Ten-Minute Field Trips*, Doubleday, Garden City, N.Y., 1970.

Samuels, Mike, and Nancy Samuels: *Seeing with the Mind's Eye*, Random House, New York, 1975.

Staterstrom, M., ed.: *Educator's Guide to Free Science Materials*, 22d ed., Educator's Progressive Service, 1981.

See also

1. Teachers' journals such as *Science and Children* and *Instructor*, in which free and inexpensive materials are identified in advertisements as well as in special lists.
2. Elementary and middle school science textbooks (teachers' editions).

CHAPTER 8

RELATING SCIENCE TO OTHER ELEMENTARY CURRICULUM AREAS

PREASSESSMENT

	AGREE	DISAGREE	UNCERTAIN
1. Limited opportunities exist for relating elementary science to other curriculum areas.	____	____	____
2. Science concepts are often the topics of children's books.	____	____	____
3. A good science curriculum could be developed using children's literature as the only reading resource.	____	____	____
4. Using reference skills is important to the scientist as well as to the elementary student.	____	____	____
5. Sets of recordings can be used to teach scientific concepts.	____	____	____
6. Copying diagrams is a valuable way to gain information.	____	____	____
7. Measuring is a mathematics skill which can be reinforced or made more meaningful in the study of science.	____	____	____

OBJECTIVES

Upon completion of this chapter you should be able to:

1. Identify several advantages of integrating study in other curriculum areas with study of science.
2. Name one way children's literature can be used to teach science and one way science can be used to encourage students to do extra reading.
3. Describe several ways to improve language skills while studying science.
4. List two ways to use music in teaching science.
5. Identify several art activities which can be related to the study of science.
6. Name several mathematical skills which can be practiced in the study of science.

INTERRELATING THE ELEMENTARY CURRICULUM

There are almost limitless possibilities for relating other curriculum areas to science. One of the real advantages of a self-contained classroom is the convenience with which the teacher can interrelate the various curriculum areas. Interrelationships can be created under other circumstances also, if the teachers work together to do the required planning. For instance, if a science teacher and a literature teacher are both aware of a reading in the students' literature book that is related to a science topic being studied, they may plan to use the reading in conjunction with the science study—or as an alternative, the reading may be used later as a reinforcement or an extension of concepts acquired in science classes. Similar plans can be evolved to take advantage of relationships between any of the other curriculum areas.

In this chapter we will present some further possibilities for interrelating science with other curriculum areas. Given a few examples, you will probably begin to think of many additional ways of relating science to other areas. Do not neglect to record and file your ideas.

One of the advantages of relating science to other curriculum areas is that such relationships can reinforce and extend concepts learned in the various areas. Another advantage is that interests students develop while studying one area can serve as motivators in other areas. Almost all students have a natural interest in science topics. A special interest in weather, for example, can motivate a student to do more reading, practice mathematical skills related to weather, consider the effect of weather on people, etc.

CHILDREN'S LITERATURE AND SCIENCE

Children's literature can be used in several ways to teach science. One way is to have students read a book (or a set of books or stories) relevant to the science topics to be studied. At appropriate times during or after the reading, you can introduce activities which relate specifically to the science topic. For example, if students read Theodore Taylor's book *The Cay*,* a story of the survival of a young boy and an old man on a raft and later on a small Caribbean cay (a reef or low island), they can do activities related to food chains, hurricanes, or marine animals. Another example is Jean Craighead George's book *My Side of the Mountain*,† which could be very easily used in a study of ecology.

A second way of relating children's literature to science—perhaps more direct and more easily used as a first effort—is first to select a science topic for study and then to seek children's literature which is related to the topic.

*Theodore Taylor, *The Cay*, Avon, New York, 1969.

†Jean Craighead George, *My Side of the Mountain*, Dutton, New York, 1975.

These are but a small sample of children's books which can be incorporated into science teaching.

An excellent example would be using the Dr. Seuss book *The Lorax** in conjunction with a study of ecology. After reading the book to the class, you could ask students to identify the ecological lessons in the book. Other examples would be using *Mickey's Magnet* by Branley and Vaughan† in studying magnets and using Margaret Graham's book *Be Nice to Spiders*‡ in a unit on animals, insects, or ecology.

Each chapter in Parts Two, Three, and Four contains a list of children's books (see the sections headed "Building a Classroom Library"). Books from those lists can be used in relating children's literature to science. Librarians in your school and at the local public library can help you to find other children's literature on science topics.

Establishing a "library corner" in your classroom, with a collection of books and magazine articles related to the topic being studied, is a good way to encourage your students to read. Books from public and school libraries make a good start for such collections—or if your students have free access to the school library, a special collection can be placed there, and the school librarian can check out books to the children. For a classroom library corner,

*Theodore S. Geisel (Dr. Seuss), *The Lorax,* Random House, New York, 1971.

†Franklin M. Branley and Eleanor Vaughan, *Mickey's Magnet,* Crowell, New York, 1956.

‡Margaret B. Graham, *Be Nice to Spiders,* Harper and Row, New York, 1967.

you can ask students to help you find additional books or other readings about the topic being studied.

Not all students will do extra reading on all subjects, but most or all children will be fascinated by topics that particularly appeal to them. If reading material at the appropriate level of difficulty is readily available, they will almost certainly do some reading that they would not otherwise have done.

LANGUAGE ARTS AND SCIENCE

Speaking, listening, writing, and doing research—all important language skills—can be readily practiced in the study of science. After children get some hands-on experience in science, it is usually not difficult to involve them in a discussion about what they did. After participating in a guided fantasy, they often find it easy to talk or write about the imaginary trip.

When you take students on a nature hike, you may want to tell them to use their senses of hearing, smell, taste, and touch—as well as the overworked sense of sight—and once again, it should be easy for them to make oral or written presentations about what they sensed. Experience with real objects or phenomena is an important basis for practicing and improving language skills.

Learning how to do research, or use library references, is especially important for elementary students. Children need to develop such skills early in order to be prepared for further study in the fields of their choice—and some of your students may be budding scientists. Use of reference materials is an essential part of scientists' work; it can also become routine practice in the study of science in elementary school. It is your responsibility, as a teacher, to see to it that students develop basic research skills. But if you help them too much with their library work, you will be robbing them of opportunities to learn. Instead, let them find the information they need, report on how they found it, and then report on the information itself.

Scientists do not obtain all their information from books, but they do not obtain all of it by direct observation either. You will sometimes need to guide students to the most efficient way to find the information they need. For example, suppose that in their enthusiasm about studying animals, your students decide they would like to watch some chicken eggs hatch. If they could bring a hen into the classroom and create a suitable environment, or if they could obtain fertile eggs and set up an incubator, their curiosity would be fully satisfied by the results. However, in this instance it would probably be more practical for you to guide them to consult library references.

MUSIC AND SCIENCE

One obvious way to relate music to science is to find songs which relate to science topics. If your school has a music teacher, he or she can help you to

Several popular and folk songs relate to science topics.

locate appropriate songs. Surveying songbooks and asking students are other ways to find useful songs.

Folk songs and some popular songs frequently relate to science topics. For example, to introduce a unit on changing weather and its effect on people, three folksongs, "Western Home" (better known as "Home on the Range"), "Lane County Bachelor," and "Dakota Land" (all written between 1870 and 1890), can be presented to or sung with students. They can then be asked to consider why the first two songs were positive in outlook while the third was quite bleak. (*"Dakota Land" was written during a period of drought.*) A study of changing weather and its affect on people can then follow.

Songs about the sea, mountains, flowers, and a large number of other topics can be used in relation to the study of science topics. To introduce a unit on mountains, for example, you might tell your students, "Next week we will begin a study of mountains. Will you please bring to class songs which relate to mountains so that we can play or sing them together." Many students, or their parents or older siblings, have record collections and will be willing to share them with the class.

When you use music in your classes, have your students share what the songs mean to them or what the songs tell them about the relevant science topic. You should also share with them the meanings that the songs have for you.

ART AND SCIENCE

Sketches or drawings can often be used to represent or interpret scientific information. Sometimes the drawings represent specific data, but often they can also be more imaginative.

One way to have students notice more about a given object—for example, a plant—is to have them make a sketch of it. Even though to others the picture may have only a faint resemblance to the object, it will contain information which is meaningful to the artist. If you give students frequent opportunities to practice drawing but do not pressure them to achieve perfection, their drawings will gradually come to resemble more closely the real objects they draw.

Copying science diagrams from a book would be boring for students, but making their own sketches of an object and then comparing them with a diagram in a book can be very educational, allowing children to notice similarities and differences between their drawings and the "ideals" illustrated in the book.

Murals which represent a science topic (e.g., the solar system, a scene from the ice age, or a habitat such as the desert) can be both educational and aesthetically appealing. Shoebox dioramas can also be created, to present information on science topics. Making paintings with chicken feathers is an unusual and interesting experience which also affords students an opportunity to learn something about feathers. Painting rocks to represent animals or using seashells to make pictures not only allows children to create art objects, but also will prompt many children to learn about the materials with which they are working.

Additional ideas for using art in science classes are given in Parts Two, Three, and Four of this text.

MATHEMATICS AND SCIENCE

Can mathematics and science be combined? Of course! Ways to relate the two areas range from counting to measuring to creating and interpreting graphs. For example, given a set of rocks, seashells, or plants, students can count the objects and divide them into subsets, according to any of several possible bases, including size, weight, and texture.

Skill in measurement can be gained in both mathematics and science classes. Students should first learn to understand the need for standard units of measurement (meters, feet, kilograms), as well as the need for multiples and divisions of the standard units. Then, practice with real objects will make students skillful in both estimating the measurements of objects and actually measuring them. Measurement of mass, volume, length, temperature, and time can be incorporated into various science activities; it can also be addressed as a topic in its own right.

Graphing skills are especially useful in science. The growth of a plant can be represented by cutting a strip of paper equal to its height each day and gluing the strips to a graph. Such activities make beginning graphing a meaningful skill with little mystery about it. Recording data on the heights or weights of students on a graph is another way to help make graphing comfortable and familiar to students.

Many educators in mathematics are urging the use of concrete experiences to teach mathematics skills. Such experiences can be especially meaningful if real examples from science are used.

SOCIAL STUDIES AND SCIENCE

The linkages between social studies and science may be more numerous than those between science and any other curriculum area; they may also be the linkages which are most identifiable in the student's real world. Interactions between the natural environment and people's lives constitute the tie between science and social studies.

The social studies which come most readily to mind are geography, history, economics, political science, sociology, and anthropology. Many interrelationships between science and the social studies are too complex for thorough exploration in elementary school, but it is easy to find interrelationships between the two which are appropriate for study by elementary school students. The influence of the physical environment—for example, the locations of rivers and large lakes—upon the settlement and development of a country has often been considered in elementary school. The use and abuse of living and nonliving natural resources is a topic in which ecology and environmental education bring science and social studies together. Weather and climate studies are also very influential in peoples' lives and have had an extensive influence upon history.

A science topic which has had profound influences upon almost all areas of social studies is energy, including such everyday applications as transportation, heating, and cooling. Petroleum resources, for example, are used as a bargaining tool in economics and politics; in addition, they have significantly affected the lifestyles and histories of countries with rich oil supplies.

HEALTH AND SCIENCE

Health and science are so closely related that in many schools health is taught as a part of the science curriculum. At other schools, health is a separate subject; but actually, any attempt to separate the two is likely to result in an artificial health curriculum which is taught only as a collection of facts.

It takes special effort to prevent health from becoming just a body of information to be memorized, repeated on a test, and forgotten. Teachers must explain health principles in terms of the human body—including psychological and emotional as well as physical factors. The connection between alcohol, tobacco, and drugs, for instance, and what is known about the human body and human behavior is usually made quite clear to students. Unfortunately, other health topics are often taught without making this connection. For example, a health unit may discuss obesity, but the probability is low that it will explain the dangers of overweight in terms of the additional miles of veins and arteries that are necessary to serve the extra pounds, and the additional work load that these miles place on the heart.

METRICS AND SCIENCE

Another area of study that is so closely related to science that it should be considered inseparable is metrics. Special effort has been made throughout this text to provide both metric and nonmetric units wherever measurements have been given. A comparison of the ease and accuracy of computing in metrics with the difficulty and inexactness of working with nonmetric units of measure will clearly demonstrate the need for teaching students to think in metrics.

USING TABLES OF SPECIFICATIONS TO RELATE SCIENCE TO OTHER SUBJECTS

In this chapter we have emphasized the need to relate science to other subjects. Furthermore, we have said that the opportunities for doing so are almost limitless—but these opportunities will not arise in your classroom unless you plan them. Careful planning is needed to ensure extensive integration of science with many subjects.

You can use a table of specifications (see Table 8-1) to ensure that science is tied to all pertinent areas of your curriculum. "Pertinent" is the key word here; each teacher must select the subjects which are to be related to science in his or her own classes.

The answer to the question "What subjects should be linked with science?" is somewhat individualistic, but all teachers should consider certain factors. Tables 8-1 and 8-2 can be used to identify the subjects which should be related to science. Use of these tables can also ensure that no pertinent subject which should be linked to science will be overlooked.

Table 8-1 is a sample table of specifications. In general, tables of specifications are used to ensure coverage of the various levels of content and objectives as measured in the educational taxonomies (see Table 8-2). A table of specifications can be designed especially to ensure adequate linkage of science concepts with other subjects. Notice that the behavioral dimension in Table 8-1 is replaced by specific subjects in Table 8-2.

TABLE 8-1 TABLE OF SPECIFICATIONS

BEHAVIORS

CONTENT	RECALL OF MATERIAL LEARNED A.0	Terminology A.1	Specific Facts A.2	Science Generalizations A.3	Science Concepts A.4	Science Principles A.5	APPLICATIONS OF KNOWLEDGE TO NEW SITUATIONS B.0	Nonquantitative B.1	Quantitative B.2	USE OF SKILLS INVOLVED IN UNDERSTANDING SCIENCE PROBLEMS C.0	Interpretation of Qualitative Data C.1	Interpretation of Quantitative data C.2	Identification of Problems C.3	Identification of Assumptions and Unanswered Questions C.4	Analysis of Scientific Reports C.5	DEMONSTRATION OF RELATIONSHIPS BETWEEN BODIES OF KNOWLEDGE D.0	Comparison D.1	Extrapolation D.2	Application to Other Areas of Science D.3	Interrelations of Facts and Principles in a New Way D.4	Development of a New Set of Interrelated Concepts D.5
Science Principle 1																					
Concept A																					
Concept B																					
Concept 3																					
Science Principle 2																					
Concept A																					
Concept B																					
Concept C																					
Science Principle 3																					
Concept A																					
Concept B																					
Concept C																					
Science Principle 4																					
Concept A																					
Concept B																					
Concept C																					

TABLE 8-2 TABLE OF SPECIFICATIONS TO ENSURE RELATIONSHIPS AMONG SUBJECTS

	ENGLISH	Generalizations (Concepts and Principles)	Facts	HISTORY	Generalizations	Facts	MATHEMATICS	Generalizations	Facts	SOCIAL STUDIES	Generalizations	Facts	LITERATURE	Generalizations	Facts	HEALTH	Generalizations	Facts
CONTENT	A.0	A.1	A.2	B.0	B.1	B.2	C.0	C.1	C.2	D.0	D.1	D.2	E.0	E.1	E.2	F.0	F.1	F.2
Science Principle 1																		
Concept A																		
Concept B																		
Science Principle 2																		
Concept A																		
Concept B																		
Science Principle 3																		
Concept A																		
Concept B																		

IN CONCLUSION: THINKING AHEAD

As you plan your teaching future, think in terms of making it suit you and your talents. Use your background and strengths to maximize the quality of your science classes. Analyze your knowledge in subjects other than science, and take time to list some science principles and concepts that can be related to concepts and principles in other subjects. You will probably be able to list and briefly describe at least one desirable link between science and each subject you have studied in college.

To plan your future, planning in depth is essential so that you can develop your teaching approach in ways that will allow your future students to benefit from the very best experiences you can provide.

SUMMARY

Almost every topic studied in elementary school science has influenced humankind, and thus science and social studies are closely interrelated. It is important to help young students to consider their values in relation to science and the social studies. In addition, educators have long held that students should be taught to make decisions and to solve problems. The links between science and social studies allow these objectives to be addressed.

Relationships between science and any other elementary school curriculum area will almost always enhance the study of both areas. Students have a natural curiosity about their world, especially their environment. The processes used in science and the attitudes important to science are not unique to science; rather, they are used in many other areas of study and thus facilitate the interrelating of science and other areas.

RECAPITULATION OF MAJOR PRINCIPLES

1. Students' natural interest in science topics can motivate them in other curriculum areas.
2. Interrelating studies in two or more curriculum areas reinforces learning in each area.
3. A selection of children's books on a given topic can serve as the core of a science unit.
4. Providing extra reading materials on a science topic can encourage students to do more reading.
5. Language skills—speaking, listening, writing, and using library references—can be readily practiced in the study of science.
6. Popular and folk songs can be used effectively in introducing or teaching science topics.
7. Skills in drawing, painting, and other artwork can be used in the study of science.
8. Among the mathematical concepts and skills that can be practiced in science are counting, grouping, measuring, and graphing.

SUGGESTED READINGS

Chenfeld, Mimi Brodsky: *Teaching Language Arts Creatively*, Harcourt Brace Jovanovich, New York, 1978.

Fuys, David J., and Rosamond Welchman: *Teaching Mathematics in Elementary School*, Little, Brown, Boston, 1979.

Jarolimek, John: *Social Studies in Elementary Education*, 5th ed., Macmillan, New York, 1977.

Kaltsounis, Theodore: *Teaching Social Studies in the Elementary School*, Prentice-Hall, Englewood Cliffs, N.J., 1979.

Moore, Karen: *NOTE: Suggested Activities to Motivate the Teaching of Elementary Music*, Educational Service, Stevensville, Mich., 1973.

Norton, Donna E.: *The Effective Teaching of Language Arts*, Merrill, Columbus, Ohio, 1980.

———: *Language Arts Activities for Children*, Merrill, Columbus, Ohio, 1980.

Striker, Susan, and Edward Kimmel: *The Anti-Coloring Book*, Holt, New York, 1978.

Wankelman, Willard F., and Philip Wigg: *Arts and Crafts*, Brown, Dubuque, Iowa, 1978.

PART 2

TEACHING THE LIFE SCIENCES

"It was the best of times, it was the worst of times, it was the age of wisdom, it was the age of foolishness, it was the epoch of belief, it was the epoch of incredulity, it was the season of Light, it was the season of Darkness, it was the spring of hope, it was the winter of despair, we had everything before us, we had nothing before us, we were all going direct to heaven, we were all going the other way."

Perhaps you recognize the source of these powerful lines: Charles Dickens used them to introduce the readers of *A Tale of Two Cities* to a time of paradoxes. And perhaps you share our perception that Dickens could have been writing about our world today. During the years that have elapsed since 1859, when *A Tale of Two Cities* was written, the world has become, if anything, even more paradoxical—partly because of what has been happening in modern-day science and technology.

Modern scientific achievements have amazed the world. Today we are continually achieving the impossible—automation, birth control, test-tube babies, space travel, organ replacement, supersonic flight. But along with our scientific advancements there has been some decline in the quality of life; and we now live with the knowledge that human beings are able to destroy the earth. The stresses of the modern world affect our well-being in many ways; for example, 1 American in 10 suffers from hypertension.

As part of the bicentennial of the United States, the National Education Association commissioned a study of goals needed to prepare our young people for the twenty-first century. The committee making the study interviewed nearly a hundred young people and found that two goals were paramount: learning to cope with a very difficult world and learning to communicate feelings. Clearly, our young people want their schools and teachers to care about them as human beings.

We believe that anyone who is planning to teach the life sciences must place a high value on life itself and must believe in the worth of the individual. There is something very precious about life—not only human and animal life but also plant life. We believe that students should be led to recognize the beauty and wonder of life. We hope that as elementary science teachers you will accept responsibility for helping children discover the dimensions of life and will yourselves be committed to protecting life and improving the quality of life. In doing so, you can help make the future "the best of times."

CHAPTER 9

PLANTS

IMPORTANT CONCEPTS

Plants are living things.

Green plants manufacture food and produce oxygen.

Cells are the "building blocks" of plants.

Most green plants have similar basic needs—air, water, soil (nutrients), and light.

Plant structures are related to their functions (e.g., roots obtain water and minerals; leaves produce food).

Many plants change through the seasons.

Plants grow where they are best suited to their environment. Different kinds of plants grow in different environments.

Plants, like other living things, have a life cycle—a series of stages through which they pass.

Plants can be classified into groups. One classification scheme divides plants with seeds from plants without seeds.

Green plants manufacture food with the aid of chlorophyll.

Plants reproduce in several ways: some produce seeds; some reproduce using other plant parts.

Nongreen plants get food from other living things or things which were once living.

People use plants for food, clothing, and beauty.

INTRODUCTION: STUDYING PLANTS

The study of plants offers many possibilities for exploring the nature and requirements of living things. A desirable goal is to have students become aware of the dependence of all living things upon plants.

When studying about the life-sustaining requirements of plants, students may be motivated to plant vegetable gardens, or to grow plants for decoration or for important functions such as keeping homes cooler in the summer. Plants come in all sizes and shapes. They range in size from among the smallest living things (algae) to the largest (redwood trees). In studying plants it is easy to include field trips, since plants can be found near the school, ranging from small plants growing in cracks in the sidewalk to trees to the great variety of plants in a vacant lot.

Much of what students learn about plants will be closely associated with the activities and learning in Chapter 12, Ecology. If study of both plants and ecology is not included in your grade level, you may nevertheless wish to use a few activities from the topic which is not included, in order to give students some understanding of the interrelationships.

Study of plants is especially helpful in acquiring skills in the processes of science. Students can *observe* and *describe* parts of plants, characteristics of whole plants, and plant communities. They can give special attention to similarities and differences at each level—which leads to *classifying.* Students can *measure* various phenomena of plants. *Communications skills,* including graphing, may be practiced. *Hypothesizing solutions* to problems can be encouraged. And there are many possibilities for *experimenting* with plants.

PLANTS FOR PRIMARY GRADES

Activities

➢ IDENTIFYING ROOTS, STEMS, AND LEAVES

Wash the soil away from the roots of several different kinds of plants which either you or the students have grown or have carefully gathered from outside. Have children identify the roots, stems, and leaves, and compare the parts of the different plants, noting similarities and differences. Ask them what they think are the functions of the different parts. Comparing real plants with drawings and photographs in a textbook is also helpful.

Figure 9-1. Radish seeds sprout quickly and are excellent plants to study.

➢ OBSERVING ROOT HAIR DEVELOPMENT

Plant some radish seeds; these are among the fastest-sprouting seeds commonly available. As shown in Figure 9-1, use either a dark-colored dish or place dark paper or cloth in a dish. Place several radish seeds on the plate, cover with paper or cloth, and moisten thoroughly; use enough water to

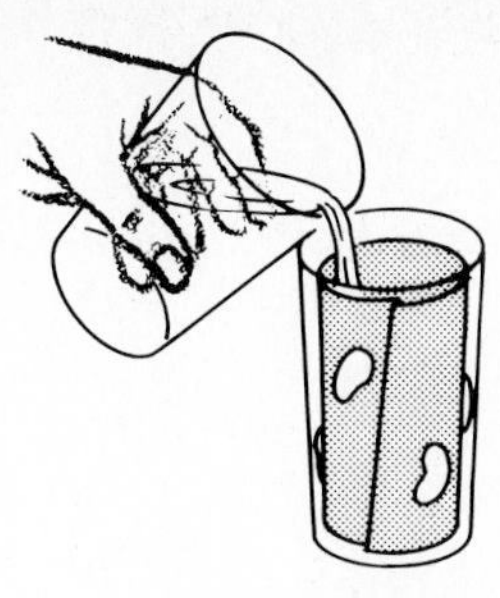

Figure 9-2. By observing lima bean sprouts, students can learn about the direction of stem and root growth. (Source: Adapted from P. F. Brandwein, *Concepts in Science*, Curie ed., Harcourt Brace Jovanovich, New York, 1980, grade 1, p. 6.)

keep the paper or cloth wet for 48 hours, but do not cover the seeds with water. Set the dish in a warm place. You may wish to cover it with an inverted plate (a throwaway pie tin will work very well) to keep it from drying out.

After 24 hours have students examine the seeds with magnifying glasses to see if they have sprouted. Ask students to describe what they see. Repeat the procedure after 48 hours. *(How is what the children see different from the previous day's observation?)*

Have the children identify the leaves and roots of the radish sprouts. Point out how fine the roots are, and especially the root hairs. Tell them that the root hairs take in the water for the plant, and ask them what they think might happen to the root hairs of a larger plant if they pulled it out of the soil. Could they replant the plant and expect it to live? Try replanting some plants, so that the students can have hands-on experience and find out for themselves what will happen.

➤ DIRECTION OF GROWTH

Place blotting paper inside a clear drinking glass or jar (preferably plastic) as in Figure 9-2. Insert three or four lima-bean seeds between the container and the blotter paper. Place the container in a warm location, and keep the blotting paper wet.

After the seed has sprouted, ask the students to note the directions in which the stem and roots are growing. Ask them what, if anything, would happen if the container were laid on its side or turned upside down. Place the container on its side and have students note the results.

➤ NAMING AND GROWING SEEDS

To familiarize students with seeds of plants which they know about but may never have seen, set out samples of such seeds as wheat, corn, soybeans, oats, and grass. Help students name each seed, and then plant the seeds on plastic sponges in aluminum pans. Add water and keep the pans in a warm place.

Have students check on the progress of the seeds each day and describe any changes. After the seeds have sprouted, ask the students to describe similarities and differences among the types of plants. Also ask them what characteristics they would use for dividing the plants into two groups. Do the groups stay the same or change as growth progresses? Have the class record which seeds grow fastest, perhaps on a graph.

➤ SEEDS FROM COMMON FOODS

Ask students to bring to school a supply of seeds from common foods such as peaches, apples, sunflowers, peanuts, squash, pumpkins, tomatoes, grapes, and prunes. Ask the class whether or not they think the seeds will grow into plants.

Plant the seeds in potting soil.* Check their progress for several days. Ask the students why they think some seeds did not produce plants.

**Note:* Plant the seeds in containers which have *drainage holes* in the base. Almost all plants that die at school do so because of overwatering.

Common seeds can be used in several activities—they can be planted, sorted, and occasionally eaten.

➢ STEMS TRANSPORT WATER AND NUTRIENTS UPWARD

Cut diagonally the stem of a white carnation. Put the carnation imme diately into a vase of water containing red food coloring. Have students observe it during several hours, describe any noticeable changes, and offer explanations for the change.

➢ FINDING THE SMALL PLANTS IN SEEDS

Ask the students whether seeds are living or dead. *(They are living, but dormant.)* Ask if they have ever seen a small plant inside a seed. Soak some lima beans (or other large beans) in water for at least 1 day. Have the students split them open and use magnifying lenses to observe the small plants attached to one half of each seed (see Figure 9-3). Can small leaves be seen?

Ask the students whether they think the seeds would grow into plants if placed in soil. Have them try growing the seeds.*

Figure 9-3. Soaked bean seeds reveal the tiny plant inside. (Source: Adapted from P. F. Brandwein, *Concepts in Science*, Curie ed., Harcourt Brace Jovanovich, New York, 1980, grade 2, p. 70.)

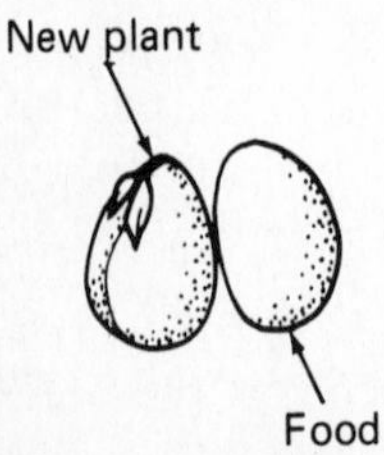

➢ PLANTS REQUIRE LIGHT

Among the most important requirements for plant life are proper conditions of light, water, nutrients, and temperature. To illustrate the requirement for light, use seedlings which have attained a height of 1 to 2 inches. Bean or potato plants work well for this experiment. Place one or more seedlings in a dark place, such as a closet or a well-sealed cardboard box. Place an additional plant or plants in normal lighting conditions.

Keep other factors equal for the two plants—i.e., water them at the

**Note:* Many seeds used for cooking have been treated so that they will not sprout.

same time and with equal amounts, use the same types of pots and potting soil, and keep the temperatures the same. Point out to students that when scientists conduct experiments, they change only one factor (such as amount of light) while holding all other variables equal. Ask the class to predict what will happen to the plants.

Have students observe and describe the condition of each plant once a day. Remove plants from the dark condition only once a day and for as short a period of time as is possible. The students should note the height and color of the plants, plus any other features that seem important to them. Many students—and many teachers—are surprised to find that plants in the dark environment often grow faster then those in normal lighting conditions. However, plants grown in the dark are not nearly as healthy-looking, and they will die unless adequate light is eventually supplied.

➤ PLANTS REQUIRE WATER

For this experiment you will need two well-established plants, such as geraniums. Place the plants side by side so that all conditions other than water are equal. Have students give one plant the amount of water recommended for it and the other no water at all. Ask the students to predict what will happen to the plants.

Each day, have students write a brief description of each plant. When the unwatered plant begins to wilt, ask the students whether or not they believe the plant would recover if water were given. Eventually, start watering the dry plant on the same schedule as the other one, and continue observing any differences.

Variations on this experiment are possible: A third plant, in a container without drainage holes; might be given 3 times as much water as the other. A fourth plant could be placed in a container of water so that the water covered the plant to 1 centimeter (½ inch) above ground level. Have students predict what will happen in each case and record their observations daily.

➤ TERRARIUMS ARE COPIES OF ENVIRONMENTS

Not only are terrariums copies of environments; they are also nice presents for parents.

Have students bring to class clear glass or plastic containers, such as food jars, from which they can make small terrariums. Use plants which you or your students collect from a local environment or which you have obtained from a generous garden shop or florist. If your students collect the plants, first tell them how to identify any endangered species which may be growing wild in their neighborhoods and warn them not to collect those species.*

Discuss with your students what kind of soil should be used in the terrarium. Ask their opinions about what kind would be best. *(Use the*

*You may have to seek the advice of a botanist or other expert—or you can do some library research to find out more about endangered plant species so that you will be able to identify them for your students.

Small terrariums not only give students firsthand experience with plants but also can become nice presents for parents.

soil the plants were growing in where they were collected, or the soil recommended by the garden shop or florist. Plants in the wild grow in soil which supplies their requirements.) Also discuss lighting conditions. How much light should the plant get? *(Again, try to duplicate the conditions in the natural environment of each plant, for those are the conditions that suit it best.*)*

In the bottom of the terrarium place a layer of small crushed rocks or sand. Ask the students why this layer is used. *(Extra water will drain into it.)* Then place soil and plants in the terrarium. Use simple, small plants—large plants will mold if the leaves touch the sides of the container. A terrarium which contains only a green moss is quite attractive. For color, add a rock or a twig with lichens.

➤ RECORDING PLANT GROWTH

As part of another plant activity or as a special activity, have students record the height of a plant first thing each morning and again just before the end of the school day. To do this, they can cut a strip of colored paper the height of the plant each day and glue all the strips to a chart. This will give the students practice in communication by making a graph which is very meaningful to them. When did the plant grow most rapidly—during the night or the day?

**Note:* Never place a terrarium in direct sunlight. Direct sunlight on a small container will cook the plants.

▶ GRASS TERRARIUM

Have students bring in containers such as disposable pie, TV dinner, or cake pans to make grass terrariums. They can place sand in the bottom with a layer of garden soil or potting soil on top, and then plant grass seeds by sprinkling them on the soil and covering them with an additional 3 millimeters (about 1/8 inch) of soil. Have them water their terrariums immediately and on a regular schedule thereafter (do not overwater).

As the grass grows, have students cut it with scissors to maintain a height of 2 to 5 centimeters (about 1 to 2 inches). If there are animals in the classroom which will eat the grass, feed it to them. Ask the students if farmers do anything similar. *(They cut hay for some farm animals.)*

Again, variations can add interest: Small dishes can be submerged in the soil to hold water, or mirrors can be placed on the soil to give the effect of a lake. Students can also make "log" (twig) cabins or figurines for their terrariums.

Do some grass types grow better than others? How much light is best? How often should a terrarium be watered? Have students investigate these and other questions which arise as they make and grow their terrariums.

▶ SPROUTING

Have students pretend they are seeds in the ground. First they roll themselves into a ball and lie on the floor. Then they sprout and use their hands to push their way through the surface of the soil. Their roots (toes) are taking up water from the soil, and their leaves (arms) begin to spread out and soak up the sunlight to produce food. The wind might blow gently, causing the plants to wave around.

Ask the students, "If you could be any plant in the world, what plant would you want to be? Why?" Have them describe what would be happening inside and around them if they were the plants they chose.

Children's literature

Read *The Seed** to your students or have them read it, and then discuss with them the concepts presented.

Find other books on plants in the school library, and identify them for the students or place them in an interest center.

Activity cards

An example of an activity card is shown in Figure 9-4. Additional activity cards could be made created for topics such as experimenting with factors that control plant growth, using seeds in art projects, finding plants that grow by means other than from seeds, learning how plants change through the seasons, and identifying food from various parts of plants.

*Ann Cameron, *The Seed,* Pantheon Books, Random House, New York, 1975.

• PRESERVE SOME PLANTS BY PRESSING THEM BETWEEN BOOKS.
• PRESERVE A FLOWER IN SOME WAY OTHER THAN PRESSING.

TAKE A FLOWER APART AND SEE IF YOU CAN IDENTIFY THE PARTS NAMED IN YOUR SCIENCE BOOK.

MAKE A DRIED-FLOWER ARRANGEMENT.

Figure 9-4. Activity card: Preserving flowers.

Bulletin boards

FOOD FROM PLANTS
Set up a bulletin board as illustrated in Figure 9-5. Have students bring to school labels from cans and boxes and magazine pictures of foods they eat which are derived from plants. They can pin their labels and pictures in appropriate spaces on the bulletin board. What parts of plants are most commonly eaten?

PARTS OF A PLANT
Make a bulletin board picturing a typical plant; in a nearby pocket place labels for names of the various parts of the plant. Send the students to the board in pairs. One student is to place the labels in the proper spots, and the other is to check on whether or not the parts are correctly labeled. The complexity of the bulletin board should of course suit the grade level of the students; at upper grade levels more parts can be identified or a picture of a flower can be added.

Field trips

Take a walk on the school grounds and in the surrounding neighborhoods, and ask the students to notice where plants grow. *(On lawns, in planter boxes, in cracks in cement, in vacant lots, in school gardens,*

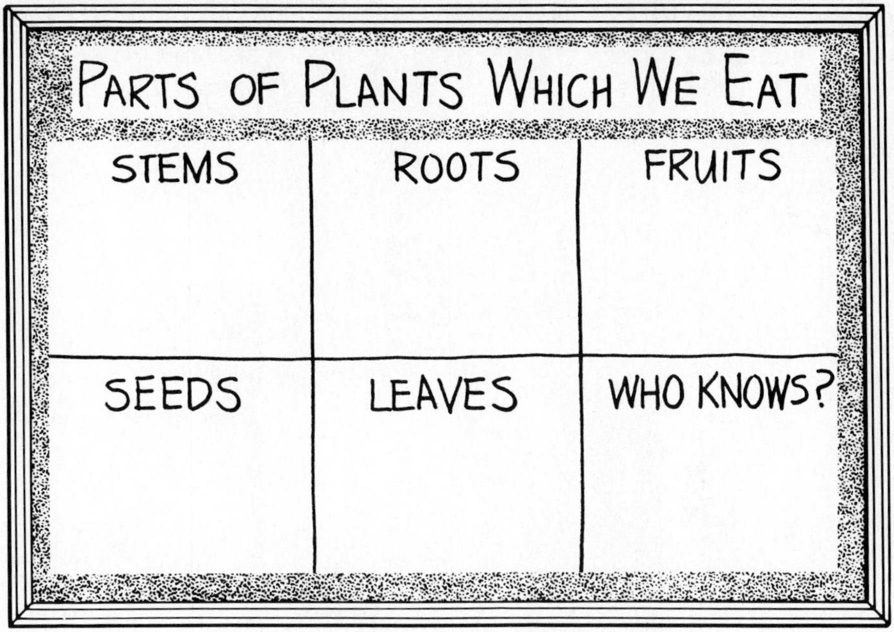

Figure 9-5. Bulletin board: Parts of plants which we eat.

etc.) Ask students to tell what kinds of plants they liked best and why. If there are trees on the school grounds, help the students to collect leaves and later to identify the trees from which they came.

Take a walk on a nature trail, if one is available, and have the children notice special features of plants which they see. Have them lie on their backs so that they can look at trees from a different perspective.

Make arrangements with the owner of a local greenhouse or garden shop to take your class for a visit and let them talk with the owner or a knowledgeable employee. Have the children pick out their favorite plants at the shop, and encourage them to ask questions about these plants. Have them also ask what trees or flowers would grow best on the school grounds. Purchase and plant a tree or some flowers, if possible.

PLANTS FOR INTERMEDIATE GRADES

Activities

➤ PARTS OF A FLOWER

Obtain enough "simple" flowers—such as gladioli or irises, which have parts that can be readily identified—to give each student at least one flower. (You can use wild flowers, or you may be able to get "throw-aways" from a local florist or garden shop.) Have students match the flowers' parts with the descriptions in their textbooks or with a drawing.

While this flower is not a "throwaway," florists do discard bruised flowers, and those can be examined and cut apart by students.

The teacher may want to use a single-edged razor blade to dissect some of the seed-producing parts of the flower.*

Also have students look at some compound flowers such as daisies. Ask them why the parts described in their books or on the drawing are not readily identifiable. If compound flowers are not being studied in the unit, ask a student who is interested to do research on compound flowers and report to the class.

➤ TREE IDENTIFICATION: A DICHOTOMOUS KEY

This activity will be explained in detail because the process used to make a tree identification key can also be used for several other classes of objects (rocks, birds, and wildflowers, to name a few).

Either supply or have students collect leaves from 12 local trees. Use one set of 12 leaves for each group of two or three students.

Step 1. Have the groups examine the specimens for 1 or 2 minutes and list the observable characteristics of the specimens.

Step 2. Ask each group to separate specimens into two categories on the basis of some major similarities or differences of the specimens.

Step 3. Compare the characteristics used by each group of students in dividing the specimens. On the blackboard, list all the characteristics used. Then list as many additional characteristics of the specimens as the students can think up.

**Note:* Never let the students use razor blades or other sharp instruments.

Step 4. The next step is to construct a dichotomous key for the specimens on hand.

Tell the students that it is possible to construct many different kinds of tree identification keys. Have on hand samples of a few different kinds of identification keys, if possible. They can be identification keys for things other than trees (flowers, seashells, rocks, etc.), and you may be able to borrow them from the library or from another science teacher. For this activity a "dichotomous" key will be developed. "Dichotomous" refers to division into two parts. For example in a dichotomous key the division "smooth edges—no smooth edges" would occur. "Sawtooth edges—lobed edges—smooth edges—etc." would not be proper division for a single stage in the key, because it has more than two choices. The example key (see Figure 9-6) shows that only a decision about possession or lack of one characteristic is made at almost every point. Very frequently the statements at the decision points are of the nature of "red—not red," "smooth—not smooth," "veins radiate from a common point—veins do not radiate from a common point," etc.

Tell the students to focus their attention on one of their specimen groups (ignoring the other group for the time) and subdivide the specimens into two smaller groups based upon similarities or differences in another characteristic. Repeat the process until they end up with one leaf in each specimen group, after making as many divisions on the key as possible.

Figure 9-6. Dichotomous key for five kinds of oak trees.

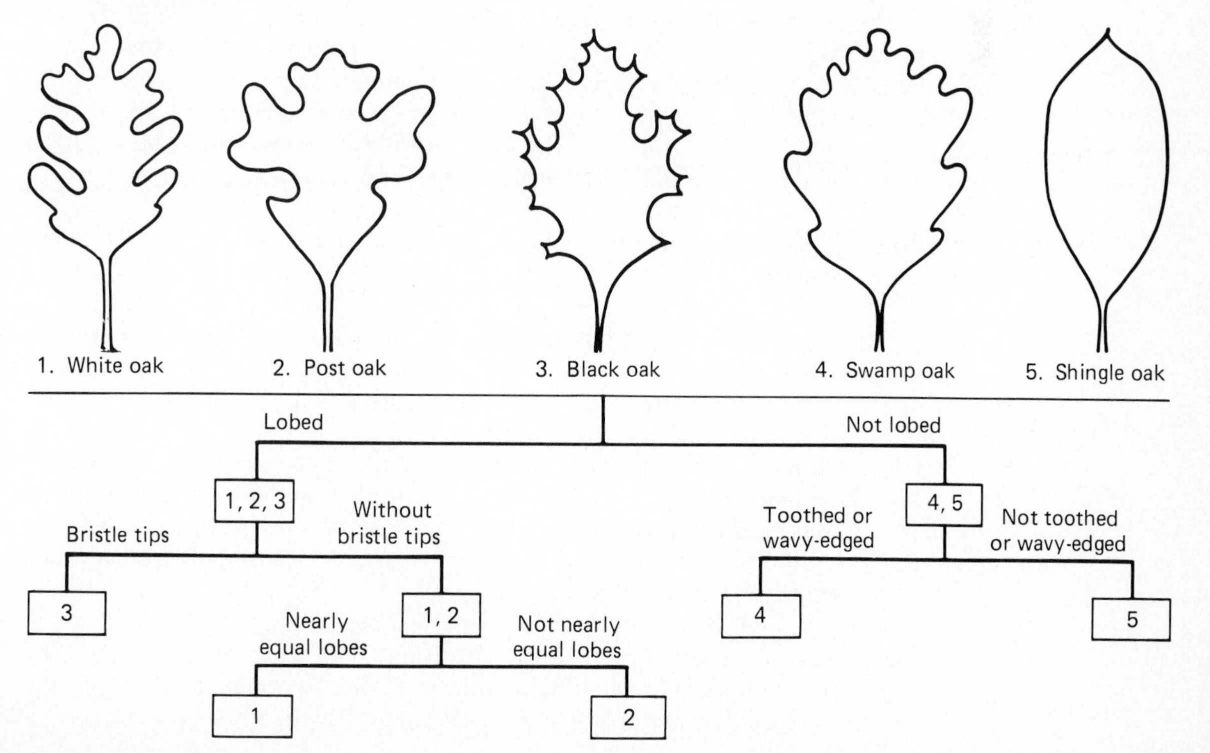

Describing the process of constructing a dichotomous key is similar to explaining the rules of a game. It usually seems confusing only until you try it. Get a set of leaves and develop a key similar to Figure 9-6.

Developing a dichotomous key has important benefits for students because it makes them focus on the similarities and differences of the specimens.

Step 5. Now have each group of students write, in sentence form, a description of one of their specimens, using the characteristics on their key. Ask one student from each group to read the description aloud, and have the other groups hold up the specimen which they think is being described.

Here is a sample description: "Find the leaf which is lobed, does not have bristle tips, and does have nearly equal lobes." *(It is the white oak.)*

Writing these descriptions and reading them to the class will show students how well they are communicating with their classmates. If the class has trouble identifying a specimen, the groups can keep working until the description is clear. Have students ask one another for suggestions on how their keys can be improved.

Step 6. After the groups of students are satisfied that their keys are usable, introduce a new specimen and have them trace it through their keys.

The students should be able to take the new specimen through the key to an end point which groups it with one other specimen. How is this similar to the way a scientist would use such a key? *(A scientist would go through a similar process in identifying a new specimen—e.g., a new species of tree.)* To complete the key, another division would be made to again generate two specimen groups each containing only one specimen.

Step 7. Have the students in each group compare their key with those of other groups.

How are the keys different? Similar? It will become apparent that useful keys can be developed in many different ways. Why are there different keys for identifying trees and many other groups of objects?

Conclusions. The students should conclude that one very important aspect of using a key is the necessity for understanding the meanings of the characteristics of the specimens in the same way that the author of the key did. Misunderstanding about meanings is the most frequent cause of problems in using identification keys.

At some point in the process the teacher should either name the specimens or invite an "expert" to class who can name them. The "expert" might be a student who was motivated enough to find out the names of the specimens (or more than one student who did this).

Many nongreen plants exist. This one's beauty rivals that of many flowering plants.

NONGREEN PLANTS

Have students research and name as many kinds of nongreen plants as they can. Suggest that they consult their textbook, encyclopedias, botany books in the library, and other references. Offer a prize for the student who names the most nongreen plants.

During a 2-week period, have the class collect samples of as many nongreen plants as possible. In what types of environments are nongreen plants found?

Invite a local collector of edible mushrooms to talk to the class.* Cook some mushrooms in class for all to sample—try something a little more unusual than cream of mushroom soup. What is the food value of mushrooms? Would one get fat eating them?

Have a contest to find the most colorful nongreen plant, or the largest. Tell the contestants to present basic information about their plants, and display the nongreen plants and information in a showcase or some other place available to the whole school.

MOLD

Give each student two pieces of bread and two plastic bags. Tell the students to see who can grow mold the fastest on one piece of bread and who can keep the other piece of bread from getting moldy for the longest period of time.

After several students have been successful at growing mold, have a

Note: Warn the students that only real experts should ever eat mushrooms which they collect.

Which items will mold? Build a large mold terrarium for your classroom and find out.

class discussion about the quantities of heat, light, and moisture that seem to speed mold growth. Try to find out whether any other factors appeared to be important. Design experiments to check the tentative answers. For example, you could keep all other factors equal and expose pieces of bread to very high, moderate, and low temperatures. On the basis of the experience of trying to keep the bread from becoming moldy, ask what the students think is the best place to store bread in the kitchen.

Make a mold terrarium out of a clean gallon jar with a lid. Place two cups of sand in the bottom, and add 100 milliliters (about ½ cup) of water. Have students bring in a few items which they believe will not mold and a few which they believe will mold. Place the items in the terrarium, and cover it with the lid. Inspect the items for mold each day. Were the students' judgments correct?

For many more ideas on mold activities, consult *Microgardening.**

➤ DIFFERENT ENVIRONMENTS—DIFFERENT PLANTS

Have students collect pictures of plants from several environments—e.g., oceans, lakes, deserts, prairie, woods. In order to collect samples of as many kinds as possible, contact a local garden club member for help.

Ask students to identify similarities and differences among the plants from different environments. Have them try to identify unusual features that help certain plants survive in their environments—e.g., quaking aspen leaves freely flutter so that the frequent strong winds that characterize their environment will not tear them from the trees.

*Elementary Science Study (ESS), *Microgardening,* Webster, McGraw-Hill, Manchester, Mo., 1976.

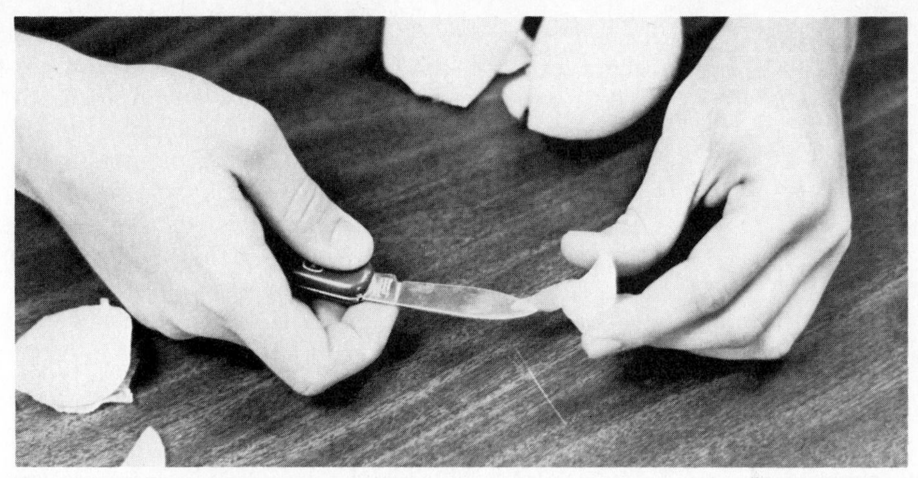

The thin "skin" between the layers of an onion can be easily stained and examined with a microscope to reveal individual cells.

➤ PLANT CELLS

To develop the concept of the cell as the building block of living things, have students observe plant cells. Plant cells which can be readily seen are found in the thin layer of skin between the layers of an onion. (Onion root-tip cells are also easy to prepare.) Remove a piece of this very thin material and stain it with a dilute solution of iodine. Place it on a microscope slide and cover it with a cover slip. (If you don't have these materials, you may be able to get them from a middle school or high school science teacher.) Have students observe the sample first under the *lowest* power of the microscope; higher powers should be used only if necessary for viewing an appropriate section of the cell. Ask students to describe the cells. What do they look like?

➤ GREEN CELLS IN ACTION

Obtain a small piece of the water plant Elodea. (A local aquarium shop may be generous and give it to you—these plants are usually very inex-

Cells from a live elodea plant (which can be easily obtained from an aquarium shop) can be examined with an inexpensive microscope.

pensive.) Keep the plant in water and relatively warm until it is needed. Take a leaf from the tip of the plant, place it on a microscope slide, and cover it with a cover slip. Using low power, focus the microscope on the edge of the leaf. Cells should be visible. Have the students describe what they see. If the plant has been kept warm enough, movement will be seen within the cells—living cells in action!

Will the movement within the cells increase if the temperature is increased? How can students check this? If the plant is cooled with an ice cube, does the motion speed up, slow down, or stay the same? Ask the students to guess what the green "dots" are and what they do. *(They are chloroplasts—the location of chlorophyl.)*

➤ CROSS SECTION OF A TREE

Suggest to the students that they try to bring cross sections of a tree, branch, or twig to school. A branch which has broken off from a tree can be sawed with a handsaw to produce a section (about 2 centimeters or 1 inch thick is a handy size).

Can growth rings be seen? Can anything be done to improve their visibility? Can the students detect other features which are identified in the textbook or an encyclopedia? What are the functions of these features?

What are the similarities and differences among sections from different kinds of trees? Which tree seems to have grown faster? Can you tell from the rings whether there has been a dry year or two? *(If a ring is smaller than other rings on either side of it, the explanation* could *be a dry year in which little growth took place.)*

➤ PLANTS NEED SUNLIGHT TO PRODUCE FOOD

To demonstrate that plants do need sunlight to produce food, cover two or three leaves on a large plant with a black plastic bag or with heavy black paper or cardboard, so that no light gets to the leaves. Ask students what they think will happen. After 3 or 4 days remove the cover and clip the leaves which were covered; mark them so that they can be identified. For comparison, clip an equal number of uncovered leaves, and ask the students to describe any differences.

Next, place all the leaves you clipped in hot (*near* boiling) water for 5 to 10 minutes. Then heat rubbing alcohol *in a beaker covered with a dish* over a hot plate until it is gently boiling.* Place the leaves in the beaker for 10 to 15 minutes. Remove the leaves and place them in a dilute solution of iodine. What are the differences between the covered and the uncovered leaves?

Plants require sunlight to produce food (starch). The presence of starch in the uncovered leaves (as indicated by the dark color produced by the

**Caution: Not* over a flame. And note that this is a demonstration to be performed by the teacher, *not* an activity for students.

iodine solution) and the absence of starch in the covered leaves (as indicated by no dark color when placed in the iodine solution) should lead students to conclude that light must be present for the plant to produce food.

➢ CAN PLANTS AIR-CONDITION A BUILDING?

Students can answer this question with an activity which is best conducted either in late spring or in early fall when plants are in full leaf and when days are relatively warm. Have students identify two buildings (houses or apartment buildings are fine) made of similar materials and located near each other. One of the buildings should be entirely or partly shaded by trees, bushes, and shrubs, while the other should be relatively unshaded. Take temperature readings around all sides of each building, and take all readings at the same distance above ground level. *(Why?)* Always shade the thermometer with your hand or a cardboard, to keep the sun from shining directly upon it. *(Why?)*

Were there any differences in the readings between the two buildings? How can the differences be explained? *(The cooler readings around the shaded building are not due solely to the effect of its being in shadow. Another cause is that water is evaporating from the plants, in a process called "transpiration," which removes heat from the air.)*

Ask the students if they think it would be financially wise to spend the money necessary to plant trees, bushes, and shrubs, considering the money which might be saved on air-conditioning costs. Would it be worth it for the comfort? Do they like the looks of plants around the buildings? Have them make a plan for shading either the school or their home.

➢ PLAN A GARDEN

Gardens can be planted on very small plots of ground, as attested to by the increasing number of them found in even very large cities. In addition to learning a great deal about plants from a garden, students can also learn lessons in economy and nutrition. The students can plan either a home garden or, if possible, a school garden. A large number of schools do have gardens.

First, obtain catalogs from seed supply companies.* Such catalogs usually include a great deal of information on planting a garden, including how far apart to space plants, when to plant the various kinds of plants in your locality, etc. Another source of help is the office of the county agricultural extension agent (yes, even in large cities).

What kinds of food plants would students like to plant? Are there certain kinds of flowers they might like to plant? Could these plants and flowers be planted in your area?

*One possible source is Gurney Seed and Nursery Company, Yankton, South Dakota 57079.

▶ DRIED-FLOWER ARRANGEMENTS

Have the students research ways to dry flowers. Check with a local florist or garden shop for sources of information. Some flowers can be dried by simply hanging them in an attic. Others can be placed in a box or can, covered (gently) with a mixture of kitty litter and borax, and left to dry for several weeks. Several wild flowers can be collected already naturally dried in many parts of the country.

Invite a florist to demonstrate to the class how dried-flower arrangements can be made, and perhaps to produce a simple cut-flower arrangement at the same time. Have the students each make a dried-flower arrangement. Most parents will be grateful for gifts of such arrangements. Have students name the various parts of the flowers. What kinds of flowers were collected?

Children's literature

Read *A Maple Tree Begins** to your students or have them read it, and then discuss with them the concepts presented. Find other books on plants in the school library and identify them for the students or place them in an interest center.

Activity cards

An example of an activity card is shown in Figure 9-7 (opposite page). Other activity cards could be made for topics such as finding out what lives in oak trees (or other trees that grow near the school), grafting plants, growing orchids, building a terrarium, attracting butterflies and birds to your yard with special kinds of plants, growing fruit trees, and growing a winter garden.

Bulletin board

Let your students make a bulletin board on methods of propagating plants other than by seeds. Then have them propagate plants by as many methods as possible.

Field trips

Arrange a field trip to a local greenhouse or floral garden. Have the manager describe the jobs of people who work there and tell briefly how plants are cared for at the site. Have individual students identify the flowers that they think are the most beautiful and the ones whose aromas they like best.

Other field trips related to plants could be made to a vacant lot, a local arboretum, a feed-and-grain store (to see what some plants are used for and what some seeds look like), or a well-landscaped commercial building or house.

*A. A. Watson, *A Maple Tree Begins,* Viking, New York, 1970.

- FIND SOME SONGS (RECORDS) ABOUT FLOWERS. WHICH DO YOU LIKE BEST?
- WHAT IS YOUR FAVORITE FLOWER? CAN YOU FIND A SONG ABOUT IT? WRITE ONE!

BECOME A DJ—
PLAN A 15-MINUTE RADIO
PROGRAM USING RECORDS
ABOUT FLOWERS. PRODUCE
IT FOR YOUR CLASS.

Figure 9-7.
Activity card:
Flower music.

RESOURCES

Sources of assistance

Of the many sources of materials, ideas, and other help in teaching about plants, several will be described in this section. Look upon these sources as a starting point only, for you should be able to think of several other sources. Let your imagination run wild for 10 minutes. Also, ask your students to help you. You may be surprised at the sources they can offer.

FREE AND INEXPENSIVE MATERIALS
Here are some good sources of free and inexpensive materials for use in teaching about plants:

1. *U.S. Department of Agriculture.** Several pamphlets on farming, plants, etc.
2. *American Forest Institute.* Posters on trees and tree products.
3. *Chevron Chemical Company.* A booklet entitled "A Child's Garden" which gives information on school and home gardens.

*The addresses of specific agencies, organizations, and businesses are listed in the Appendix.

4. *Garden Clubs of America.* An education packet on the environment, including materials on plants.
5. *Green Giant Company.* Posters about vegetables.
6. *State department of agriculture.* Information on agriculture, including plant products.

LOCAL RESOURCES

Local resources which can help in teaching about plants include the following:

1. *County extension agent.* A large amount of information on plants, especially ones which grow in your area (trees, agricultural crops, and so on.)
2. *Garden club.* Members may be willing to help you with materials and information on plants.
3. *Garden shop, greenhouse, or florist.* Information on plants. May also be able to give you throwaway plants and flowers for your students to study.
4. *Special-interest groups.* African violet organizations, maple-syrup producers, cotton growers, peanut growers, etc.

ADDITIONAL RESOURCES

1. Magazines such as *Ranger Rick, Better Homes and Gardens,* and *National Geographic.*
2. Books from your school or municipal library on botany, ecology, horticulture, and gardening.

Strengthening your background

1. Learn how to dry flowers. Then make a dried-flower arrangement for your home.
2. Build a terrarium for your home or classroom, using only plants that you can find in your local area (within about 50 kilometers or 30 miles of home).
3. Learn how to grow African violets (or some other plant of your choice) and keep a couple of them in your home or classroom.
4. Go to a nearby vacant lot and take pictures of six beautiful wildflowers.
5. Attend a meeting of a local garden club.
6. If you have access to a small plot of ground, plant a flower or vegetable garden.
7. Purchase and plant an inexpensive tree that will grow well in your yard or on the school grounds.
8. Take a trip to some place you would like to visit. Notice what kinds of plants live there, and take pictures (slides if possible) of several of them.

Getting ready ahead of time

1. Find three pieces of children's literature which you can use with your students on the topic of plants.
2. Collect magazine pictures of plants from three different environments (for example, ocean, desert, prairie) which you could use for a bulletin board.
3. Obtain at least six bulletins from your county extension agent on the growing of plants in your area.
4. Prepare six activity cards on plants.
5. Make a collection of leaves from trees common in your area. Identify the trees which they come from.

Building your own library

Baker, Samm: *The Indoor and Outdoor Grow-It Book,* Random House, New York, 1966.

Brandwein, Paul, et al.: *Life: A Biological Science,* Harcourt, Brace, Jovanovich, New York, 1975.

Elementary Science Study (ESS): *Budding Twigs,* Webster, McGraw-Hill, Manchester, Mo., 1970.

———: *Growing Seeds,* Webster, McGraw-Hill, Manchester, Mo., 1969.

———: *The Life of Beans and Peas,* Webster, McGraw-Hill, Manchester, Mo., 1976.

———: *Microgardening,* Webster, McGraw-Hill, Manchester, Mo., 1976.

———: *Starting From Seeds,* Webster, McGraw-Hill, Manchester, Mo., 1976.

Gale, Frank, and Clarice Gale: *Experiences with Plants for Young Children,* Pacific, Palo Alto, Calif., 1975.

Garden, John F.: *Book of Nature Activities,* Interstate, Danville, Ill., 1967.

Hillcourt, William: *The New Field Book of Nature Activities and Hobbies,* Putnam, New York, 1970.

Klein, Richard, and Deana Klein: *Discovering Plants: A Nature and Science Book of Experiments,* Natural History Press, Doubleday, Garden City, N.Y., 1968.

Rahn, Joan: *Grocery Store Botany,* Atheneum, New York, 1974.

———: *How Plants Travel,* Atheneum, New York, 1973.

Ramsey, William, et al.: *Life Science,* Holt, New York, 1978.

Russell, Helen R.: *Foraging for Dinner: Collecting and Cooking Wild Food,* Nelson, Camden, N.J., 1975.

Silverstein, Alvin, and Virginia Silverstein: *Beans: All about Them,* Prentice-Hall, Englewood Cliffs, N.J., 1975.

Smallwood, William, et al.: *Biology,* Silver Burdett, Morristown, N.J., 1977.

Smith, Herbert, et al.: *Exploring Living Things,* Laidlaw, River Forest, Ill., 1980.

Walsa, Anne: *A Gardening Book: Indoors and Outdoors,* Atheneum, New York, 1976.
Zim, Herbert: *Flowers,* Golden Press, New York, 1950.
——— and Alexander Martin: *Trees,* Golden Press, New York, 1952.

Building a classroom library

Anderson, Lucia: *The Smallest Life around Us,* Crown, New York, 1973.
Anderson, Margaret: *Exploring City Trees and the Need for Urban Forests,* McGraw-Hill, New York, 1976.
Beck, B.: *The First Book of Fruits,* F. Watts, New York, 1967.
Bentley, L.: *Plants that Eat Animals,* McGraw-Hill, New York, 1968.
Blough, Glenn: *Plants round the Year,* Harper and Row, New York, 1959.
Borland, Hal: *The Golden Circle: A Book of Months,* Thomas Y. Crowell, New York, 1977.
Branley, Franklyn: *Roots Are Food Finders,* Thomas Y. Crowell, New York, 1975.
Cameron, Ann: *The Seed,* Pantheon, New York, 1975.
Cobb, Vicki: *Lots of Rot,* Lippincott, Philadelphia, 1981.
Cole, Joanna: *Plants in Winter,* Thomas Y. Crowell, New York, 1973.
Cooper, Elizabeth, and Padraic Cooper: *A Tree Is Something Wonderful,* Golden Gate Junior Books, Chicago, Ill., 1972.
Craig, M. Jean, and William Grimm: *Wondrous World of Seedless Plants,* Bobbs-Merrill, Indianapolis, Ind., 1973.
Davis, Bette: *Winter Buds,* Morrow, New York, 1973.
Day, Jennifer: *What Is a Flower?* Western Publishing, New York, 1975.
———: *What Is A Fruit?* Golden Press, New York, and Western Publishing, Racine, Wis., 1976.
Earle, Olive, and Michael Kantor: *Nuts,* Morrow, New York, 1975.
Edwards, Joan: *Caring for Trees on City Streets,* Scribner, New York, 1975.
Farb, Peter: *The Story of Life: Plants and Animals through the Ages,* Harvey House, New York, 1962.
Gallob, Edward: *City Leaves, City Trees,* Scribner, New York, 1972.
Goldin, Augusta: *Grass: The Everything, Everywhere Plant,* Nelson, Camden, N.J., 1977.
Hathaway, Polly: *Backyard Flowers,* Macmillan, New York, 1965.
Heady, Eleanor: *Plants on the Go: A Book about Seed Dispersal,* Parents Magazine, New York, 1975.
Hogner, Dorothy C.: *Endangered Plants,* Thomas Y. Crowell, New York, 1977.
Hopf, Alice L.: *Animal and Plant Life Spans,* Holiday, New York, 1978.
Johnson, Hannah L.: *From Apple Seed to Applesauce,* Lothrop, New York, 1977.
Jordan, Helene: *How a Seed Grows,* Thomas Y. Crowell, New York, 1960.

Krauss, Ruth: *Bouquet of Lillies,* Harper and Row, New York, 1963.
———: *Everything under a Mushroom,* Scholastic Book Services, 1973.
Lauber, Patricia: *Seeds: Pop – Stick – Glide,* Crown, New York, 1981.
Lewis, Richard: *In a Spring Garden,* Dial, New York, 1976.
McMillan, Bruce: *Apples, How They Grow,* Houghton Mifflin, Boston, 1979.
Millard, Adele: *Plants for Kids to Grow Indoors,* Sterling, New York, 1975.
Newton, James R.: *Forest Log,* Thomas Y. Crowell, New York, 1980.
Norris, Louanne, and Howard K. Smith: *An Oak Tree Dies and a Journey Begins,* Crown, New York, 1979.
Overbeck, Cynthia: *Sunflowers,* Lerner, Minneapolis, 1981.
Parker, Bertha: *Dependent Plants,* Harper and Row, New York, 1957.
Petie, Harie: *The Seed the Squirrel Dropped,* Prentice-Hall, Englewood Cliffs, N.J., 1976.
Pringle, Laurence: *Water Plants,* Thomas Y. Crowell, New York, 1975.
———: *Wild Foods: A Beginner's Guide to Identifying, Harvesting and Cooking Safe and Tasty Plants from the Outdoors,* Four Winds, New York, 1978.
Rahn, Joan Elma: *More about What Plants Do,* Atheneum, New York, 1975.
———: *Plants up Close,* Houghton Mifflin, Boston, 1981.
———: *Seven Ways to Collect Plants,* Atheneum, New York, 1978.
Raskin, E.: *The Fantastic Cactus: Indoors and in Nature,* Lothrop, New York, 1968.
Ross, Wilda: *Who Lives in This Log?,* Coward-McCann, New York, 1971.
Selsam, Millicent E.: *The Amazing Dandelion,* Morrow, New York, 1977.
———: *Eat the Fruit, Plant the Seed,* Morrow, New York, 1980.
———: *Maple Tree,* Morrow, New York, 1968.
———: *Plants We Eat,* Morrow, New York, 1981.
———: *Vegetables from Stems and Leaves,* Morrow, New York, 1972.
——— and Joyce Hunt: *A First Look at Flowers,* Walker, New York, 1977.
——— and ———: *A First Look at Leaves,* Walker, New York, 1976.
Silverberg, R.: *Vanishing Giants: The Story of the Sequoias,* Simon and Schuster, New York, 1969.
Sterling, D.: *The Story of Mosses, Ferns, and Mushrooms,* Doubleday, Garden City, N.Y., 1955.
Tarsky, Sue.: *The Prickly Plant Book,* Little, Brown, Boston, 1981.
Waters, John: *Carnivorous Plants,* F. Watts, New York, 1974.
Welch, Martha McKeen: *Sunflower!,* Dodd, Mead, New York, 1980.
Wilson, Jean: *Useful Plants,* Addison-Wesley, Reading, Mass., 1969.
Wilson, Ron: *How Plants Grow,* Larousse, Paris, 1980.
Woodside, Dave: *What Makes Popcorn Pop?,* Atheneum, New York, 1980.

CHAPTER 10

ANIMALS

IMPORTANT CONCEPTS

Animals are one group of living things.

Animals require proper amounts and kinds of food, air, and water to live.

Animals live in habitats which satisfy their needs.

Animals have special adaptations which help them survive.

Scientists who study animals classify them into groups.

Animals are first classified as vertebrates or invertebrates.

Invertebrates (animals without backbones) are the most numerous animals.

Insects are complex invertebrates.

Fish, amphibians, reptiles, birds, and mammals are the major groups of vertebrates (animals with backbones).

People who have pet animals are obligated to provide proper care for them.

INTRODUCTION: STUDYING ANIMALS

Animals, especially exotic animals, capture the attention of almost everyone. The next time you visit a museum which includes live-animal displays, notice where most of the people are congregated—almost certainly it will be around the live animals.

Among the goals in studying animals in the elementary school are increasing the students' awareness of the diversity of animals and learning something about the groups into which animals are classified for study. Students will enjoy applying several of the processes of science to their study of animals—especially observing, classifying, using numbers, measuring, communicating, inferring, formulating hypotheses, and experimenting.

People often overlook the importance of "small" animals, especially invertebrates. Small animals are not only fascinating but also abundant in almost all environments. You probably will not be able to name all the animals your students encounter, but what you can do is make sure they understand that, if they want to know the names of animals, they can find them. Then you can encourage them to develop their research skills by looking up specific animals that interest them.

Elementary school children often have pet animals, or wish they had them. Studying animals helps them to become more knowledgeable about the needs of their pets. Keeping wild animals as pets should be discouraged, since we often do not fully understand their requirements. Domesticated animals are better suited to life as pets.

Some activities related to animals are included in this chapter, and more can be found in Chapter 12, Ecology.

ANIMALS FOR PRIMARY GRADES

Activities

➤ ANIMAL PICTURES

Have each student cut out at least four pictures of animals from old magazines or obtain them from other sources. Then, working in groups of three or four, the students can classify their animal pictures. Have each group explain to the class how they classified their animals. Classifications such as "dogs," "cats," "pets," "farm animals," and "wild animals" will probably be most prevalent. A few students may have cut out pictures of fish and insects, but most children (and many adults too) think first of mammals.

Next, tell the students, one at a time, to place each picture in one of these six groups: "invertebrates," "fish," "amphibians," "reptiles," "birds," and "mammals." Have the students name as many of the animals as they can. Explain the difference between vertebrates and invertebrates at this time, but let the students do the best they can to differentiate the other groups. After the pictures have all been placed in groups, tell the students a little about each group and name a few examples of animals in each group. Then ask them if any of the animals that have been placed in the various groups should be changed.

Finally, have the students search for additional pictures, especially for classifications for which not many pictures have yet been collected. If some of the pictures are small enough, have students paste them onto cards and write the appropriate classification on the back of each card. The cards can then be used as flash cards to help students learn the animal classifications.

➤ ANIMALS IN THE CLASSROOM

Animals in the classroom, either permanently or as visitors, are not only fascinating and enjoyable but also very educational. For a description of how to set up and maintain aquariums and terrariums, see Chapter 12. In this section the keeping of hamsters, gerbils, mice, and rats in the classroom will be described.

Caring for classroom animals can help students learn about the behavior and living requirements of the animals. An additional, very important lesson is the responsibilities of an animal's caretaker. Students should gain an increased appreciation for life and living things, and should translate that appreciation into their treatment of any animals with which they come in contact.

Let the students help you make plans for bringing an animal into the

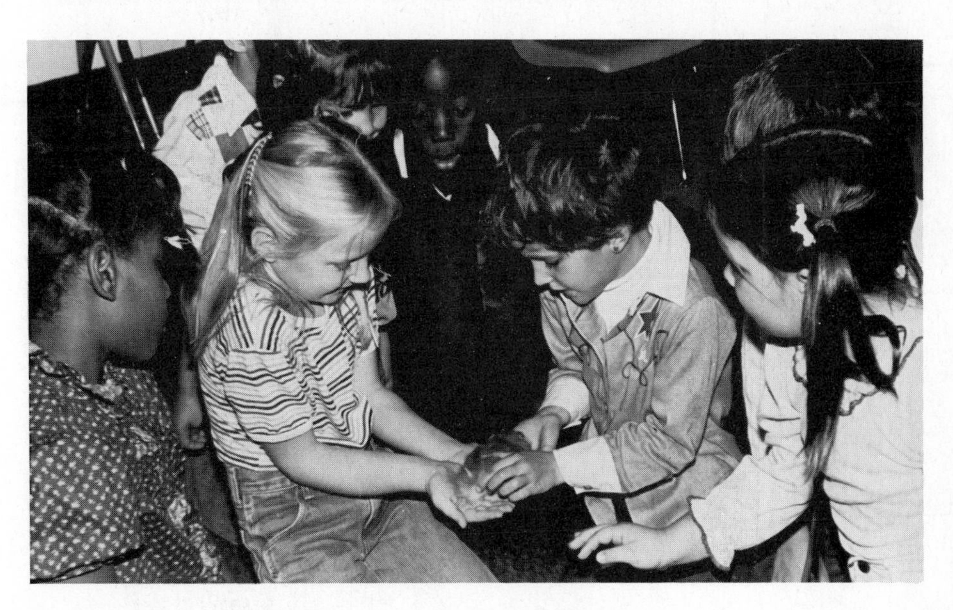

Students enjoy and learn from having animals in the classroom.

classroom. If only one kind of animal is available, help the students become familiar with some of the characteristics and requirements of that animal. But if a choice is to be made about which of several animals is to be adopted, let them help choose. Before making the decision, study the characteristics and requirements of each of the possible choices. Invite someone who has kept most or all of the animals as pets to talk to the class and to answer the students' questions. You and the class can then choose an animal together.

Next, work with the students to make plans for the animal's housing and care. This valuable learning situation would be lost if the teacher simply brought the animal in and told the students how to care for it. A trip by the class or a group of students to a local pet shop or the home of someone who keeps the same kind of animal as a pet can be very informative and interesting. Make sure that information on the care of the animal is available to students in several books, including books from a pet shop. Additional information can be obtained from a humane society (see "Free and Inexpensive Materials").

After the students have agreed on guidelines for feeding, watering, and handling the animal and for cleaning the cage, prepare the cage. The animal which you obtain will usually have the best chance of being a good

Care charts not only inform students of their responsibilities but also let the teacher know that the animal has indeed received proper care. With a chart like this one, students can remove their names when they have performed their jobs.

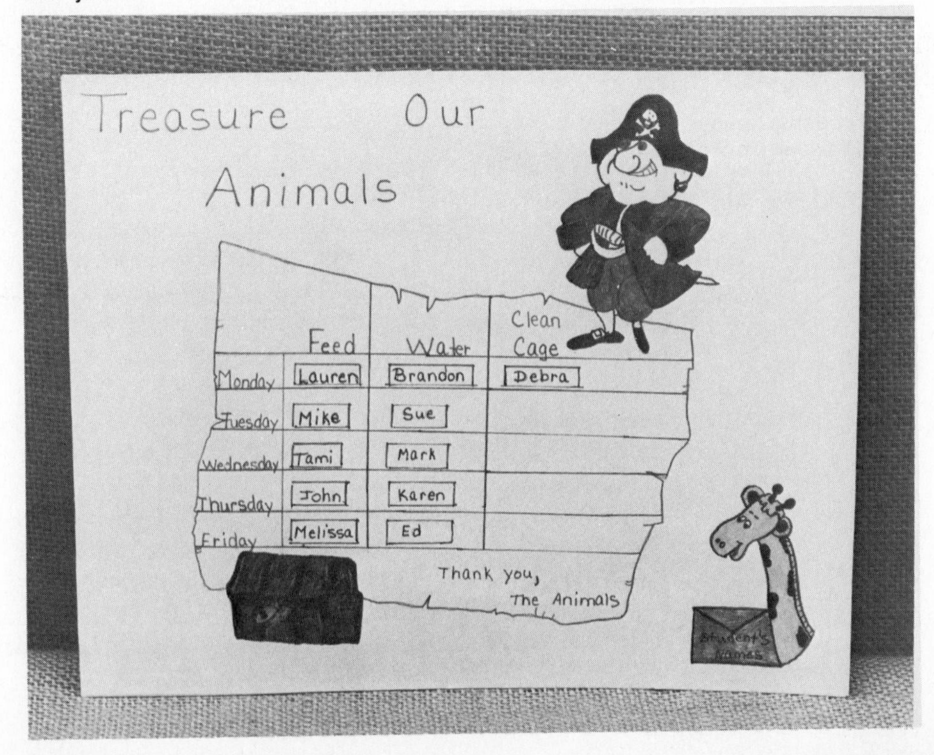

classroom pet if it is obtained shortly after weaning. Most animals should be handled frequently during their first several weeks of living in the classroom; but always follow recommendations for care of specific animals.

On the basis of many years of keeping animals in classrooms, the authors would like to share some recommendations. Hamsters, gerbils, mice, and rats all make fine classroom animals. Most people initially judge the rat to be repulsive, but it may be the best classroom pet.* It is a very gentle animal if handled and petted enough when young—and it is large enough that it is not likely to be injured if accidentally held too tightly by small hands. An aquarium (10-gallon or about 40-liter capacity) with a wire lid is a very adequate cage. Clean wood shavings (from a local cabinet shop) are an appropriate litter material for the cage bottom. Food is simply placed directly on the cage bottom. Water should be placed in a suspended water bottle, for open containers on the cage floor are messy, smelly, and unhealthy for the animal. An adequate food is a mixture (equal parts by volume) of unsalted sunflower seeds, chopped corn, and rabbit pellets, all of which can be obtained at a feed store or a pet shop. Water-soluble vitamins (these are important for all animals kept as pets, especially birds) can be placed in the drinking water. The cage should be cleaned once a week, and—to reiterate—the animal should be handled frequently.

Many educational opportunities arise when a pet is kept in the classroom. Students can investigate questions such as: At what times of day is the animal most active? *(Charts and graphs can be used to record amounts or times of movement, or distance moved during each hour of the day.)* What food does the animal consume first? Is the pet's activity affected by a change in room temperature? How long does it take for the animal to learn to come to a selected corner for food in response to a single tap on the cage? Does the animal learn more quickly if we change the feeding location to a second corner? If we change to a third corner? Can the animal be trained to go to one corner in response to one tap, to another in response to two taps?

Library research can help students find the answers to questions such as "What other animals are related to our classroom animal?" To help students learn more about the group to which their animal belongs, you can ask them to tell how the animal exhibits some of the visible characteristics of that group.

Creative activities can also be useful and interesting. For example: "Pretend you are our classroom animal. If you escaped from the cage, what would you do? Write about what your relatives are doing in the wild. How do you like being our classroom animal?"

The enjoyment and educational possibilities of keeping a live animal in the classroom far outweigh the slight amount of extra work involved. In fact, there really is very little extra work for the teacher, who just has to help the students get organized to do the necessary tasks.

**Note:* Warn students never to approach or play with rats that have not been domesticated.

A field trip to a local pet shop can be a good learning experience.

➤ PET CARE

Many young children have or would like to have pets. The most common pets are dogs and cats, though many others are possible. Students who do not have pets of their own nevertheless often interact with other people's pets. For these and other reasons, it is desirable to help students learn about pets.

In studying pets in elementary school, dogs and cats should receive primary consideration. Other animals which students have or would like to have may be included also. Information on pet care can be obtained from pet shops, libraries, and humane societies (see "Free and Inexpensive Materials").

After the students have studied pet care, invite the owners of several kinds of pets to visit the class with their animals. Ask them to share with the class how they care for their pets and why they have these particular pets.

At the end of the unit on animal care, have students form groups. Each group should choose a different pet and make a presentation (either written or oral) to the whole class on caring for that kind of pet.

➤ VISITORS' DAY

Seeing live examples of various animals will give students a better understanding of the different groups of animals and their characteristics. Identify groups of animals for your students—fish, amphibians, reptiles, birds, mammals, and so forth—and name one or two examples of animals in each group. Ask the students to name more examples. Then have them

Guest lecturers from a local museum or pet shop can often be found. They can bring several animals to visit the classroom.

do some library research in order to find some unusual animals in each group, as well as additional well-known examples.

Tell the students that you would like their help in arranging a "Visitors' Day" (or days). Ask them to think of people (themselves, their parents, pet-shop owners, zoo keepers, etc.) who could be invited to bring some of the animals they have been talking about. You can then write letters of invitation to the people you and the class select, explaining "Visitors' Day" and asking for their help. The letters can either be mailed or delivered personally by the students who suggested the names.

On "Visitors' Day" have the students name the characteristics of the animal visitors that differentiate them from animals in other groups. Students might also point out special characteristics of the animals which make them better able to live in their natural environments.

➤ ANIMALS THROUGH THE SEASONS

While studying ways animals adapt that help them survive, students can investigate seasonal behavior and characteristics of animals. Some animals—ermine, arctic hares—have seasonal variation in coloration, which helps them survive by making it difficult for predators to see them. Other animals—geese, some butterflies—migrate north in the summer and south in the winter. Others—groundhogs, some kinds of bears—hibernate or reduce their activity in winter.

Have students make drawings of animals whose characteristics or behavior change from season to season. They can use library resources to find some of these animals, and they should try to show the changes in their drawings. After the drawings have been completed, have the students tell, if possible, what special advantages the animals they drew gain from the changes. If the students do not know the reasons for the changes, ask them to think of the best explanations they can. Then they can do some research to check out their guesses.

ANIMALS AS FOOD

To help students appreciate how often animals are used for food for humans, create an "animal food store" in the classroom. Ask students to collect pictures of animal foods, as well as containers and labels; then they can help you stock the shelves of the store with their collections. They can also collect pictures of the live animals from which the food came.

Ask the students to determine where the animals that are hunted or gathered lived. Also ask them to explain where the animals that are raised as food (e.g., cows, pigs, sheep, chickens) came from originally. Do not tell them where the animals came from—let them do the research and make a report to the class.

INVESTIGATE AN ENVIRONMENT

Tell the students to close their eyes, relax, and pretend that they are small mice living in a vacant lot or a field just outside town. Give them time to think about each of these questions and others you may think of: What is it like as you move about in your environment? What sort of home do you live in? What do you eat? How do you protect yourself? What other animals live near you? What animals do you have to avoid so that you will not be eaten? What do you do on warm, sunny days? On cold, rainy days?

After the guided fantasy is over, ask the students to share their thoughts. Some students may later wish to verify some of their thoughts—encourage them to use the library. Some of the students' experiences may also suggest art projects and creative writing—encourage them to carry out any such projects.

Children's literature

Read *Animals that Hide, Imitate, and Bluff** to your students or have them read it, and then discuss with them the concepts presented.

Find other books on animals in the school library, and identify them for the students or place them in an interest corner.

Activity cards

An example of an activity card is shown in Figure 10-1. Additional activity cards can be made for topics such as: investigating animals (inverte-

*Lilo Hess, *Animals that Hide, Imitate, and Bluff,* Scribner, New York, 1970.

· USE 1 OR MORE ROCKS TO MAKE AN INSECT.

· THE ROCKS CAN BE MARKED WITH PAINTS OR MARKERS.

· IF TWO OR MORE ROCKS ARE USED, GLUE THEM TOGETHER.

IS YOUR "INSECT" LIKE A REAL ONE? HOW?

MAKE SOME ANIMAL OTHER THAN AN INSECT.

Figure 10-1. Activity card: Rock animals.

brates, fish, amphibians, reptiles, birds, mammals) in their natural environments, finding examples of camouflage in animals, writing or telling adventure stories about animals, and keeping pets.

Bulletin boards

ANIMALS WORK FOR US
Make a bulletin board illustrating work that animals have done for people in the past, as well as work they still do—or better yet, have your students help you make the bulletin board. Examples include horses, some dogs, and oxen. Students could also identify laboratory animals and make a bulletin board showing them.

ANIMALS MOVE
Label sections of a bulletin board (see Figure 10-2) with ways in which animals move—flying, walking, swimming, creeping, etc. Have the student cut pictures of animals from magazines or other sources and place each picture in a space that identifies one way each animal moves. The bulletin board will not only make students aware that animals use various means of locomotion but also help them understand how several specific animals move.

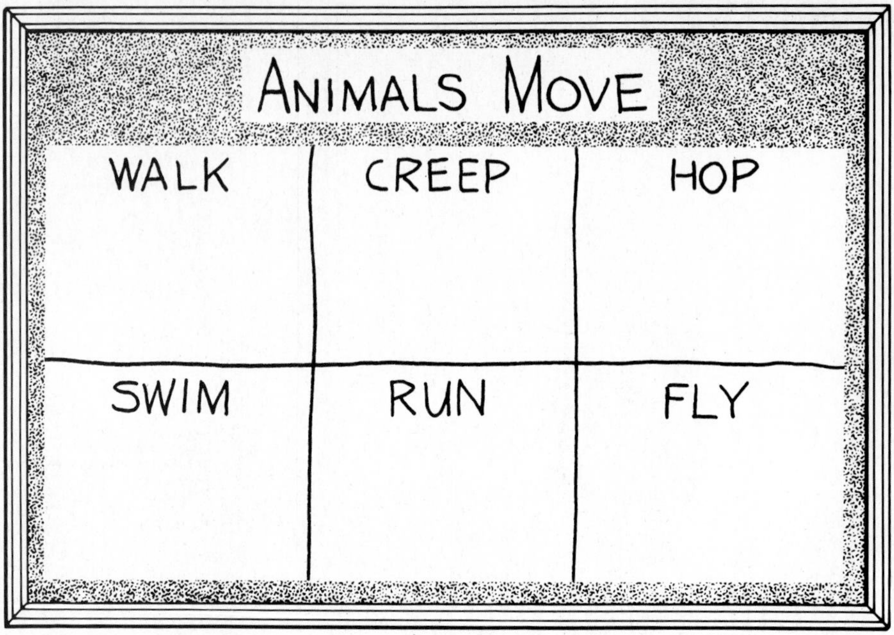

Figure 10-2. Bulletin board: Animals move.

Field trips

A local zoo is an especially good place to go on a field trip while studying animals. Many kinds of animals will probably be found in the zoo. Prepare the students ahead of time to look for certain things—for example, names of animals from each group they are studying. The students can also find out what the various animals are fed, and they can search for similarities among animals which live in a certain habitat.

If a zoo is not available, you could plan a visit to a natural history museum or a pet shop. In addition, a field trip to almost any outdoor location will reveal a sizable assortment of animals. A visit to the school grounds, a vacant lot, a field outside town, or a local forest should provide a rich experience in studying animals.

ANIMALS FOR INTERMEDIATE GRADES

Activities

➤ ANIMAL FAIR

To acquaint your students with the various major groups of animals used by scientists, plan an animal fair. Include invertebrates as well as fish, amphibians, reptiles, birds, and mammals. Ask the students to name five animals each for the fish, amphibian, reptile, bird, and mammal groups and fifteen animals for the invertebrate group. (Encourage the students to identify invertebrates from several subgroups within the invertebrates,

Animal skins, such as these bird skins, can often be borrowed from a local museum or university to make the study of animals more realistic.

even though the subgroups themselves will not be identified as part of the activity.) They can use biology textbooks and natural history guides to help them find the animals. Once all the animals have been named, the students can help you obtain a specimen of each animal. The specimens may be live or preserved. One good source of assistance is science teachers in grades 7 to 12.

After specimens have been located, form the class into six groups—one group of students for each animal group listed above. Each group of students should then prepare a brief information sheet for their animal group, identifying the major characteristics which differentiate their animals from those in other groups. Each group of students should then present their information sheet to the whole class and point out the visible characteristics which have been listed for each of their specimens. Have the students post their information sheets near the specimens to which they relate and leave them on display for a few days after the fair is over.

➤ POND WATER

A very good source of small living animals is pond water. A sample of pond water from a good location—rich in plant life and with a rich, mucky bottom—will produce many interesting creatures. But clear water over a clean sandy bottom will not provide such an abundance of animal life. You may find it more practical to collect the samples of pond water yourself; but if you are able to let your students do it, have them collect samples from several locations in the pond and label each sample carefully. Then they will be able to compare the numbers and kinds of organisms present in the different samples.

Figure 10-3. Microscopes can be useful for observing small living organisms.

While some organisms from pond water may be visible with the unaided eye, most will have to be viewed with a microscope. Many good and inexpensive microscopes are available from supply houses which specialize in science supplies for elementary schools (see the Appendix); or a secondary school science teacher may be able to help you borrow microscopes for this study.

Tell the students to watch one organism carefully (see Figure 10-3). Have them make a rough sketch and try to identify the organism. Reference books or a resource person may be very helpful in identification. Can students see their chosen organism eating? What does it eat? In turn, what might eat the various organisms found in the pond water? Do fish depend upon such organisms?

If there are noticeable differences among samples, have the students propose reasons for the differences. Ask the students where they would go fishing in that pond if they wanted to catch the most fish. Why there?

➤ AQUATIC INSECTS

Students are seldom aware of the abundance of animals in a small area and are often surprised when they discover that an apparently quiet place is teeming with animals. That is certainly true about life on or near the bottom of a pond, and a field trip to a pond can be very rewarding. Dredge a fishnet (at least 15 centimeters or 6 inches in diameter) quickly along the bottom of the pond (see Figure 10-4). Then hold the opening of the net above water to allow the water and much of the soil to drain away.

Figure 10-4. A fishnet can be used to discover a variety of life forms.

Wearing a rubber glove, carefully pull your finger through the remaining material. Several insect larvae and other animals will probably be present. Using a book such as George Reid's *Pond Life,** try to identify some of the animals.

Sample the bottom in several locations. If differences are noted, ask students to try to explain them. There are many ways to approach the study of aquatic animals. For instance, would the same kinds and numbers of animals be found at all times during the daytime? How about at night? Would the kinds and numbers of animals present remain the same throughout the year? Would it be possible to watch some of the larvae develop into their adult forms? What kind of environment would have to be provided in order to make this observation possible? Try to set up an appropriate environment in the classroom, and encourage students to make drawings of the animals as they go through various stages. Ask them how what they have learned about aquatic insects would be useful to a person who enjoys fishing.

➤ BIRD MODELS

Students may have been told that eagles are large, majestic birds. But just how large are they? Without seeing birds at close range, it is difficult to comprehend their sizes. This activity will help students gain a perspective on the actual sizes of birds that are often seen from a distance or discussed.

Use a bird guide or a good encyclopedia to find out the sizes of several birds that you think may interest your students. Have students sketch freehand outlines of the birds, in flying position, on clean, discarded cardboard boxes; or the outlines can be traced from opaque projections. After cutting the shapes, the students may either decorate the cutouts to resemble the birds being represented or leave them as silhouettes. (See Figure 10-5, page 208.)

Hang the models in the room or in a hallway, and compare the sizes of the birds. Are the students surprised? Is the hummingbird the size they thought it would be? How about the eagle? Did some of the students make cutouts of any birds larger than eagles? Is a blackbird larger or smaller than a robin?

Have the students also make larger-than-life models of the beaks and feet of several birds and place the beaks and claws in groups based on their shapes. Is there any special similarity among the birds with similar-shaped beaks? Claws? What do the different kinds of birds eat? Look at pictures of some unfamiliar birds. Have the students guess what those birds eat, on the basis of their beak and claw shapes, and then do some research to find out what they really eat. How accurate were the students' predictions?

*George Reid, *Pond Life,* Golden Press, New York, 1967.

Figure 10-5. Silhouettes of birds.

➤ MEASURING GROWTH

How rapidly does a baby rat (or mouse, gerbil, or hamster) grow? This activity will not only answer that question but also give students valuable practice in measuring and graphing.

Adopt a pregnant rat as a classroom pet (see "Animals in the Classroom"). Starting the second day after the rats are born, weigh them each day at the same time. Handling the young animals is permissible—just be sure to avoid unduly upsetting the mother or exposing your hand to her teeth. Since mothers are very protective of their litters, it would be a good idea to use a large plastic cup to transfer the mother to another cage while the babies are being weighed. Each day, record the weight of each rat in grams, and chart their growth on a graph (see Figure 10-6). Continue measuring at least until 2 weeks after the rats are weaned and separated from their mother.

Consider the following questions: During which week did the most rapid increase in weight take place? Did all members of the litter have their most rapid growth during the same week? Did males and females grow at the same rate? Did the rats which were originally smallest remain the smallest?

Interested students should be encouraged to do similar measurements of other animals, if possible. Are the growth graphs similar for all types of animals? Does litter size seem to make a difference? Does a second litter from the same mother grow at the same rate as the first litter? If there are differences, what may have caused them?

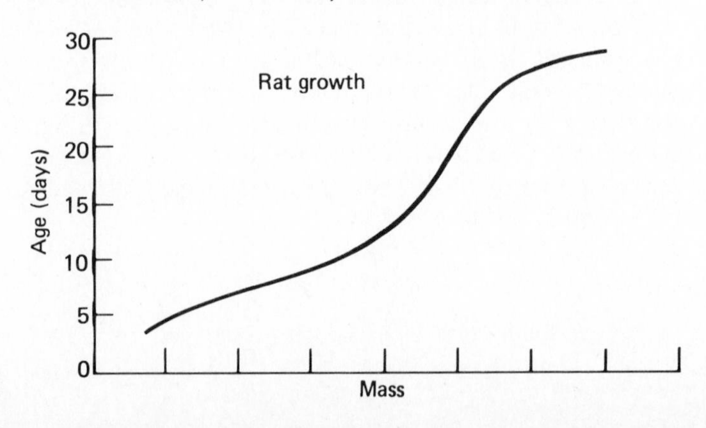

Figure 10-6. A graph can be used to show the growth rates of animals.

➤ MEALWORMS

Mealworms are readily available, and many experiments can be performed with them. They can be obtained from local pet shops, where they are sold as pet food. They are clean animals and are easily kept for classroom use.

Each student can be given two or three mealworms in a small container to care for and observe. Some students will note that their mealworms undergo dramatic changes. Mealworms pass through the following stages: egg–mealworm–pupa–adult beetle. (The adult does not fly or otherwise easily escape from its container.) Discuss life cycles of insects with students. Ask them to find out whether or not all insects go through the same stages as the mealworm. *(The grasshopper is an example of one insect that does not.)* Have the class prepare a chart depicting the different types of life cycles which insects go through.

Next have the students design and conduct experiments for answering these and other questions: Do mealworms prefer bright light, darkness, or something in between? Which of four breakfast cereals do mealworms prefer? Do mealworms prefer dry or moist places? What temperatures do they prefer? Many other questions can be posed.*

Encourage the students to think carefully about the methods they select to find out the answers. Have they controlled enough of the possible variables so that their experiments really do answer the questions?

➤ ANIMAL HOMES

Have students collect pictures or make drawings of animal homes, then display them on a bulletin board or incorporate them into a scrapbook. Each picture or drawing should be accompanied by a brief information card which tells the location of the home, the animal which lives in it (if these facts are not obvious from the picture or drawing), and how the home functions as protection for the animal. What is the relationship of the home's location to the animal's source of food?

➤ SKELETON

Challenge interested students to try to obtain a complete skeleton of some animal. The skeleton of a broiled fish is relatively easy to obtain. To save the skeleton of a cooked chicken is a slightly more difficult challenge, but it can be done. With a little luck, other skeletons can be found on farms, on ranches, or in forests. Sometimes an animal dies and the remains are disturbed only by scavengers, which remove most of the remains except the bones and possibly the skin. (See Figure 10-7 for examples of skeletons.)

To learn how to treat and prepare the skeletons, the students will have to do a considerable amount of research. A well-displayed skeleton can become a matter of pride. Skeletons can also become puzzles: give the students a box of bones which constitute the complete skeleton of some

*An excellent source of ideas is the Elementary Science Study unit *Behavior of Mealworms* (see "Building Your Own Library").

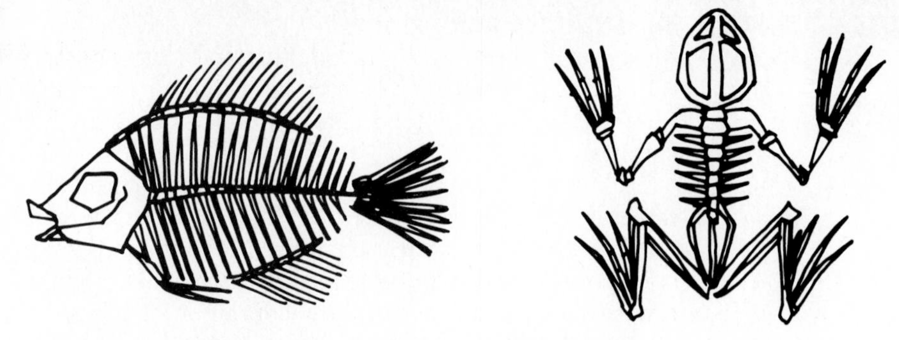

Figure 10-7. Animal skeletons can be used as puzzles.

animal and only the instructions, "Reconstruct the skeleton." The students will be amazed at how well they can reconstruct the skeleton without directions and will learn a great deal about bones and joints.

You can mark the individual bones of another skeleton with letters, to make discussing them convenient. Here are some questions to talk over with students: What advantage is there in the size and shape of the leg bones? What do you think is the function of the bone with the letter A painted on it? Why is the joint near the letter B better suited to be in that location than the joint near the letter C?*

ANIMAL TRACKS

When studying animal tracks is mentioned, students usually expect to learn to identify the animals which made specific tracks—and identification is certainly a legitimate track activity. If you put tempera paint on the feet of classroom or "borrowed" animals (hamsters, mice, gerbils, dogs, cats, ducks, etc.), they will make visible track prints when they walk across paper of a contrasting color.† Students can examine the prints for several characteristics, including number of toes and symmetry. They can also relate the size of the animal to the size of the tracks.

Students can be challenged to obtain other information from real animal tracks. What does the animal eat? In what direction is the animal moving? How fast is it moving? What is the animal's size?‡

Students can also cut "tracks" out of potatoes or carrots. Using stamp pads or tempera paint, they can create stories involving one or more animals by making track prints on paper. This activity leads quite nicely into language experiences.

*Additional suggestions for the study of skeletons may be found in the Elementary Science Study unit *Bones* (see "Building Your Own Library"). Inexpensive skeletons to be used with the *Bones* unit can be purchased from a science supply house (see the Appendix).

†*Note:* Be sure to wash or wipe the residue off the animals' feet as soon as the prints have been made.

‡For several more teaching ideas for animal tracks, refer to the Elementary Science Study unit *Tracks,* listed in "Building Your Own Library."

Figure 10-8. Mississippi migratory flyway—the migratory path of many ducks and geese.

ANIMAL MIGRATION

It is widely known that ducks and geese migrate annually (see Figure 10-8), and another fairly well-known migration is that of the monarch butterfly. However, there are several other animals that also migrate. Challenge students to identify as many specific migrating animals as possible.

Mark the migration routes of the animals on one or more maps. (Separate maps for birds and for other animals may be appropriate.) Find out the location of wildlife preserves controlled by the U.S. Fish and Wildlife Service (see the Appendix). Is there any relationship between the times when game birds migrate and the hunting seasons along the migratory routes?

Why do animals migrate? Why don't they just stay in one place where the weather is moderate all year? Does a given individual animal go to the same two places each year? Have students propose answers to these questions and to others which you and they think of. The answers can then be verified either by library research or by talking with a wildlife biologist.

Have students plan a wildlife refuge for migrating wildlife. Where would the refuge be located? Would arrangements have to be made for ensuring the safety of the wildlife? What about arrangements for food for the animals?

BECOMING A PET

Tell your students, "Imagine that you are some small animal out in the woods." Then go on: "A child who is walking through the woods notices you. You are such a fascinating animal that the child catches you in an empty pickle jar. What are your thoughts and concerns?" Have the students think for 3 or 4 minutes before responding out loud. They may write down their areas of concern if they wish.

Then have a class discussion in which the students are free to express their thoughts and concerns. Bring up the problems involved in keeping wild animals as pets. Encourage the students to conclude that wild animals should not be kept as pets—that, at *most,* wild animals might be observed for a day and then released exactly where they were originally found. Why exactly? *(Consider habitat requirements and the territorial nature of many animals.)*

Children's literature

Read *Gifts of an Eagle** to your students or have them read it, and then discuss with them the concepts presented.

Find other books on animals in the school library, and identify them for the students or place them in an interest center.

*Kent Durden, *Gifts of an Eagle,* Simon and Schuster, New York, 1972.

Figure 10-9.
Activity card:
Design an animal.

Activity cards

An example of an activity card is shown in Figure 10-9. Additional activity cards could be made for topics such as investigating the pill bug (or any other specific animal), caring for a cocoon, finding examples of one group of animals (e.g., reptiles), identifying special adaptations which help animals survive, collecting insects, and caring for a pet.

Bulletin boards

LOCAL ANIMALS

Label a bulletin board "Local Animals" and have students who have seen and identified local wild animals post pictures or drawings of the animals along with short statements about where they were seen. After several have been posted, group them as invertebrate, fish, amphibian, reptile, bird, or mammal. Tell students to look especially for members of groups which have not yet been seen or have been seen only seldom.

ANIMAL CLASSIFICATION

Divide a bulletin board into six areas and label the areas "Invertebrates," "Fish," "Amphibians," "Reptiles," "Birds," and "Mammals." Paste pictures or drawings of animals on one side of index cards and write the group to which they belong on the reverse side—or have the class help you make the cards. Tell the students to post the pictures in what they think are the proper groups. After all the cards have been posted, the students can check the reverse sides of the cards to see if the animals were properly classified.

Field trips

Especially appropriate for intermediate students would be a field trip to a local museum which has study collections of animals. Ask the educational director of the museum to help you arrange for a visit which will allow students to see several animals from each group which they have studied. Attempt to have several local animals included.

Other field trips could be taken to a local wildlife refuge, a zoo, and any outdoor environment. Tell the class to observe similarities and differences within and among groups of animals.

RESOURCES

Sources of assistance

FREE AND INEXPENSIVE MATERIALS

Here are some good sources of free and inexpensive materials for use in teaching about animals:

1. *American Humane Association.** Pet care information. A unit plan and several other aids.
2. *National Wildlife Federation.* Reprints from *Ranger Rick.* Information on endangered species.
3. *Pendleton Woolen Mills.* An information kit with samples on wool processing.
4. *U.S. Department of the Interior, Fish and Wildlife Service.* Conservation notes which describe at least 20 different animals.
5. *Animal Protection Institute of America.* Information on the responsibilities of pet ownership.

LOCAL RESOURCES

Local resources which can help in teaching about animals include:

1. Zoos, museums, and universities
2. Members of a local humane society
3. Audubon Club members
4. Pet shops
5. Professional or serious amateur fishers or hunters, who often know a great deal about animal behavior

*The addresses of specific agencies, organizations, and businesses are listed in the Appendix.

ADDITIONAL RESOURCES

1. Magazines such as *Ranger Rick, National Geographic, National Geographic World, Audubon, National Wildlife, International Wildlife,* and *3–2–1, Contact.*
2. Many state departments of natural resources (which sometimes have other titles, such as fish and wildlife department) prepare teaching materials, including pictures and printed information for use by teachers.
3. State departments of agriculture can provide information on many domestic animals.
4. Boy Scout, Girl Scout, and 4-H Club leaders usually have available handbooks of activities on many science topics, including various kinds of animals.

Strengthening your background

1. Visit a large city museum of natural history and a zoo. Make photographs (preferably slides) of several fish, amphibians, reptiles, birds, and mammals. Also attempt to photograph or obtain slides of invertebrates.
2. Keep a pet rodent for at least 2 months. Consult frequently with pet shop personnel (or someone you know who keeps an animal of the same kind) about proper care for the animal. If your rodent is female, try to raise a litter.
3. Sit quietly by yourself or with one other person near a small pond for an hour in the morning or evening and observe animals.
4. Visit a chicken or fish hatchery and ask the manager or a staff member to show you the entire operation and explain it.
5. Learn to identify and name 10 local birds.
6. Touch an amphibian and a reptile. Do they feel different?
7. Have a student help you find a cocoon.

Getting ready ahead of time

1. Collect at least 50 pictures of different animals from magazines and other sources.
2. Prepare a guided fantasy about children visiting a zoo.
3. Prepare at least five "shining light" cards for each of the following topics:
   a. Animal classification (e.g., "Which animal is not a reptile?" "Which animal is not a mammal?")
   b. Caring for pets
   c. Characteristics of animals in a particular group.
4. Obtain free or inexpensive teaching materials about animals from at least 10 different sources.
5. Obtain the skull of some animal or—preferably—a whole skeleton.
6. Prepare materials for a bulletin board on animals.

ANIMALS

Allen, William H., Jr.: *Animals in the Classroom,* Elementary Science Study (ESS), Webster, McGraw-Hill, Manchester, Mo., 1970.

Axelrod, Herbert R., and Rolf Bader: *The Educational Aquarium,* Tropical Fish Hobbies, Jersey City, N.J., n.d.

Brin, Ruth F.: *Butterflies Are Beautiful,* Lerner, Minneapolis, 1974.

Buckstaum, Ralph: *Animals without Backbones,* University of Chicago Press, Chicago, 1975.

Carr, Archie: *The Reptiles,* Young Readers Edition, Life Nature Library, Time-Life, New York, 1977.

Carrington, Richard: *The Mammals,* Young Readers Edition, Life Nature Library, Time-Life, New York, 1977.

Clarkson, Jan Nagel: *Tricks Animals Play,* National Geographic Society, Washington, 1975.

Cooper, Elizabeth K.: *Science in Your Own Backyard,* Harcourt Brace Jovanovich, New York, 1958.

Day, Jennifer: *What Is an Insect?* Golden Press, New York, and Western Publishing, Racine, Wis., 1976.

Eimerl, S., and I. DeVore: *The Primates,* Young Readers Edition, Life Nature Library, Time-Life, New York, 1977.

Elementary Science Study (ESS): *Animals in the Classroom,* Webster, McGraw-Hill, Manchester, Mo., 1970.

———: *Behavior of Mealworms,* Webster, McGraw-Hill, Manchester, Mo., 1976.

———: *Bones,* Webster, McGraw-Hill, Manchester, Mo., 1968.

———: *Brine Shrimp,* Webster, McGraw-Hill, Manchester, Mo., 1976.

———: *Butterflies,* Webster, McGraw-Hill, Manchester, Mo., 1970.

———: *Crayfish,* Webster, McGraw-Hill, Manchester, Mo., 1968.

———: *Earthworms,* Webster, McGraw-Hill, Manchester, Mo., 1970.

———: *Eggs and Tadpoles,* Webster, McGraw-Hill, Manchester, Mo., 1974.

———: *Mosquitoes,* Webster, McGraw-Hill, Manchester, Mo., 1971.

———: *Tracks,* Webster, McGraw-Hill, Manchester, Mo., 1971.

Engel, Leonard: *The Sea,* Young Readers Edition, Life Nature Library, Time-Life, New York, 1977.

Naden, Corinne: *Let's Find Out about Frogs,* Watts, London, 1972.

National Geographic: *Creepy Crawley Things: Reptiles and Amphibians,* National Geographic Society, Washington, 1974.

Nickelsburg, Janet: *Nature Program for Early Childhood,* Addison-Wesley, Reading, Mass., 1976.

Ommanney, F. D.: *The Fishes,* Young Readers Edition, Life Nature Library, Time-Life, New York, 1977.

Orlans, Barbara: *Animal Care from Protozoa to Small Mammals,* Addison-Wesley, Reading, Mass., 1977.

Peterson, Roger T.: *The Birds,* Young Readers Edition, Life Nature Library, Time-Life, New York, 1977.

Rinard, Judith E.: *Wonders of the Desert World,* National Geographic Society, Washington, 1976.

Ross, Wilda: *Cracks and Crannies: What Lives There,* Coward-McCann, New York, 1975.
Timgergen, Niko: *Animal Behavior,* Young Readers Edition, Life Nature Library, Time-Life, New York, 1977.
Whitlock, Ralph: *Spiders,* Raintree, Milwaukee, Wis., 1976.
Zim, Herbert S., and Clarence Cottam: *Golden Nature Guide to Insects,* Golden Press, New York, 1951.
———and Ira Gabrielson: *Golden Nature Guide to Birds,* Golden Press, New York, 1956.
———and Donald Hoffmeister: *Golden Nature Guide to Mammals,* Golden Press, New York, 1955.
———and Lester Ingle: *Golden Nature Guide to Seashores,* Golden Press, New York, 1955.

Building a classroom library

Arnosky, Jim: *Crinkleroot's Book of Animal Tracks and Wildlife Signs,* Putnam, New York, 1979.
Behrens, June: *Look at the Forest Animals,* Children's Press, Chicago, Ill., 1974.
Blough, G. O.: *An Aquarium,* Harper and Row, New York, 1959.
Bonners, Susan.: *Panda,* Delacorte Press, Dell, New York, 1978.
Branley, Franklyn M.: *Big Tracks, Little Tracks,* Thomas Y. Crowell, New York, 1975.
Brenner, Barbara: *On the Frontier with Mr. Audubon,* Coward-McCann, New York, 1977.
Brown V.: *How to Follow the Adventures of Insects,* Little, Brown, Boston, 1968.
Carrick, Carol: *The Blue Lobster: A Life Cycle,* Dial, New York, 1975.
Cole, Joanna: *Find the Hidden Insect,* Morrow, New York, 1970.
Compere, M.: *Dolphins,* Scholastic Book Services, New York, 1970.
Conklin, Gladys: *Tarantula: The Giant Spider,* Holiday, New York, 1972.
Dallinger, Jane: *Grasshoppers,* Lerner, Minneapolis, 1981.
Davis, Burke: *Biography of a King Snake,* Putnam, New York, 1975.
Earle, O.L.: *Praying Mantis,* Morrow, New York, 1969.
Eberle, Irmengarde: *Pandas Live Here,* Doubleday, Garden City, N.Y., 1973.
Facklam, Margery: *Wild Animals, Gentle Women,* Harcourt Brace Jovanovich, New York, 1978.
Ford, Barbara: *Black Bear: The Spirit of the Wilderness,* Houghton Mifflin, Boston, 1981.
Freedman, Russell: *Farm Babies,* Holiday, New York, 1981.
———: *Tooth and Claw: A Look at Animal Weapons,* Holiday, New York, 1980.
Friskey, Margaret: *The True Book of Birds We Know,* Children's Press, Chicago, Ill., 1981.
Garelick, May: *About Owls,* Four Winds, New York, 1975.
George, Jean Craighead: *The Wounded Wolf,* Harper and Row, New York, 1978.

Goudy, Alice E.: *Here Come the Bears,* Scribner, New York, 1954.
Gray, R.: *Children of the Ark,* Grosset and Dunlap, New York, 1968.
Gross, Ruth B.: *What Do Animals Eat?* Scholastic Book Services, New York, 1973.
Haley, Neale: *Birds for Pets and Pleasure,* Delacorte Press, Dell, New York, 1981.
Hartman, Jane E.: *Animals that Live in Groups,* Holiday, New York, 1979.
———: *How Animals Care for Their Young,* Holiday, New York, 1979.
Hess, Lilo: *Animals that Hide, Imitate, and Bluff,* Scribner, New York, 1970.
Hinshaw, Dorothy: *Beetles and How They Live,* Holiday, New York, 1978.
Hoover, H.: *Animals at My Doorstep,* Parents Magazine, New York, 1966.
Hopf, Alice L.: *Biography of a Snowy Owl,* Putnam, New York, 1979.
Horsburgh, Peg: *Living Light: Exploring Bioluminescence,* Messner, New York, 1978.
Hutchins, Ross E.: *Hop, Skim, and Fly: An Insect Book,* Parents Magazine, New York, 1970.
———: *A Look at Ants,* Dodd, Mead, New York, 1978.
Jacobs, Francine: *Coral,* Putnam, New York, 1980.
Kane, H. B.: *Wings, Legs, or Fins: How Animals Move,* Knopf, New York, 1965.
Lauber, Patricia: *Earth Worms: Underground Farmers,* Garrard, Champaign, Ill., 1975.
———: *What's Hatching out of That Egg?* Crown, New York, 1979.
Leinwoll, Stanley: *The Book of Pets,* Messner, New York, 1980.
Lowrey, L.F., and A. B. Carr: *What Can an Animal Do?* Western, New York, 1969.
——— and ———: *What Does an Animal Eat?* Golden Press, New York, 1969.
Mason, George F.: *Animal Feet,* Morrow, New York, 1970.
Mason, Robert G.: *The Life Picture Book of Animals,* Time-Life, New York, 1969.
May, Charles Paul: *A Book of Insects,* St. Martin's, New York, 1972.
McClung, Robert M.: *Bufo: The Story of a Toad,* Morrow, New York, 1954.
———: *Green Darner: The Story of a Dragonfly,* Morrow, New York, 1980.
———: *Peeper, the First Voice of Spring,* Morrow, New York, 1977.
McGovern, Ann: *Shark Lady: True Adventures of Eugenie Clark,* Four Winds, New York, 1979.
McNulty, Faith: *Whales, Their Life in the Sea,* Harper and Row, New York, 1979.
Mellanby, Kenneth: *Talpa: The Story of a Mole,* Collins, William, Cleveland, Ohio, 1977.
Milne, Margery, and Larus Milne: *Gadabouts and Stick-at-Homes: Wild Animals and Their Habitats,* Sierra, and Scribner, New York, 1980.
Parker, Bertha: *How Animals Get Food,* Harper and Row, New York, 1959.

Patent, Dorothy H.: *Butterflies and Moths, How They Function,* Holiday, New York, 1979.

———: *The Lives of Spiders,* Holiday, New York, 1978.

———: *The World of Worms,* Holiday, New York, 1978.

Pfadt, Robert: *Animals without Backbones,* Follett, Chicago, Ill., 1967.

Prince, J. H.: *How Animals Hunt,* Elsevier, Amsterdam, and Nelson, Camden, N.J., 1980.

———: *How Animals Hunt,* Elsevier, Amsterdam, and Nelson, Camden, N.J., 1980.

Ricciuti, Edward R.: *Sounds of Animals at Night,* Harper and Row, New York, 1977.

Rinard, Judith E.: *Zoos without Cages,* National Geographic Society, Washington, 1981.

Scott, Jack D.: *City of Birds and Beasts,* Putnam, New York, 1978.

———: *Little Dogs of the Prairie,* Putnam, New York, 1977.

Selsam, Millicent E.: *How to Be a Nature Detective,* Scholastic Book Services, New York, 1975.

———: *Night Animals,* Four Winds, New York, 1980.

——— and Joyce Hunt: *A First Look at Animals without Backbones,* Walker, New York, 1977.

——— and ———: *A First Look at Sharks,* Walker, New York, 1979.

Shapiro, Irwin: *Darwin and the Enchanted Isles,* Coward-McCann, New York, 1977.

Shapp, Charles, and Martha Shapp: *Let's Find Out about Animal Homes,* F. Watts, New York, 1962.

Silverstein, Alvin, and Virginia B. Silverstein: *Mice: All about Them,* Lippincott, Philadelphia, 1980.

Simon, Seymour: *Animals in Your Neighborhood,* Walker, New York, 1975.

———: *Poisonous Snakes,* Four Winds, New York, 1981.

———: *What Do You Want to Know about Guppies?* Four Winds, New York, 1977.

Snyder, Gerald, S.: *Is There a Loch Ness Monster? The Search for a Legend,* Messner, New York, 1977.

Steiner, Barbara: *Biography of a Polar Bear,* Putnam, New York, 1972.

Stemple, David: *High Ridge Gobbler: The Story of the American Wild Turkey,* Collins, London, 1979.

Stonehouse, Bernard: *A Closer Look at Reptiles,* F. Watts, New York, 1979.

Stoutenberg, A.: *Vanishing Thunder,* Natural History Press, Doubleday, Garden City, N.Y., 1967.

Strong, Arline: *Veterinarian, Doctor for Your Pet,* Atheneum, New York, 1977.

Waters, John: *Creatures of Darkness,* Scholastic Book Services, New York, 1977.

Weber, William J.: *Care of Uncommon Pets,* Holt, New York, 1979.

Zim, Herbert S.: *Bones,* Morrow, New York, 1969

CHAPTER 11

THE HUMAN BODY

IMPORTANT CONCEPTS

Cells, the "building blocks" of all living things, make up tissues which in turn make up organs which are parts of body systems which finally make up an organism.

The skeletal system is the body's support and protective system.

Muscles create movement in and of the body.

The digestive system prepares foods for use by the body.

The respiratory system provides oxygen for the body.

The circulatory system distributes food, oxygen, and water throughout the body and removes waste materials to appropriate organs.

The nervous system is the controlling system of the body.

Several body actions are automatic.

People are products of both heredity and their environment.

People, like other living things, have a life cycle.

For the body to function properly (be physically fit), it must receive the appropriate kinds and amounts of food, exercise, and rest.

Our bodies combat microorganisms which might cause diseases by creating conditions unfavorable for those particular microorganisms.

A balanced diet includes several types of foods.

The misuse of drugs and certain other materials harms people's bodies.

INTRODUCTION: STUDYING THE HUMAN BODY

The topic "the human body" is often studied in a separate health curriculum in the elementary schools, since it is so important to students. In the activities sections of this chapter, only a few of the many topics related to the human body have been included. If you judge it desirable to study topics not included in this section (the reproductive system, for example), you can adapt the ideas presented here for use with them. It is especially important to cover a broader range of topics if your students do not have a separate health curriculum. (Dental care is another important topic; however, it is likely to be part of a health curriculum.)

Activities for this chapter focus upon the component parts of the body, their functioning, and major aspects of their care. Many of the activities are included because they offer good opportunities for students to sharpen their application of the scientific processes.

THE HUMAN BODY FOR PRIMARY GRADES

Activities

➤ SKELETAL SYSTEM

Make outlines of the bodies of four or five students by having them lie on large pieces of paper and using a pencil to trace around their bodies. Then in small groups (one body outline to a group) have students feel their own bodies to detect bones. Each group should draw as many bones into their body outline as they can feel (see Figure 11-1). After carefully identifying

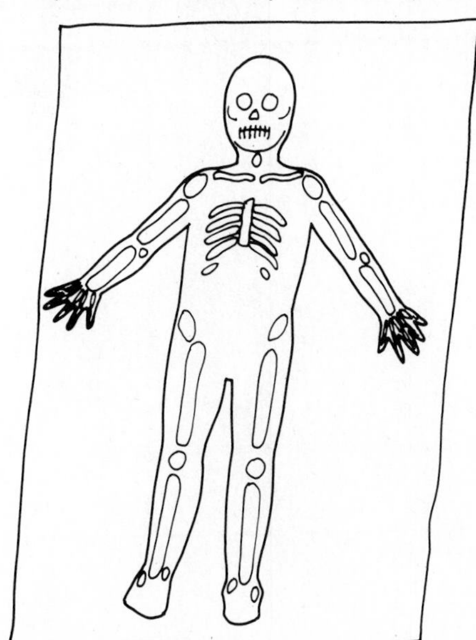

Figure 11-1. Children can draw their own skeletons.

and drawing as many bones on the outlines as possible, the groups can compare their drawings. Students can then feel for and add to their drawings additional bones which they saw on other groups' drawings. Finally, allow the students to compare their drawings with drawings of a skeleton in some reference book.

Ask the students to suggest what function the bones in our body serve. What might we look like if we had no bones? What if we all awakened some morning and had no bones? How would we and other people move around? What would a "day without bones" look like?

▶ X-RAYS

Ask your doctor or a bone specialist to give or lend you a few x-rays which show a significant number of the bones of a human body. Try to obtain some x-rays which show rather common bone fractures and others which show arm or leg joints.

Show your students the x-rays and have them attempt to locate the same bones in their bodies. Have them especially notice bone ends—often joints. What do the joints look like? Can they find similar joints in other parts of their bodies?

X-rays permit us to see our bones.

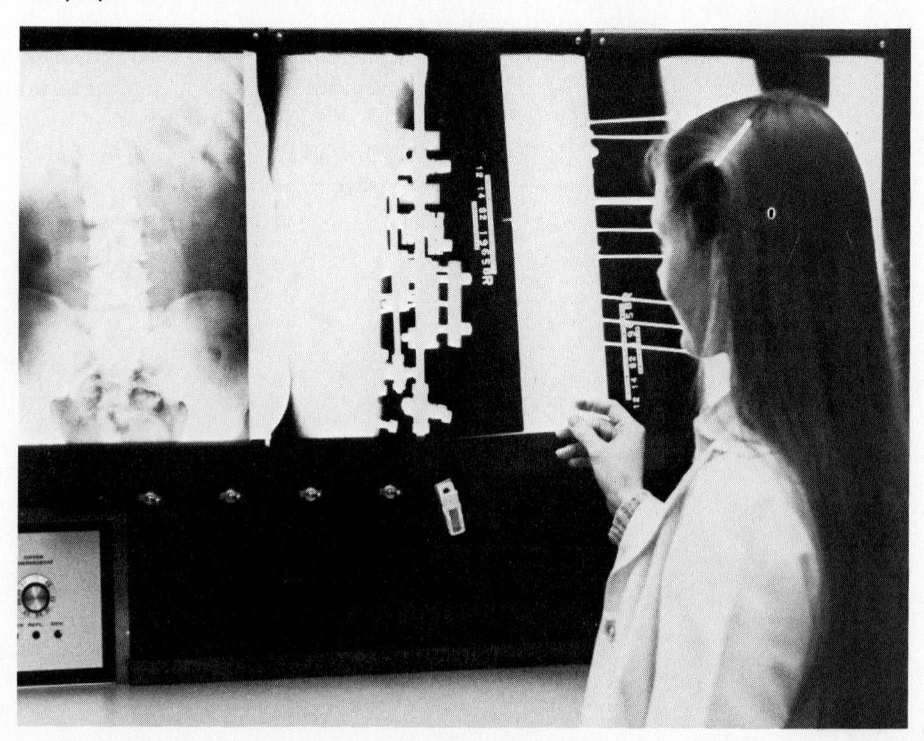

➤ BONE JOINTS

Use a whole raw chicken as a vivid illustration of the structure of a skeleton, and especially to demonstrate different types of bone joints. Let students move the various parts of both the legs and wings. What types of motion are possible at each joint? Do some joints allow a circular motion? Do some move only back and forth, like a hinge?

Cut into a chicken leg* and let the students look at the joint. Point out that this joint allowed a certain type of motion. Ask the students to guess which if any of the wing joints are similar in shape. Remove the meat from the wing joint and check the prediction. Now examine another type of leg joint. Are there any wing joints like this one? Is it likely that the joints in the chicken's neck are similar to any of the types examined so far?

Have students name joints in their bodies which they think may be similar to those in the chicken's leg. Are they located in similar places?

Let the students watch you while you carefully remove as much meat as possible from the skeleton. Note the muscles: How are the muscles connected to the bones? Boil the skeleton so that you will be able to remove the remaining meat. Set all the bones aside in a convenient place until they dry thoroughly. The students will then be able to examine a complete skeleton—and they will also have a bone puzzle to put together.

Your collection of bone joints can be enlarged by visiting a local butcher.

➤ FINDING AND LISTENING TO YOUR HEARTBEAT

Ask the school nurse to visit your class, bringing one or—preferably—several stethoscopes, and to show the students how to listen to their heartbeats. Ask the students to describe the sounds of their heartbeats. Does a person's heartbeat sound the same standing up as lying down? Sitting? What happens if the students run in place for 1 minute and then listen again?

Have students count the number of times their own hearts beat in 30 seconds, after they have been relatively inactive for at least 15 minutes, then multiply by 2 to obtain the number of heartbeats per minute. Record all the students' pulse rates on the chalkboard. What was the fastest rate? The slowest? What appears to be a common rate for students in the class?†

➤ BREATHING RATE

Have students count the number of times they inhale in 1 minute, after being relatively inactive for at least 15 minutes. (Only complete breaths should be counted—i.e., make sure they do not count inhaling and exhal-

**Note:* Do the cutting yourself; don't let children use sharp knives or scissors.

†*Note:* Students with pulse rates lower than 60 or higher than 110 (resting pulse rate) should be unobtrusively referred to the school nurse for examination.

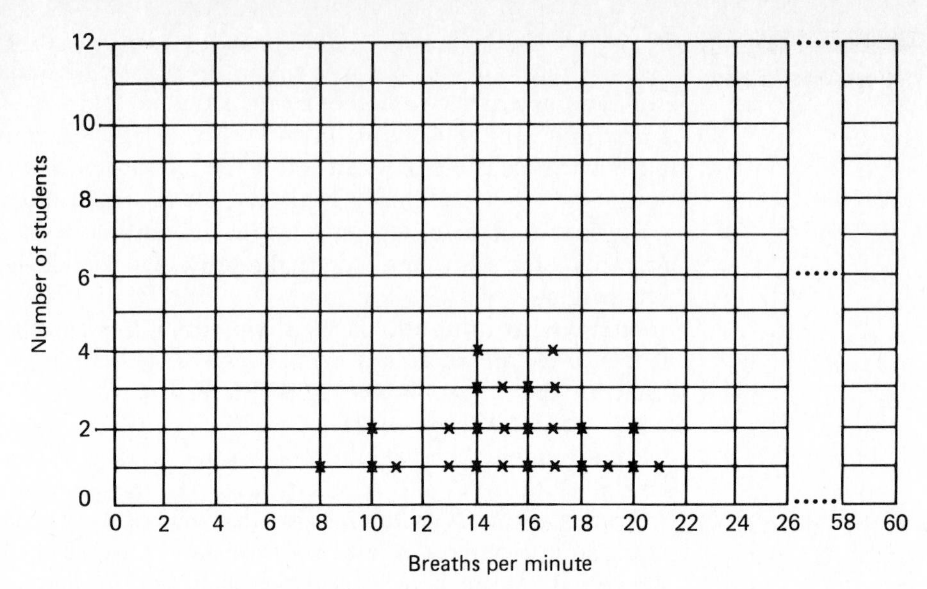

Figure 11-2.
Breath histograms.

ing as separate breaths.) Breathing rates of most elementary school students fall between 10 and 25 breaths per minute. Draw a grid on the chalkboard similar to the one illustrated in Figure 11-2, and have each student record her or his own breath rate, making a histogram. Each student should place an X on a horizontal line above the appropriate breath-rate number. The first student with a given rate places an X on the first line, the second student with the same rate places an X on the second line, etc.

After all rates are recorded, ask the students to identify the lowest rate and the highest rate. Ask the children what they think would happen to their breathing rates if they exercised. Then have them run in place or jump up and down for 1 minute and determine their breathing rates again. This time, plot the rates using 0's. What can students conclude from comparing the X's and the 0's?

CELLS: BUILDING BLOCKS

To introduce the concept of the cell as the building block of all living things, have your students examine some living cells of the plant elodea, which can be easily and inexpensively obtained at most tropical fish shops and pet stores. Place one leaf from the plant on a microscope slide and let students look at it at 100X magnification, which will make the cells easily visible. Make sure the students know that the little "boxes" which they see in the leaf are individual cells and that the whole leaf is made up of cells. What can be seen if the cells are watched closely? *(Movement of material in the cells.)* Is anything especially visible? *(Students will probably notice the green pieces of matter—chlorophyll, the material used in food production by green plants.)*

Try to obtain some prepared slides of cells from the human body; you may be able to borrow them from a high school or college biology teacher. Ask the students to describe similarities and differences among human cells, and also between elodea calls and human cells. Tell the students that our whole body is made up of different kinds of cells.

➤ PEOPLE'S LIFE CYCLES

People, like other living things, have life cycles. We begin life as a fertilized egg in our mother's body, we are born, we grow and mature, we may reproduce, and we finally grow old and die. Have students collect pictures from magazines or make drawings which represent the human life cycle.

Students can do sequencing activities with each other's sets of pictures. One thing which children should begin to think about is that just as birth and growing up are natural processes, so is death. Ask students whether all people go through all the stages in the human cycle. Why might some not? *(Accidental death, disease.)* In what ways is it nice to be young? Middle-aged? Old?

➤ CLEANLINESS

To help us stay healthy, we practice cleanliness. Ask students to name some things we do to keep our bodies clean, and some ways we try to make sure the things we eat and drink are clean. Post pictures on a bulletin board to illustrate how we practice personal cleanliness *(brushing teeth, washing, grooming, etc.)* and how we make our food and beverages clean *(cooking, ensuring cleanliness in food-processsing plants, purifying water, etc.).*

We can help students to be aware of proper practices for personal cleanliness and encourage them to use those practices.

➤ FOOD GROUPS

Students should learn to understand that their bodies need a diet which includes a balanced intake of carbohydrates, fats, proteins, vitamins, minerals, and water. Have them cut out magazine pictures of foods representative of each type. Label four boxes "Minerals and Vitamins," "Proteins," "Fats," and "Carbohydrates," and have the students place each picture in the box they consider appropriate.

Then remove some of the pictures, one at a time, and ask the students if each one really does belong in the box in which it was placed. Could it also be placed in one or more other boxes? *(Almost always yes.)* Do any of the pictured foods contain water? *(Almost all foods do.)* Minerals and vitamins are seldom the food group with which a food is primarily identified—because minerals and vitamins, like water, are found in virtually all foods.

Stress that our bodies function best when we receive appropriate amounts of each kind of food. A balanced diet is attained by eating foods from each of the four food groups: meats, breads, dairy foods, and fruits and vegetables. Ask the students to write down what they ate on the previous day and determine whether or not foods from each group were included. As an alternative, ask them to keep a record for a week, showing the foods from each group that they eat each day.

Have the students plan meals, using the food pictures they collected. After one child plans a breakfast, lunch, or supper, another child can check to see if the four groups are represented.

➤ CREATE AN ANIMAL

Show the students leg bones from two or three different animals—say, a chicken, a pig or sheep, and a cow (you may be able to get them from a local butcher). Tell them these are leg bones, and ask them to try to picture what kind of animal each bone came from. Then have them create an imaginary animal from which each bone was taken; they should describe the size of the animal and tell a little about what it might usually be seen doing.

Make sure that the students understand, by the end of the activity, what animal each bone actually did come from.

Children's literature

Read *The Skeleton Inside You** to your students or have them read it, and then discuss with them the concepts presented.

Find other books on the human body in the school library, and point them out to the students or place them in an interest center.

*Philip Balestrino, *The Skeleton Inside You,* Thomas Y. Crowell, New York, 1971.

- ASK A DOCTOR OR A NURSE WHAT KINDS OF FOOD WOULD MAKE HEALTHY SNACKS FOR YOU WHEN YOU GET HOME IN THE AFTERNOON. MAKE A LIST.
- WHICH OF THOSE "HEALTHY" SNACKS DO YOU LIKE?
- ASK YOUR PARENT(S) TO GET YOU SOME OF THE HEALTHY SNACKS THAT YOU LIKE.

Figure 11-3. Activity card: Identifying healthy foods for snacks.

Activity cards

An example of an activity card is shown in Figure 11-3. Additional activity cards should be made for topics such as checking pulse rates at different times of the day; taking pictures which illustrate the human life cycle; getting proper exercise; making a skeleton; designing muscles for a wooden leg *(two boards with a hinge connecting them)*; and poems about bones, muscles, cells, and other body parts.

Bulletin boards

SKELETONS
Post drawings of a human skeleton and the skeleton of some familiar animal. Tell students to connect strings between bones or joints which are similar on the two skeletons. Ask them to explain the functions of the bones they selected. How do the joints work? Is the movement similar for the animal and the human?

FOOD GROUPS
As described in the activity "Food Groups," collect pictures of foods in the various food groups and post the pictures in the proper spaces. (See Figure 11-4.)

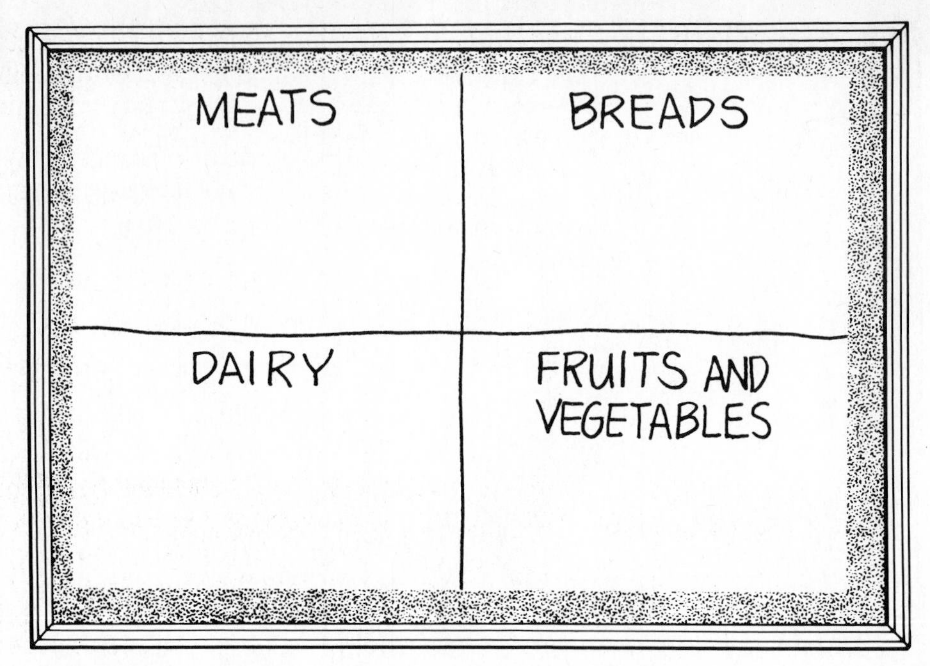

Figure 11-4.
Bulletin board:
Food groups.

Field trips

A field trip to the nurse's office in your school can be an opportunity for your students to get to know the nurse and what he or she can do to help them, and can also give the nurse a chance to make a presentation on why and how we keep our bodies clean. If you take your class on a field trip to the school cafeteria, the dietician can explain how nutritious foods are planned and prepared for school lunches.

Other informative field trips can be made to a local museum; a hospital; a grocery store (to identify nutritious foods); a doctor's or dentist's office; and a gymnasium, a health club, or some other place that provides exercise facilities and training.

THE HUMAN BODY FOR INTERMEDIATE GRADES

Activities

▶ CELL STUDY

In addition to observing elodea cells (see the activity "Cells: Building Blocks"), students can easily examine another plant cell—that of an onion. Slice an onion and pull off a piece of the very thin "skin" between the layers. Place the skin on a slide and examine it under a microscope. The cells of this skin will not have living materials moving around in them, unlike the elodea cells. If the students have difficulty seeing the cells, apply a drop of diluted iodine to the onion skin.

Obtain prepared slides of human cells—if possible, include at least

There are several sources of free or inexpensive teaching materials related to the cell and other areas of study of the human body. (Refer to the section at the end of this chapter.)

bone, skin, three kinds of muscle, nerve, and blood. These slides may be purchased from a science supply house (see the Appendix), or you may be able to borrow them from a high school or college biology teacher.

Ask the students to describe similarities and differences among the human cell types. Where in the body would each of the types of cells be found? Would it be possible for the students to prepare their own slides of animal cells? *(In most cases, it would not be easy; but if they really want to, they can.)* How about using a *very* thin slice of meat? And could they see their own blood cells? If your students are eager to do this kind of activity, refer them to a reference book for directions, or a junior or senior high school life science or biology teacher.

➤ BODY SYSTEMS

Tell the students that, if they want to, they can bring a T shirt to school and paint one of the body systems on it. Invite an art teacher to demonstrate how to paint the shirts and to show them what kinds of paint are safe to use. Systems which can fairly easily be painted include the skeletal, digestive, circulatory, and respiratory systems.

Have the students find library books which show the locations of the various parts of the systems they select to paint. The systems should be sketched and then painted as realistically as possible. (One student teacher, as a joke, painted a circulatory system minus the heart for her science education professor—claimed he did not have one anyway.)

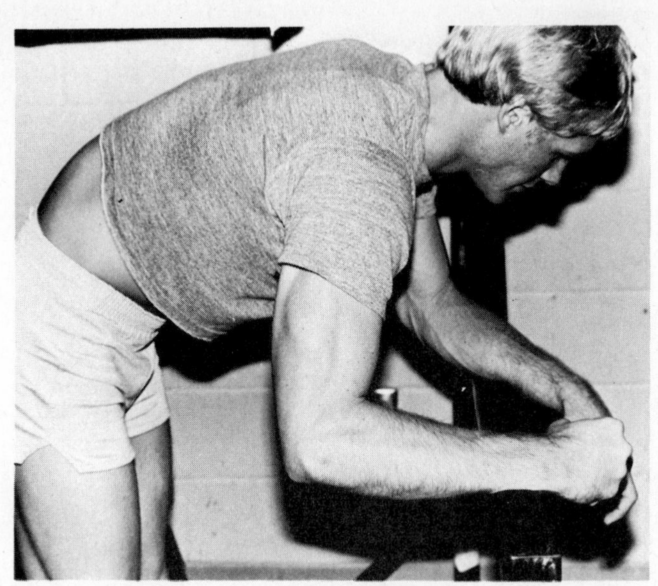

Placing the arms on a solid surface and pressing down tightens the tricep muscle. Placing the arms below the surface and lifting tightens the bicep muscle.

MUSCLES

By feeling their own muscles, students can learn much about the muscular system. One thing they can learn is the concept of pairs of opposing muscles. By first placing one hand, palm up, under their desk and lifting, they can feel the larger, front-arm muscle (biceps) tighten. Then by placing a hand palm up on the top of the desk and pushing down, they can feel the smaller, behind-the-arm muscle (triceps) become firm. Those two muscles move the arm in opposite directions. Only one works at a time. Why is the front muscle so much larger? *(We use it for more and harder work. How often do you push down on something with the back of your hand?)*

Have the students find another muscle which they can feel moving some part of the body and then try to find the muscle which moves that part in the opposite direction. They can do this with several arm and leg muscles, and also with fingers and toes. And how about muscles in their torsos?

To enable students to see some muscles, bring to class a raw chicken leg and thigh (not separated). Remove the skin and bend the leg backward and forward. Can the muscle pairs be identified? The students wil be able to see that the meat on the chicken leg is not simply a single mass of meat, but rather several separate muscles. Can the muscles be separated?

NERVOUS SYSTEM

One interesting aspect of the nervous system is the area of automatic reflexes. To introduce the activity, use a nonautomatic action as an

These students are lining up to do an activity which will help them calculate a person's reaction time.

example. Have all your students form a line and hold hands, and tell them that when the student ahead of them squeezes their hand, they are to "pass it on." All students except the first one in the line are to close their eyes. Visually signal the first student to squeeze the second person's hand. Either you or a student who is not part of the line can determine the total elapsed time between the starting signal and the last hand squeeze. Divide the total time by the number of students to determine time per student (a single student's reaction time). Ask the students whether or not they think they could reduce the time if they repeated this activity several times. Then repeat the activity several times until it is apparent that the time is no longer being reduced. Enter the data obtained on a graph, with one axis representing time per student and the other axis representing trial number.

Now demonstrate an automatic reflex—something which you cannot usually stop even if you try. One such action is blinking your eyes when someone's hands are clapped near your face. How much time elapses between the hand clap and the blink? *(Far too little to have had time to decide to blink the eyes—just try not to blink!)* Another automatic reflex is the knee-jerk reaction when the leg is tapped gently just below the kneecap. Reflex actions are directed by a nerve function that does not carry the message to the brain for a decision. The decision is automatically made much closer to the eye (or knee), and the motion is signaled from that point.

Ask students to name other automatic reflex actions. *(How about closing the eyes when something approaches them? Or removing a finger from a hot surface?)* What are the advantages of such actions?

DIGESTION

The human digests food by several specific actions. The enzymes in saliva, for example, act upon starches. If a small piece of soda cracker is chewed for approximately 1 minute, allowing saliva to act on it, it will develop a sweet taste, indicating that the startches are being broken down into sugars. If any of your students wonder whether the sweet taste is really sugar that was mixed into the cracker before it was cooked, refer them to a health textbook which describes tests for starch and sugar (both are simple—an iodine test for starch and Tes-Tape for sugar) and have them verify the change.

LUNG CAPACITY

How much air can you move into and out of your lungs in one breath? How could you determine the amount? One way to do this is to calibrate a 4-liter (1-gallon) jug by placing a strip of tape down the side, from top to bottom, pouring measured amounts of water into the jug (100 milliliters or ½ cup at a time is a reasonable amount), and placing a mark on the tape to indicate the level of the water after each increment of water has been added. Then fill the jug completely full of water, hold your hand over the top, and invert it into a dishpan which contains enough water to insert the jug into it at least 3 centimeters (1½ inches).

Insert a rubber hose into the inverted jug as shown. Inhale deeply and then blow as much air as possible through the rubber hose. The air will displace water in the jug, and its volume can be determined by reading the calibrations on the tape. (See Figure 11-5.)

Is everyone in the class able to exhale nearly the same maximum amount of air? Record the results for each student and make a graph to illustrate the information. What was the least amount exhaled? The most? What appears to be the most common amount exhaled?

Figure 11-5. Measuring the capacity to inhale and exhale. (Source: Adapted from V. N. Rockcastle et al., *Science: Level 5*, Addison-Wesley, Reading, Mass., 1980.)

These students are exercising to increase their pulse rates.

➤ RECORDING PULSE RATE

The purpose of this activity, and the next two activities, is to cause students to think about physical fitness and to encourage them to design a personal exercise program. First, each student's normal resting pulse rate must be determined. Normal rates for elementary school students generally range from 60 to 110 heartbeats per minute. Students should understand that there is no single "best" pulse rate. Each person's body is unique and has its own normal pulse rate.

For this activity it is necessary for students to become proficient in checking their pulse rates. Most students can readily check their pulse rates on the arm near the wrist. A few students may have to find another pulse point (temple, top of foot, behind ear, throat). Practice pulse taking several times with your students. After a rest period or other quiet time, have students count their pulses for 15 seconds, record the count (allow 15 seconds recording time), count another 15 seconds, record the second count, count another 15 seconds, and make a final record. Have the students multiply each 15-second count by 4 to determine the pulse count per minute. Conduct this practice procedure several times until the students become proficient at counting and feel at ease doing it. Then have the students record their resting pulse rates on a permanent record sheet.

Next, have the students record a sample exercise pulse rate. Have the students run in place or jump vigorously for 1 minute.* Immediately after the exercise, have them count their pulses for 15 seconds, record, count another 15 seconds, and record. If necessary, practice this procedure also until the students become proficient.

At this point, several questions can be posed to students: "Is your resting pulse rate the same at all times during the day and night? When do you think it might be different? Would it be different just before an important test? How about while you are watching a suspense movie? Before you make a speech to your whole class? Are older persons' rates in the same ranges as younger persons'?

RECOVERY RATE

The pulse "recovery rate" is the length of time it takes for the pulse rate to return to the resting rate after vigorous exercise. In general, a long recovery time indicates poor physical fitness and a rapid recovery time indicates good physical fitness. In this activity the students will determine their pulse recovery rates and try to change them.

The exercise in this activity will be to trot around the school grounds or other appropriate place. Either you or a designated student should set the pace (do not run rapidly—just move a little faster than jogging). Keep the students in a close group during the trotting, and continue to trot until all are breathing rapidly.† Then have the students immediately sit down and begin counting their pulses for 15 seconds, recording them on paper which they have carried with them. Repeat the counting and recording procedure at 15-second intervals for 5 minutes.

After returning to the classroom, have the students calculate their pulse rates per minute for each of the readings by plotting the data on a graph, using pulse rate and time as the coordinates.

Ask the students to determine their recovery rates. How long did it take for their pulses to return to the resting rate? Ask them to talk about factors that could have slowed their recovery rates. *(There are many possibilities, including physical fitness, present health, anxiety, and tension.)* Ask the students to predict whether or not trotting for 5 minutes each morning and each afternoon for a week would affect their recovery rates. Beginning the following Monday, have them trot 5 minutes each morning and each afternoon. On Friday afternoon check their recovery rates. Was there any change?

Physical fitness cannot be greatly improved by 1 week of moderate exercise, but 1 week will often make a noticeable change in recovery rate. Ask students to predict how much they think they could shorten their

*If any students should not or cannot do this for reasons of health or physical handicap, get the advice of the school's physical education instructor or the children's own doctors or therapists and devise other exercises which they can do.

†As with the preceding activity, be sure to make adjustments in the exercise if any students have health problems or physical handicaps.

recovery time if they continued their exercise for a month. Invite someone who is involved in a good fitness program and who is in very good physical condition to visit your class and check her or his recovery rate.

➤ EXERCISE PROGRAM

Have the students check several books or pamphlets on physical fitness activities and programs to identify exercises which can improve physical fitness. One program which they should consider is the President's Physical Fitness Program. Invite a physical education teacher, athletic trainer, or exercise specialist (e.g., from a local "health club") to visit your class and describe or demonstrate exercises. Before the visit, prepare the students to ask questions about different exercises or physical fitness programs.

Ask each student to design a personal physical fitness program in which he or she would be willing to participate. You design one too! Ask your students to write you a note telling you whether or not they really will do the exercises that they have selected and also telling you when you can check with them about how well they are doing. Challenge some other class at your grade level to see which class will have the best percentage improvement in recovery time by the end of the year. (Remember to use your students' initial recovery rates as determined in the previous activity before they began the exercise program.)

➤ LIFE EXPECTANCY

This activity will focus students' attention on the reasons for the great increase in life expectancy in this country during the last 80 years. Table 11-1 indicates the average number of years of life expected for a child born in the United States in the years indicated.

Have the students graph the information. Why has the life expectancy in the United States increased so much? Is life expectancy in other countries the same? *(A world almanac should give them the information they need to answer this question.)*

TABLE 11-1 LIFE EXPECTANCIES

DATE OF BIRTH	LIFE EXPECTANCY
1910	47
1920	54
1930	60
1940	63
1950	68
1960	70
1970	71
1980	74

▶ DRUGS

While drugs serve many good purposes, their abuse has created many problems. Preaching about the "evils" of drugs does little good, and knowledge about the effects of drugs by itself will not stop the problem. (If you are not sure this is true, consider that doctors and nurses certainly know about the problems drugs may cause, and yet many of them do misuse drugs.) However, a more positive approach may help. Using value-clarifying activities, help students consider and evaluate alternatives other than drugs which will help them achieve the effects they might want from drugs.

Ask students to name as many reasons as they can for misuse of drugs. For each reason named, have them list as many other things as possible that people could do to achieve the indicated goals. Give them your own ideas too. Then have them form small groups. Do not supervise the groups closely. Tell them to evaluate each of the suggested means of achieving the goal (including misusing drugs) briefly, in writing.

▶ A BLOODY TRIP

Have the students imagine they are blood cells. Guide them through the circulatory system on three trips to three different parts of the body. Describe not only the parts of the system through which they are traveling but also how they are changing along the way. (You may want to have the students help you write the script for this imaginary trip before you begin the activity itself.)

Children's literature

Read *The Food You Eat** to your students or have them read it, and then discuss with them the concepts presented.

Find other books on the human body in the school library, and identify them for the students or place them in an interest center.

Activity cards

An example of an activity card is shown in Figure 11-6. Additional activity cards could be made for topics such as exercise programs; the senses; dental care; investigating or making models of body systems (including kits which are commercially available); different people's need for rest and rest habits; diseases; and drugs.

Bulletin boards

EXERCISE

Describe, with drawings and explanations, an exercise program which your students could use. Make up forms that the students can, if they

*John Marr, *The Food You Eat,* M. Evans, Philadelphia, 1973.

- PLAN MEALS FOR THREE DAYS WHICH WILL SUPPLY YOU WITH THE NUTRITIOUS FOOD YOU NEED AND WHICH INCLUDE FOODS YOU WOULD LIKE TO EAT.
- WHAT KIND OF FOODS DO YOU NEED?

IDENTIFY SOME KIND OF FOOD THAT YOU HAVE NEVER EATEN BEFORE AND TRY IT.

PREPARE ONE OF THE MEALS FOR YOUR FAMILY.

Figure 11-6. Activity card: A balanced diet.

choose, fill out to define a program that they personally intend to use. List on the bulletin board the advantages of participating in an exercise program.

THE DIGESTIVE SYSTEM
As illustrated in Figure 11-7 (page 238), post a drawing of the digestive system on the bulletin board and place in a nearby pocket labels for each of the major parts (mouth, salivary gland, esophagus, liver, stomach, pancreas, small intestine, large intestine, anus). Students are to tack each label near the part it identifies.

Field trips

Several of the field trips described for primary-grade students would also be appropriate for intermediate grades. Some other possibilities are a health food store, a physical therapy facility, the exercise room or gymnasium of a secondary school or college athletic department, and a blood bank.

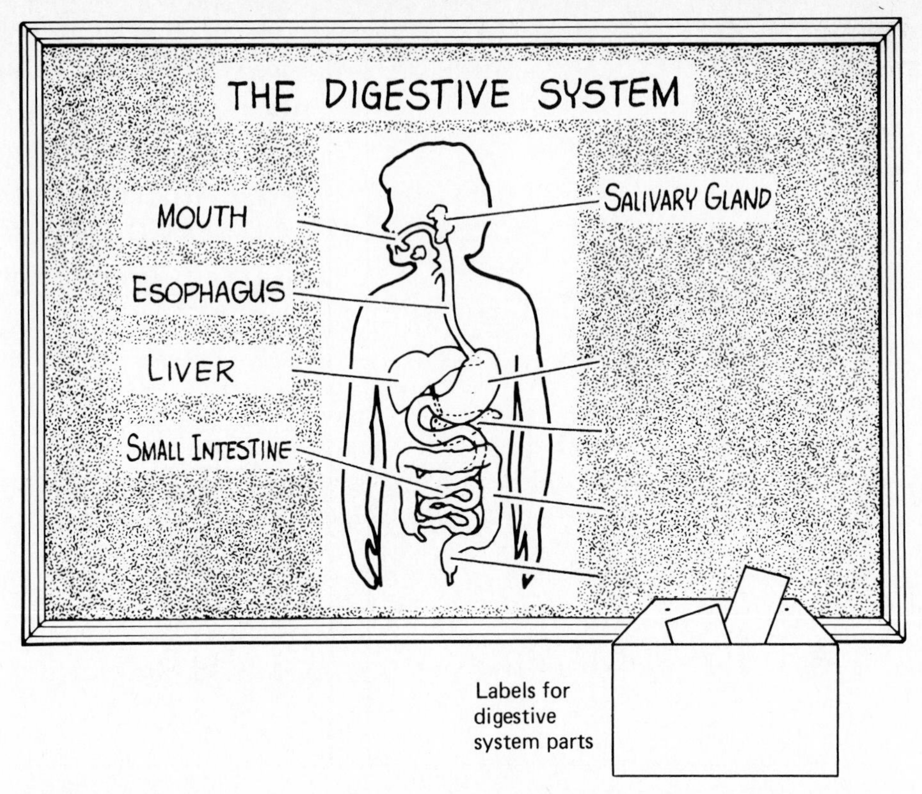

Figure 11-7. Bulletin board: The digestive system. (Source: Adapted from V. N. Rockcastle et al., *Science: Level 5*, Addison-Wesley, Reading, Mass., 1980, p. 115.)

RESOURCES

Sources of assistance

FREE AND INEXPENSIVE MATERIALS

Here are some good sources of free and inexpensive materials for use in teaching about the human body:

1. *American Dental Association.** Booklets and posters.
2. *American Institute of Baking.* A food mobile which illustrates the components of a balanced diet.
3. *American Medical Association.* Many booklets.
4. *Council on Family Health.*
5. *Del Monte Kitchens.* Posters on the food groups.
6. *Department of Citrus, State of Florida.* Nutrition posters.
7. *Johnson & Johnson Products, Inc.* "First Aid for Little People."

*The addresses of specific agencies, organizations, and businesses are listed in the Appendix.

8. *National Dairy Council.* Many good aids and materials.
9. *U.S. Department of Agriculture.* Many aids. (Ask for a list.)
10. *California Raisin Advisory Board.* A teaching kit, "Selective Snackology."

LOCAL RESOURCES
Local resources which can help in teaching about the human body include:

1. *Medical society.* Booklets on health and body function, plus resource persons you could invite to visit your class.
2. *Hospitals, physical therapy centers, doctors' offices.*
3. *Health and exercise center.*
4. *County extension agent.* Publications on nutrition and food. May have a nutrition specialist whom you could invite to visit your class.
5. *Museums.*
6. *Funeral homes.* May distribute publications on helping children deal with death.
7. *Law-enforcement agencies.* Often have drug information programs.
8. *School nurse.*
9. *Fire departments.* May have an emergency medical technician who would be a good resource person to visit your class.

ADDITIONAL RESOURCES
1. Magazines such as *Today's Health, Psychology, World* (published by the National Geographic Society), and *Reader's Digest.*
2. State departments of health and human resources; state departments of agriculture.
3. American Red Cross.
4. Medical and nursing schools.

Strengthening your background

1. Take a cardiopulmonary resuscitation (CPR) course and become certified.
2. Find out what classes the local Red Cross chapter offers, and attend one or more which are of special interest to you. If you do not have first-aid training and certification, begin to work toward getting it.
3. Dissect a chicken.
4. Have a fish fry and save a few of the skeletons.
5. Design your own personal exercise program—and use it. Not only will you feel better physically, but also you will be mentally more alert.
6. Do some research on the cause of the bubonic plague. Could such a plague occur in modern times? What is being or could be done to prevent such a disaster?
7. Make a drawing of the nervous system.
8. Find out what circadian rhythms are. What are the implications of a knowledge of circadian rhythms for teaching elementary school?

Getting ready ahead of time

1. Prepare 20 game-board cards for use in teaching about the human body.
2. Collect free and inexpensive materials for use in teaching about the human body.
3. Have your school order several prepared slides of body cells.
4. Obtain information on several physical fitness programs for children.
5. Write for information on the Health Activities Project (Hubbard, P.O. Box 104, Northbrook, IL 60062), and order several of the modules.
6. Prepare 10 activity cards based upon activities suggested in *Blood and Guts: A Working Guide to Your Own Insides* by Linda Allison (see "Building Your Own Library").

Building your own library

Allison, Linda: *Blood and Guts: A Working Guide to Your Own Insides,* Little, Brown, Boston, 1976.
Asimov, Isaac: *The Human Body: Its Structure and Operation,* Houghton Mifflin, Boston, 1963.
———: *The Human Brain: Its Capacities and Functions,* Houghton Mifflin, Boston, 1964.
Gilbert, Sara: *Feeling Good: A Book about You and Your Body,* Scholastic Book Services, New York, 1979.
Randal, Judith: *All about Heredity,* Random House, New York, 1963.
Silverstein, Alvin, and Virginia Silverstein: *Alcoholism,* Lippincott, Philadelphia, 1975.

Building a classroom library

Aho, Jennifer J., and John W. Petras: *Learning about Sex,* Holt, New York, 1978.
Aliki: *My Five Senses,* Thomas Y. Crowell, New York, 1962.
Andry, Andrew: *How Babies Are Made,* Time-Life, New York, 1968.
Asimov, Isaac: *How Did We Find Out about Germs?* Walker, New York, 1974.
———: *How Did We Find Out about Vitamins?* Walker, New York, 1974.
Balestrino, Philip: *Fat and Skinny,* Thomas Y. Crowell, New York, 1971.
———: *The Skeleton Inside You,* Thomas Y. Crowell, New York, 1971.
Berger, Melvin: *Disease Detectives,* Thomas Y. Crowell, New York, 1978.
Berry, James: *Why You Feel Hot, Why You Feel Cold: Your Body Temperature,* Little, Brown, Boston, 1973.
Blochman, Lawrence: *Understanding Your Body,* Macmillan, New York, 1968.
Burns, Sheila L.: *Allergies and You,* Messner, New York, 1980.
Calder, R.: *The Wonderful World of Medicine,* Doubleday, Garden City, New York, 1964.
Cavallaro, Ann: *Physician's Associate: A New Career in Health Care,* Nelson, Camden, N.J., 1978.

Cohen, Daniel: *Medicine: The Last Hundred Years,* M. Evans, Philadelphia, 1981.
Curtis, Patricia: *Cindy, A Hearing Ear Dog,* Dutton, New York, 1981.
Daly, Kathleen N.: *Body Words: A Dictionary of the Human Body, How It Works, and Some of the Things that Affect Its Health,* Doubleday, Garden City, N.Y., 1968.
Day, B., and H. M. Liley: *The Secret World of the Baby,* Random House, New York, 1968.
Elgin, Kathleen: *The Digestive System,* F. Watts, New York, 1973.
———: *The Skin,* F. Watts, New York, 1970.
Epstein, Beryl, and Sam Epstein: *Dr. Beaumont and the Man with the Hole in His Stomach,* Coward-McCann, New York, 1978.
Farber, Norma: *How Does It Feel to Be Old?* Unicorn, London, and Dutton, New York, 1979.
Fenton, Carroll, and E. F. Turner: *Inside You and Me: Introduction to the Human Body,* John Day, New York, 1961.
Freedman, Russell, and James Morriss: *Brains of Animals and Man,* Holiday, New York, 1972.
Freese, Arthur S.: *The Bionic People Are Here,* McGraw-Hill, New York, 1979.
Gallant, Roy A.: *Memory: How It Works and How to Improve It,* Four Winds, New York, 1980.
Gardner-Loulan, Joann, Bonnie Lopez, and Marcia Quackenbush: *Period,* Down There Press, Burlingame, Calif., 1979.
Goldreich, Gloria, and Esther Goldreich: *What Can She Be? A Veterinarian,* Lothrop, New York, 1972.
Goldsmith, Ilse: *Anatomy for Children,* Sterling, New York, 1964.
Gross, Ruth B.: *A Book about Your Skeleton,* Hastings House, New York, 1979.
Haines, Gail Kay: *Brain Power: Understanding Human Intelligence,* F. Watts, New York, 1979.
Harris, Robie H., and Elizabeth Levy: *Before You Were Three,* Delacorte Press, Dell, New York, 1977.
Howe, James: *The Hospital Book,* Crown, New York, 1981.
Kalina, Sigmund: *Your Blood and Its Cargo,* Lothrop, New York, 1974.
———: *Your Nerves and Their Messages,* Lothrop, New York, 1973.
Kelly, Patricia: *Mighty Human Cell,* John Day, New York, 1967.
Kipnis, Lynne, and Susan Adler: *You Can't Catch Diabetes from a Friend,* Triad Scientific Publishers, Gainesville, Ill., 1979.
Klein, Aaron: *You and Your Body: A Book of Experiments to Perform on Yourself,* Doubleday, Garden City, N.Y., 1977.
Knight, David C.: *Viruses: Life's Smallest Enemies,* Morrow, New York, 1981.
Krementz, Jill: *How It Feels when a Parent Dies,* Knopf, New York, 1981.
Lamb, Mina, and Margarette Harden: *The Meaning of Human Nutrition,* Pergamon, New York, 1973.
Lauber, Patricia: *Your Body and How It Works,* Random House, New York, 1962.

Limburg, Peter: *The Story of Your Heart,* Coward-McCann, New York, 1979.

Marr, John: *The Food You Eat,* M. Evans, Philadelphia, 1973.

Mintz, Thomas, and Lorelie M.: *Threshold, Straightforward Answers to Teenager's Questions about Sex,* Walker, New York, 1978.

Nourse, Alan E.: *The Body,* Time-Life, New York, 1970.

Patterson, Lillie: *Sure Hands, Strong Heart, the Life of Daniel Hale Williams,* Abingdon, Nashville, Tenn., 1981.

Paul, Aileen: *The Kids' Diet Cookbook,* Doubleday, Garden City, N.Y., 1980.

Rutland, Jonathan: *Human Body,* F. Watts, New York, 1977.

Schick, Alice: *Serengeti Cats,* Lippincott, Philadelphia, 1977.

Seixas, Judith E.: *Alcohol: What It Is, What It Does,* Greenwillow, New York, 1977.

Showers, Paul: *A Drop of Blood,* Harper and Row, New York, 1972.

———: *No Measles, No Mumps for Me,* Thomas Y. Crowell, New York, 1980.

Silverstein, Alvin, and Virginia Silverstein: *Cells: Building Blocks Of Life,* Prentice-Hall, Englewood Cliffs, N.J., 1969.

——— and ———: *Circulatory Systems: The Rivers Within,* Prentice-Hall, Englewood Cliffs, N.J., 1969.

——— and ———: *Itch, Sniffle, Sneeze: All about Asthma, Hay Fever, and Other Allergies,* Four Winds, New York, 1978.

——— and ———: *The Muscular System: How Living Creatures Move,* Prentice-Hall, Englewood Cliffs, N.J., 1972.

——— and ———: *Nervous System: The Inner Networks,* Prentice-Hall, Englewood Cliffs, N.J., 1970.

——— and ———: *The Respiratory System: How Living Creatures Breathe,* Prentice-Hall, Englewood Cliffs, N.J., 1969.

——— and ———: *Sense Organs: Our Link with the World,* Prentice-Hall, Englewood Cliffs, N.J., 1970.

——— and ———: *The Skeletal System: Frameworks for Life,* Prentice-Hall, Englewood Cliffs, N.J., 1972.

Simon, Seymour: *Exploring with a Microscope,* Random House, New York, 1969.

Stwertka, Albert, and Eve Stwertka: *Marijuana,* F. Watts, New York, 1979.

Thompson, Brenda, and Rosemary Giesen: *Bones and Skeletons,* Lerner, Minneapolis, 1977.

Wilson, Ron: *How the Body Works,* Larousse, Paris, 1979.

Zim, Herbert S.: *Your Skin,* Morrow, New York, 1979.

———: *Your Stomach and Digestive Tract,* Morrow, New York, 1973.

CHAPTER 12

ECOLOGY

IMPORTANT CONCEPTS

An "ecosystem" is a group of organisms and their physical environment.

Living things are interdependent with each other and their environment.

Living things receive energy from their environment.

Food chains and food webs are tools which assist in studying the energy flow in an ecosystem.

Living things have characteristics which enable them to live in their specific environments.

When an environment is changed, organisms which live in that environment must adapt to the change or perish.

People affect environmental processes.

An environment becomes polluted when matter or energy is placed into it in abnormal amounts.

People have caused the extinction of some organisms.

By considering the consequences of what we do, we can reduce pollution.

INTRODUCTION: STUDYING ECOLOGY

In teaching about ecology, a great deal of emphasis has been placed on environmental quality and problems of pollution. Those topics are within the realm of ecology and certainly are important; however, ecology has many other components which are equally important. If people were more knowledgeable about the entire realm of ecology, they would probably take more care to avoid problems in the environment; further, there would be a better basis from which to maintain a quality environment with tolerable limits of pollution.

Children—and teachers—enjoy being in environments where animals and plants abound. There is a real fascination in such places: so much can be learned there, and they are close at hand. We might all enjoy going to a forest or to the seashore; but rich environments can also be found on almost any school grounds, and if they are not already present, very little work is required to establish them.

In order to give the wide range of ecology its proper perspective, this chapter presents ideas in three categories: ecology for primary grades, ecology for intermediate grades, and environmental quality. One important component of science in the elementary school is the consideration of the interaction of science, people, and the environment. Consideration of environmental quality is not an "extra"—it should be an integral part of ecology. In this book we have chosen to separate environmental-quality teaching activities from other ecology teaching activities in the hope that this arrangement will help teachers and students become more aware that a sound knowledge of ecological concepts is important because it enables people to make wise decisions about the quality of the environment.

The study of ecology and environmental quality is rich in potential for presenting interesting questions and research possibilities for elementary school students. The area deals with objects and phenomena seen by stu-

An elementary school campus like this one offers rich possibilities for activities outside the classroom. The same is true of almost any school grounds if a little creative thought is applied.

dents—and teachers—every day. During your daily life you should think about questions which you can pose to your students. Just be careful to keep the scope of the problems you present realistic for your students. Depending on where they live, they may not relate closely to a forest or marine ecosystem, but they may be able to visit a vacant lot environment. Also remember that your students cannot solve the air pollution problems of the world, but they can think about what they personally can do to reduce air pollution.

ECOLOGY FOR PRIMARY GRADES

Activities

ECOSYSTEM DIORAMAS

Making shoebox dioramas of several different ecosystems (see Figure 12-1) will help students form a concept of "ecosystem." They can work in small groups to represent ecosystems such as forests, grasslands, ponds, rivers, saltwater estuaries, and deserts. If possible, while they are creating the dioramas they should also prepare a fact sheet for each ecosystem—a concise listing of a few of the animals and plants present in the ecosystem, as well as a short description of some of the interactions of those plants and animals. If you think that your students are not ready to prepare fact sheets, you may be able to work with an elementary school teacher whose students are preparing fact sheets on the same ecosystem.

While the students work on the dioramas, ask them questions: "Is there food present for each of the organisms represented? Who eats what? What is the weather like in such an environment? What would happen if all the frogs (or other organism) were removed from your ecosystem?"

After the dioramas are completed, display them and have each group tell the class about their ecosystem. Tell the students that each diorama represents something which we call an "ecosystem," and ask them to explain what they think an ecosystem is. (*An ecosystem is a group of plants and animals linked by a common need—food—and by the physical environment in which they live.*)

Figure 12-1. Shoebox diorama of an ecosystem.

Finally, have the class as a whole construct an ecosystem diorama of a vacant lot. Either during or after school, the students can explore a vacant lot and look for animals and plants living there. (*Insects and other small animals such as rodents, birds, lizards, and snakes will probably be present. There will probably be a wide variety of plants.*) Indirect evidence may have to be used—a snakeskin, feathers, tracks. If possible, dig into the ground to find animals just below the surface—insects, worms. Ask questions: "What do the birds eat? What do the insects eat? Does that plant-eating insect eat all the plants or just certain ones?" For comparison, you might wish to have your students construct a lawn ecosystem diorama. (*It will contain fewer organisms. Why?*)

➤ A UNIQUE HABITAT

A habitat is a place where organisms—plants and animals—live. You and your students can create a new and special habitat for study. Locate a relatively undisturbed and secluded place on your school grounds or a nearby vacant lot (get permission from the owner, of course). Attach a handle to one edge of a 50-centimeter-square (20-inch-square) piece of plywood and place it in the secluded spot. (A smaller piece of wood will work, but will accumulate a less interesting collection of organisms; the handle is needed because of the remote possibility that the board may attract stinging or biting animals.)

Before you and the class return to observe the new habitat, discuss it with them. Help them realize that the organisms in this environment, like organisms in any environment, have special requirements for staying alive. If the board is removed or disturbed too often, the habitat will be destroyed and the organisms will either move or perish. Work with the class to establish guidelines for the number of visits to the board (one visit a day might be reasonable), but do not be too restrictive—the students may be very eager to show the new habitat to their friends and parents. Make sure that they can explain the life requirements of the organisms in the habitat to visitors.

After 2 or 3 weeks, carefully use the handle to life the edge of the board. Unless there is a major disturbing factor in the environment, you should find a number of interesting creatures under the board. The animals in this new habitat will vary with the location; but there should be insects and worms, and there may even be a lizard or a small snake.

Challenge interested students to identify the organisms in the habitat. Ask them: "What special things about these animals help them to live in the habitat which we have created? How are these animals alike? How are they different? Is there any evidence that some of these animals eat the others? What are the characteristics of this special environment? Where do you think these animals could be found in the natural environment?" (*Under rocks, along foundations of buildings, in the soil.*)

➤ TRACK DETECTIVES

A relatively simple and easily constructed device for obtaining samples of tracks of animals living in many environments is a sandbox. The sandbox can be quite small; but if it is 1 meter square (about 1 yard square) or larger, it will be especially useful. It can be constructed on the school grounds in a place that receives relatively little travel, or at some other site with the owner's permission. Mark off the size of the sandbox and dig a square hole approximately 10 centimeters (or 4 inches) deep. Place 10-centimeter-wide (4-inch-wide) boards around the sides of the hole; the tops of the boards should be at soil level or just a little above it. Fill the box with very fine sand—the finest you can get, for very fine-grained sand will reveal insect tracks as well as tracks of larger animals. Use another board to smooth the surface of the sand so that it is even with the top of the boards and very flat.

Visit the sandbox the following morning and examine it carefully for tracks; look closely for insect tracks. If you find tracks, have the students guess what animals made them. Would they have expected to find those kinds of animals in this environment?

Ask the students what could be done to attract more animals to the sandbox. (*How about supplying various kinds of food?*) Have them predict from day to day what kinds of animals will come to the sandbox, as they try out their various suggestions. If the tracks are well defined, try making plaster casts. (If you need to, do some research to find out how to work with plaster.)

Have the students predict what tracks might be found in sandboxes in other specific environments. If possible, make some sandboxes in other environments—or as an alternative, look for natural tracks in different environments.

➤ FOOD CHAINS

To help students understand the interdependence between animals and their environment (which includes other animals), have them construct food chains, either with mobiles (see the activity "Food Web Mobiles" later in this chapter) or by drawing pictures.

First they can illustrate a simple food chain such as grass–mice–snake–hawk or algae–water insect–small fish–large fish–osprey (see Figure 12-2). Next the class can take a field trip on the school grounds and observe some animal—probably an insect. (*What does it eat? Does anything else eat the animal which they are observing?*) Have the students construct a two- or three-member food chain based upon their direct observations. Also, have them hypothesize what would be higher up on the food chain.

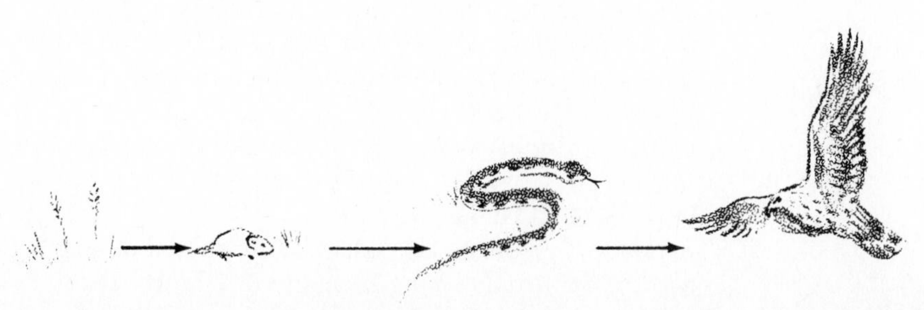

Figure 12-2. Simple food chain.

Present the concept that all food relationships are energy relationships. Explain that each organism receives energy from the organisms which it eats and provides energy for organisms which eat it. People too are part of the food chain; we use the energy from our food to build and maintain our bodies and to carry on our activities.

▶ CLASSROOM AQUARIUM ECOSYSTEM

One enjoyable way to help students understand the components of an ecosystem and the requirements of its organisms is to set up a classroom aquarium. The students should be involved with all the planning and work. You may help them—but do not do it all yourself, or you will be the only one to benefit from the experience. You will need the following basic supplies: a 40-liter (approximately 10-gallon) aquarium, an air pump, an underground filter system, a lid for the aquarium, and very fine gravel (pea-sized or slightly smaller). Often most or all of these materials are sold as a unit by a local pet shop or discount store for a very reasonable cost. The gravel can be obtained very inexpensively from a local sand and gravel company. You will need enough gravel to make a 5-centimeter (2-inch) layer over the entire bottom of the aquarium.

Place the materials in the classroom. When the students notice them and begin asking questions, tell them they can use the materials to set up an aquarium. Have them make a list of the living requirements for fish in an aquarium. (*The list should include water, air, food, proper temperature, an appropriate number of fish, plants and other objects to provide hiding places for young and small fish, and protection from outside dangers, such as cats.*)

Tropical fish are a popular choice for classroom aquariums, but native fish from a nearby pond will help students learn about their own environment—and besides, native organisms are free. Perhaps they would like to make a native aquarium first and then change it to a tropical fish aquarium. A game warden or local high school biology teacher may be able to help you secure native organisms.

Two weeks before obtaining any living materials for the aquarium, have the students place the filters, gravel, and water in the aquarium and then connect and start the air pump, following this procedure: First place the filter in the bottom of the aquarium. Then wash the gravel thoroughly and spread it over the filter. Make the gravel deeper at the back of the aquarium and shallower at the front, to facilitate cleaning when necessary. Put a plate over the gravel and carefully pour water onto the plate until the aquarium is approximately two-thirds full. Remove the plate, hook up the pump, and turn it on. After you are sure that the pump is in good working order, add water to within 7 centimeters (about 3 inches) of the top.

Allow the aquarium to stand for 2 weeks before doing anything other than adding water to maintain the original level. Ask the students to note any changes. (*There will probably be some cloudiness at first due to small soil particles in the water—they should settle or be filtered out.*) If the water becomes slightly green, ask the students for possible explanations. (*There are algae in the water.*) If a microscope is available, have them examine samples of the water.

After 2 weeks, help the class collect water plants from the location where the fish will be obtained and plant them in the aquarium. If possible place a few snails in the aquarium also. (*A possible problem with snails is that they may devour the plants. However, a fresh supply of local aquatic plants can easily be obtained.*)

Wait another week and then take the students to collect fish for the aquarium; try to obtain several different kinds of fish. Make sure the students understand that, upon the advice of a game warden, biologist, or biology teacher, you have duplicated the natural environment of the fish as closely as possible so that all their living requirements will be met. Place the fish in the aquarium and tell the students to note the behavior of the fish and other organisms.

Each student should select one organism (either animal or plant) to observe closely and should keep records on it. Suggest that the students consider such questions as "Where does my fish or snail spend most of its time? Where and how does it eat? Does my organism have its own territory? Does it chase other organisms out of its territory? Can I find snail eggs? What is happening to my kind of plants? Did the situation change when fish were introduced to the aquarium? Do snails eat or avoid my kind of plant?" Tell your students to think of their own questions also, and to find the answers to them.

In feeding the fish, try to use both native foods (a good research topic for an interested student) and prepared fish food from a pet shop.

After the students have satisfied their curiosity about native fish, you may want to establish a tropical fish aquarium. If so, carefully return the native organisms to the environment from which they were collected. Following the advice of a pet shop owner or a book on tropical fish aquariums (several very inexpensive ones are available at pet shops), set up the tropical fish aquarium.

➤ TRAVELS THROUGH A VACANT LOT

A good guided-imagery activity is to have the students imagine they are ants traveling through a vacant lot. What do they see? What do they have to watch out for so they will not be destroyed? What kinds of food do they search for?

Children's literature

Read *Look at Pond Life** to your students or have them read it, and then discuss with them the concepts presented.

Find other books on ecology in the school library, and identify them for the students or place them in an interest center.

*Rena Kirkpatrick, *Look at Pond Life*, Raintree, Milwaukee, Wis., 1978.

Figure 12-3. Activity card: Adopt an animal.

Activity cards

An example of an activity card is shown in Figure 12-3. Additional activity cards could be made for topics such as life in local habitats, looking for special characteristics of animals that live in a given habitat, identifying food chains, discovering the diversity of life in local habitats (a lawn, for example), and finding out how zoos provide for the needs of animals.

Bulletin boards

WHICH HABITAT?
As illustrated in Figure 12-4, identify selected habitats on a bulletin board. The background area for each habitat should be a distinctive color—green for forest, yellow for desert, and so forth. Have the class help you collect magazine pictures and make drawings of many animals and plants which live in the habitats on the bulletin board. Glue a piece of paper to the back of each picture, color-keyed to the appropriate habitat. Attach an envelope to the bulletin board and place the pictures in it. Ask students to match the pictures with the habitats and then to check the colors on the backs of the pictures to see whether they have identified the correct habitats for the various organisms.

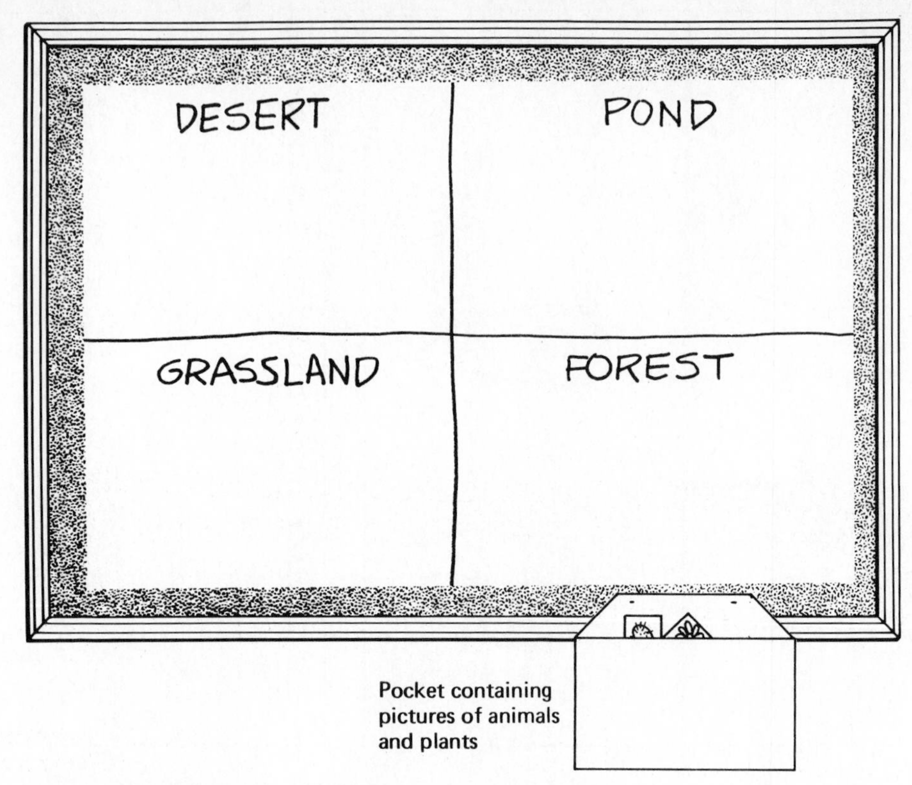

Figure 12-4. Bulletin board: Animal habitats.

SCHOOL GROUNDS HABITAT
Have students identify and draw plants and animals which live on the school grounds. They can then post their drawings on the bulletin board.

Field trips

Field trips are almost imperative when studying ecology, but you do not have to go far. Many living things and their interrelations can be studied on the school grounds, on a local vacant lot, at the school arboretum, on a nature trail, in a forest, at a pond or grassland, or at a local environmental education center. Trips to a local zoo are often especially rewarding. Ask someone on the zoo staff to explain how zoo workers determine and satisfy the life requirements of organisms. Children are usually very interested in learning about the special requirements of familiar and unfamiliar animals.

ECOLOGY FOR INTERMEDIATE GRADES

Activities

➤ FOOD WEB MOBILES
Constructing food webs is one way to learn about the various organisms in a habitat and the interrelationships among them. To construct a food

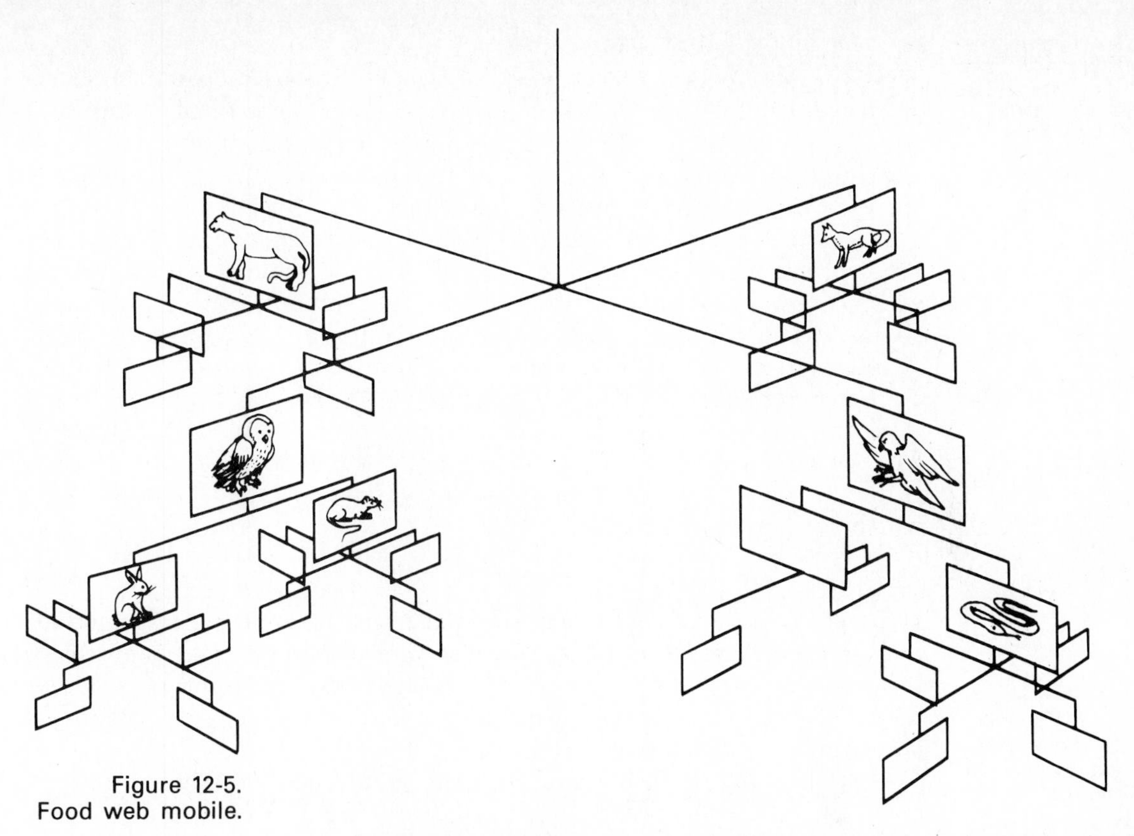

Figure 12-5.
Food web mobile.

web, you list the organisms in a habitat and then draw lines between each organism and what it eats.

In a slightly modified procedure, a mobile which represents the food web can be constructed (see Figure 12-5). First identify animals in a selected habitat which are not eaten by predators—examples are eagles, cougars, wolves, some kinds of sharks, and some large freshwater fish. Put the names, sketches, or cardboard silhouettes of these animals at the top level of the mobile and—using wooden dowels or clothes hangers and string—suspend below each animal two to four organisms which it eats. Then, below each of the second-level organisms, suspend two to four organisms which it eats. Add as many levels to the food web mobile as possible.

Divide the class into several groups and have each group make a mobile for a different habitat. Possible habitats include desert, mountain, hardwood forest, pine forest, prairie, freshwater pond, saltwater estuary, and river. After completing the mobiles, each group can describe the energy flow in their particular habitat.

Ask each student to choose two habitats that he or she would like to visit. Then have the class do some research together to locate within their state and within the nation examples of favorite habitats. If one or more habitats were not chosen, ask the class why they might not consider those habitats attractive places to visit. Also, ask them for reasons why those habitats might, after all, be interesting to visit.

➤ HABITAT FACT SHEET

Have your students work in groups of two or three to prepare fact sheets on these habitats: hardwood forest, pine forest, grassland, freshwater pond, river, desert, saltwater estuary, mountain, vacant lot, and lawn. Explain to the class that they will be giving the fact sheets to third-grade students to help them learn about different kinds of habitats.

Each sheet should include information on weather and climate, the kinds of organisms present, special adaptations of animals in the habitat which allow them to survive in their environment, examples of food chains within the habitat, and one or more unusual and interesting pieces of information about the habitat or its organisms.

After the sheets are finished, have all your students review them and make any necessary revisions. Then arrange with a third-grade teacher to have your students visit the class to present the fact sheets, explain their purpose, and tell the younger students about the particular habitats which they investigated.

➤ INSECT WATCH

Although we seldom pay much attention to insects unless they annoy us, they are very numerous and are an important part of all ecosystems. Have the students go outside and look for insects which they normally do not notice. Each student should select one insect to study closely; they need not know the names of the ones they select.

After choosing an insect, each student should begin to observe it carefully. The students should plan to find out the answer to the question "What does it eat?" (*Where did they find it? It was probably near its food source.*) Also ask them, "Where is your insect most commonly found? What eats it? Can you find evidence of its life cycle? What is the name of your insect?"

Encourage students to learn as much as possible about insects by direct observation. They should identify their insects by using a book such as *A Field Guide to the Insects** from the Peterson Field Guide Series or *Insects*† from the Golden Nature Guide series (see "Building Your Own Library"). Then have them look for answers to other questions: "What other insects is yours closely related to? What does your insect eat in addition to what you observed it eating? (*Have them verify this, if possible, by observing the insect in the presence of their foods.*) Where in the United States is your insect found? What does your insect do that is judged good? Bad? Overall, is your insect judged useful or harmful to human beings? What is the life cycle of your insect? What eats your insect?"

*Donald Borror and Richard White, *A Field Guide to the Insects of America North of Mexico*, Houghton Mifflin, Boston, 1970.

†Herbert Zim and Clarence Cottam, *Insects: A Guide to Familiar American Insects*, Golden Press, New York and Western, Racine, Wis., 1956.

MAKE AN INSECT DIFFICULT TO FIND

To help students understand the survival advantage of camouflage for insects and other animals, have them construct model insects (using colored paper and marking pens) which are difficult to see when placed in selected locations—on the side of the school building and other different-colored buildings; on leafy plants; in decaying plant matter; on certain kinds of trees or flowers; on different-colored soils.

After the students have constructed their model insects, have them search for real insects with similar characteristics. Tell them to determine what eats the insects they find. How do each insect's characteristics help keep it from being eaten by its predator?

If they like, the students can also construct models of other animals which use camouflage for protection. Ask them what they might do to make themselves able to observe insects or other animals closely. (*Camouflage themselves, of course.*)

BIRD LOCATOR

To introduce this activity, ask your students, "Where are birds found? How about robins? Are they found more often in some locations than others—on the ground, high in trees, etc? How about nuthatches? Flycatchers? Eagles? Hawks?"

You can help your students to realize that not all birds (or fish, insects, mice, or other animals) are found in random locations in the environment by instructing them to select one kind of bird and make several observations of it. They should note their bird's most frequent locations. Considering what the bird usually eats, does it surprise them where they most frequently see the bird?

Choose a particular bird yourself, and do not tell the class its name; instead, describe its eating habits, and ask them where they think it would most commonly be seen. Then tell them the name of the bird and challenge them to verify their predictions. Students can help make a challenging game of describing the eating habits and frequent locations of local birds (and other animals) and having classmates guess their names.

MUSIC

Obtain a copy of Pete Seeger's album, *God Bless the Grass** and play the song "The People Are Scratching." The words and music appear in Figure 12-6. Ask the students, "What is an important lesson to be learned from the song?" (*If we do one thing in an ecosystem, other things—often unexpected—happen.*)

Have the students find songs about specific kinds of environments that give some information about the environment and convey how people feel in that environment, or how they feel about it. Songs about mountains and the seashore are common, but there are also songs about other

*Columbia CL2432.

THE PEOPLE ARE SCRATCHING*

1. Come fill up your glasses and set yourselves down,
I'll tell you a story of somebody's town.
It isn't too near and it's not far away
And it's not a place where I'd want to stay.

Chorus:

The people are scratching all over the street
Because the rabbits had nothing to eat.

2. The winter came in with a cold icy blast,
It killed off the flowers, and killed off the grass.
The rabbits were starving because of the freeze
And they started eating the bark on the trees.

3. The farmers said, "This sort of thing just won't do,
Our trees will be dead when the rabbits get through;
We'll have to poison the rabbits, it's clear,
Or we'll have no crops to harvest next year."

4. So they bought the poison and spread it around
And soon dead rabbits began to be found.
Dogs ate the rabbits, and the farmers just said,
"We'll poison those rabbits 'til the last dog is dead."

5. Up in the sky there were meat-eating fowls
The rabbits poisoned the hawks and the owls,
Thousands of field mice the hawks used to chase
Were multiplying all over the place.

6. The fields and the meadows were barren and brown,
The mice got hungry and moved into town.
The city folks took the farmers' advice
And all of them started to poison the mice.

7. There were dead mice in all the apartments and flats,
The cats ate the mice and the mice killed the cats.
The smell was awful, and I'm glad to say
I wasn't the man hired to haul them away.

8. All through the country and all through the town
There wasn't a dog or a cat to be found;
The fleas asked each other, "Now where can we stay?"
They've been on the people from then till this day.

9. All you small creatures that live in this land,
Stay clear of the man with the poisonous hand!
A few bales of hay might keep you alive,
But he'll pay more to kill you than to let you survive.

*from The Sierra Club Survival Songbook, pp. 106-107

Figure 12-6. "The People Are Scratching." (Source: THE PEOPLE ARE SCRATCHING, music by Pete Seeger, words by Ernie Marrs and Harold Martin, © copyright 1963 by FALL RIVER RIVER MUSIC, INC.; all rights reserved; used by permission.)

THE PEOPLE ARE SCRATCHING

Words by Ernie Marra and Harold Martin. Music by Peter Seeger.

environments—e.g., "Home on the Range" ("Western Home") for prairies. Books on folklore often include folk songs about local environments. Have students also identify currently popular songs which deal with specific environments.

SPECIAL RELATIONSHIPS

Occasionally very special relationships, called "symbiotic" relationships, exist between two plants, between a plant and an animal, or between two animals. The yucca plant and the yucca moth have one such relationship, and the lichen actually is a combined alga and fungus. In these special relationships, one organism cannot live without the other. Challenge interested students to identify other symbiotic relationships, especially in local organisms.

BUILD A TERRARIUM

A terrarium, or a small copy of a land-based environment, is relatively easy to build. Any container that will not leak can be used to create miniature replicas of environments such as grasslands, woods, deserts, moss-covered banks, and lawns. In this activity, your students can build a number of different types of terrariums. The basic procedure is the same in each instance. Students can collect plants and animals from the environment to live in a terrarium. Only very small animals can live in a terrarium, and they must have special care. However, it is usually all right to have an animal as a "visitor" to a terrarium only for a few days.

Directions are given below for building a terrarium in a 4-liter (1-gallon) container. With slight adjustments, you can use other sizes. A 40-liter (approxmately 10-gallon) aquarium is a very good size.

To start making a terrarium, place approximately 1.5 centimeters (½-inch) of charcoal in the bottom of a 4-liter container. The charcoal acts as a "purifier" in the system.* Next, place 2 to 5 centimeters (1 to 2 inches) of crushed rock or gravel (pea-sized) on top of the charcoal. One function of the gravel is to provide a place for excess water to drain off, in case the terrarium is ever overwatered.

Now come the important aspects of making terrariums: Divide the class into several groups, and have each group decide what kind of terrarium they wish to make. After the decisions have been made, ask each group, "How can we decide what kind of soil should be placed in the terrarium?" (*Soil from where the plants are collected will be the right kind.*) Have the students gather the soil and place 2 to 5 centimeters (1 to 2 inches) of it on top of the gravel. The next task is to choose and collect plants. Use small (even tiny) plants; they will usually be successful, whereas larger ones may touch the sides of the container and finally become moldy and die. When you collect the plants, keep as much of the original soil around the roots as possible; make holes in the terrarium soil and insert the plants. For "scenery," add little extras—decaying wood

*Personally, I never use the charcoal and have had great results, but the experts recommend it.—D. Janke.

with brackets or fungi, pretty rocks, and petrified wood. If animals are to be used, wait until the terrarium has been established for at least 3 weeks before introducing them, and then add only very small animals.

A terrarium needs a minimum of care. A terrarium in a sealed aquarium can go for weeks without any special attention. If the top is not sealed, however, the aquarium will require watering—but remember that the most common reason for failure of terrariums is overwatering. Watch the plants, and water only when the leaves begin to droop, shrivel up, or show other signs that water is needed. Another indicator is the gravel, if it is visible; note whether it has water standing in it, is dark and damp, or is lighter-colored and dry. In a 4-liter terrarium, 100 milliliters (½ cup) of water as needed should be sufficient. A good research project is to have your students give terrarium plants varying amounts of water to determine optimum amounts.

The best lighting conditions for each terrarium can be another subject of investigation. Ask your students where in the classroom they think the terrariums should be placed. (*As nearly as possible, the plants' natural lighting conditions should be duplicated.*) Experiment by placing terrariums in various locations.*

Constructing a terrarium teaches students about environmental factors—lighting, moisture, soil type, temperature—which affect an ecosystem. Children also learn what plants live in the environment and some of the interaction among various organisms. The class should become aware of special adaptations of organisms which make them well suited to survive in their environment. Another lesson is that an oversupply or undersupply of some component of the environment—water, air, heat, light, plants, animals—has significant effects on the total environment.

➤ MAKE A CARD GAME

To help students learn about the organisms in several ecosystems, and about the interdependence of all organisms in a given ecosystem, have them work in groups to make cards similar to the ones described in the activity "Who's for Dinner" in Chapter 16. Each group of four or five students can make a deck of 20 (or fewer) organism cards for an ecosystem. A good deal of library research will probably be needed. Books from the Golden Nature Guide series (see "Building Your Own Library") are especially helpful in determining what organisms exist in selected ecosystems and what foods each animal eats.

Separate sets of cards can be made for deserts, hardwood forests, pine forests, grasslands, and tundra ecosystems. By making the cards, the students in each group will learn about one ecosystem. After the cards are finished, the groups can exchange cards and play the game several times, until every group has played with all the decks. Thus the entire

**Caution*: Never place a terrarium—even a desert terrarium—in direct sunlight; it will get much too hot.

class will become familiar with organisms in all the ecosystems. Playing the game also makes them aware of levels of consumers and predator-prey relationships.

▶ ECOLOGICAL POETRY
Much poetry has been written about various ecosystems and their components. Find some examples and read them aloud or have the students read them. Be sure to include one or two of Robert Frost's poems.

Students can also write poetry about ecosystems or their parts. They can write haiku about an ecosystem which interests them, then make an appropriate drawing for each poem. The drawings and poems make nice posters to hang in the classroom or in a hallway.

▶ A TRIP TO THE MOUNTAINS
Have the students take a guided-imagery trip to the mountains. After the trip is over, ask questions about it.

Have the students work in groups to create imaginary trips to other ecosystems.

Children's literature

Read *Once upon a Sidewalk** or *Year on Muskrat Marsh*† to your students or have them read one of the books, and then discuss with them the concepts presented.

Find other books on ecology in the school library, and identify them for the students or place them in an interest center.

Activity cards

An example of an activity card is shown in Figure 12-7. Additional activity cards could be made for topics such as investigating a local habitat, constructing a terrarium, studying a specific animal, identifying and drawing food chains or webs from selected ecosystems, describing an imaginary trip to an ecosystem, and drawing a cartoon about an organism escaping one of its predators.

Bulletin boards

ECOLOGICAL ZONES OF A MOUNTAIN
Construct a bulletin board showing the ecological zones which you might pass through when climbing a mountain (see Figure 12-8). Ask students to cut out magazine pictures of wild plants and animals and place them in the proper spaces on the bulletin boards.

*Jean George, *All upon a Sidewalk*, Dutton, New York, 1974.
†Bernice Freschet, *Year on Muskrat Marsh*, Scribner, New York, 1974.

Figure 12-7. Activity card: Inspect a tree.

Figure 12-8. Bulletin board: Ecosystems on a mountain. (Note: To give the altitudes in meters, divide feet by 3.3.)

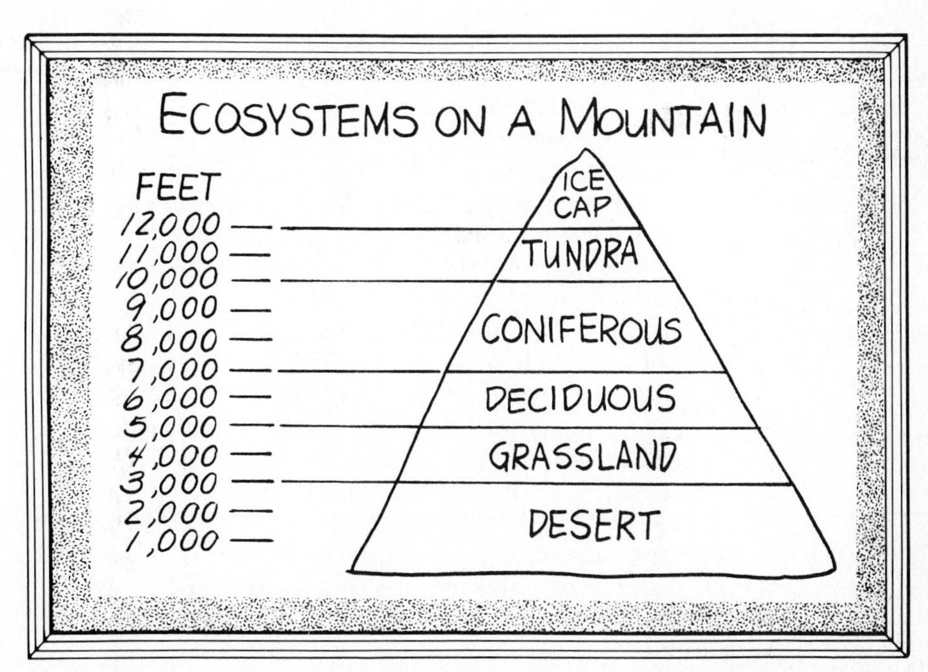

SEASONS IN AN ECOSYSTEM
Select a local small ecosystem, such as a pond in the park, which the class can observe throughout the school year. Have student "reporters" and "photographers" visit the pond every 6 or 8 weeks. For each visit, the reporter can write a brief article describing the ecosystem and the photographer can take three or four photographs of the ecosystem, including its plants and animals.

The pictures and articles should be posted throughout the year. Save all of them, and at the end of the year have the class examine the articles and photographs and note the changes through the seasons. Include the articles and photographs in an almanac; the next year's students can study the almanac as they construct a bulletin board for another ecosystem.

Field trips

The field trips described for primary grades would also be appropriate for intermediate grades. If there is a national or state forest or wildlife refuge nearby and if transportation can be arranged, contact the person in charge of education and arrange for a visit. Ask for information on the purpose of the forest or wildlife refuge, and share the information with your students before the visit.

ENVIRONMENTAL QUALITY

The condition of an environment is of vital concern to people and other organisms which live in that environment. Students should learn to think about the environmental consequences of their decisions. We need to maintain a healthy environment, and we want to maintain an environment which we can enjoy.

The environment is complex and has many components. In studying environmental quality, these topics should be considered: land, air, water, solid wastes, sound, population, wildlife, and mineral resources.

A good environmental education program makes students more knowledgeable about their environment, helps them to recognize problems in the environment, and moves them to do something about creating and maintaining a quality environment. Environmental education encompasses much more than science. This section deals with environmental activities which focus especially on science.

Activities

➤ PRESENT ENVIRONMENTAL CONCERNS
Ask students to cut out newspaper articles related to environmental quality or pollution. You might find it helpful to save your newspapers and news magazines for 2 weeks and then have the students search

through them for articles. Group the articles into areas such as water quality, wildlife management, and air quality.

The purpose of this activity is to make students familiar with problems in the environment. If any of the articles contain proposals for or descriptions of actions designed to treat the problems, use a marking pen to point out those actions. Ask the students these and other questions: "Would you like to help with the proposed actions? How could *you* help reduce the problem which was presented? Can we as a class take any actions?"

AIR QUALITY

No one person or group can tackle the overwhelming problem of global air pollution. Giving up is a temptation, considering the immensity of the job. But a more productive approach is to consider what contributions one person can make to an overall effort to maintain and improve air quality.

Air quality and energy consumption are very closely related. In general, the more energy individuals consume, the more air pollutants we are responsible for creating. If we use more electricity than we need, unnecessary fuel is used to generate that electricity and unnecessary air pollution is created by burning that fuel—for most electricity in the United States is generated by burning fuels. If we burn extra gasoline by unnecessary travel, again unnecessary pollutants are added to the air. Thus, a useful learning activity is to examine how we can use less energy.

Have students keep a 1-week record of every trip, which they know of, that is taken in the family automobile. To avoid embarrassment, have them turn in their records without adding their names. Have the whole class examine your own travel record (you do not have to tell them that it is yours) and recommend as many ways as possible that the amount of travel could have been reduced. (*Walking on trips less than about 1½ kilometers or 1 mile, doing several errands in one trip instead of in several trips, sharing rides with a neighbor to do shopping, riding a city bus if available, riding a bicycle, using buses or trains for longer trips if available.*) Divide the class into small groups and have each group examine one or two of the students' travel records and make further suggestions for reducing automobile travel.

Use of electricity in the home, or school, could be similarly examined.

SOLID WASTE

To help students realize what enormous amounts of solid wastes are produced by people in their town or city, have them examine the amount of one kind of solid waste generated by their families. Each student should count and save all the cans used by his or her family for 1 week.*

**Note:* Tell the students to remove labels and rinse out cans before placing them in the collection container.

An analysis of what a family throws out each week will yield much valuable information.

At the end of the week have them bring the cans to school, and weigh the cans brought by each student.

A very practical application of mathematics is now in order: First add the total number of cans and the total weight of the cans for the whole class. Then determine the average number and the average weight of cans from each family by dividing the totals by the number of families represented in the classroom. If the assumption is then made (be sure to tell the class that this assumption may not be correct) that every family in your town or city uses the same number and weight of cans, calculate the total number and weight of cans thrown away in 1 week. One question which should occur to your students is, "Hey, how can we find out how many families live here?" Have them solve that problem if they can. (*One way would be to assume that the average number of persons per family in your city or town is the same number as the number of persons per family represented in your class; this assumption is inaccurate because it does not take into consideration families without children, but for various reasons the number is nevertheless adequate for reasonable estimates. Then you can have students look up the local population and divide the population by the number of persons per family.*)

Similar studies of glass containers or paper products could be made.

➤ POSTERS AND BUTTONS

Have your students create posters or "campaign" buttons with slogans about environmental quality. An example is given in Figure 12-9. If the students want to make buttons, give them round pieces of paper approximately 10 centimeters (4 inches) in diameter.

Figure 12-9. Environmental poster.

SOUND

We enjoy some sounds and dislike others—and there are times when we want to get away from all sounds and relax, or, if that is not possible, at least to reduce the total level of sound so that we can relax.

Make tape recordings of a street in your town at various times during the day—perhaps the last 2 or 3 minutes of each hour. Play the tape for your class, and ask the students to identify the recorded sounds. Then have them attempt to identify the times at which the various segments were recorded.

Have the students make recordings of sounds they like and sounds they dislike. Then other students can try to figure out where the recordings were made.

Discuss unpleasant sounds with the class, and have them list several sounds which they dislike. Then ask them to propose ways of reducing or eliminating those sounds. Select a small committee of students to visit a local architect to learn how sound is controlled in buildings. Also, ask the architect for information on planting trees, bushes, and shrubs to reduce sound levels around buildings. Have your students design a planting scheme to reduce noise around the school. Present the plan to the principal and if possible arrange to implement parts or all of the plan.

WILDLIFE

It has been said that wildlife is like the "miner's canary." Ask your students what they think that means. (*In earlier times, miners would take a canary into the shaft with them. If there were poisonous gases present, the canary would die, thus warning the miners to evacuate quickly because the air was unhealthy.*) How might wildlife in the environment serve as an indicator of the healthfulness of the environment?

Have students do some research to identify plants or animals which are endangered species. One famous example is the whooping crane, and there are many others. Has humankind caused the extinction of organisms? What lessons can we learn from plant and animal extinction?

One of humankind's most potent capabilities is the ability to think ahead of time about the consequences of our actions. If we do not take advantage of this ability, we may suffer. Ask students to explain how we can use this unique ability to avoid environmental problems. How are we already using this ability? In what other ways could we use it?

➤ SELF-POLLUTING EARTH

Present the following figures to your students:

Carbon monoxide. Trees and other green plants produce 93 percent of this poisonous gas (which is found in automobile exhausts), whereas only 7 percent of it comes from people's activities.

Oxides of nitrogen. Nature produces approximately 99 percent of the oxides of nitrogen (which produces smog) in the atmosphere; humankind produces only 1 percent of these gases.

Particles in the atmosphere. It has been calculated that the eruption of just three volcanoes—Krakatoa in 1883, Katmai in 1912, and Hekla in 1947—put more particles and gases into the air than all of humankind's air pollution during thousands of years of life on earth.

Water pollution. The Mississippi River, which many people consider "dirty," carries over 2 million tons (about 1.8 million metric tons) of natural sediment into the Gulf of Mexico each day.

Ask the class these questions: "Does this information indicate that people do not have to be concerned about environmental quality, since we really do not do very much polluting? How is the way people add pollution to the environment different from the earth's self-polluting activities?" (*Humankind concentrates pollutants in relatively small areas.*)

One important point students should learn is that the earth is self-polluting and self-cleansing. Have students find local instances of the earth's self-pollution. (*Wind, water erosion.*) How much particulate matter has been added to the atmosphere as a result of the eruptions of Mount St. Helens in Washington State? What could humankind do to reduce the effect of its polluting activities?

➤ ENVIRONMENTAL MUSIC

Find some records which relate to environmental quality and play them for your students. There are several songs on Pete Seeger's album *God Bless the Grass,** and another possibility is Ed Ames's song "Leave Them

*Columbia CL2432.

a Flower." Ask students to find other appropriate recordings, especially currently popular ones.

A very good source of folk songs which teach environmental lessons is the publication *Folksongs*.*

Have the students talk about lessons they learn from the recordings. Is there something which they as individuals can do to help resolve the problems mentioned in the music?

▶ ECOBIRDS

Tell students to imagine they are ecobirds. Their purpose in life is to fly about the country identifying pleasant and polluted areas. When they find polluted areas, they are to offer suggestions for resolving the problems. The students can tell or write about their travels, findings, and suggestions.

Children's literature

Read *The Lorax*† to your students or have them read it, and then discuss with them the concepts presented.

Find other books on environmental quality and identify them for the students or place them in an interest center.

Activity cards

An example of an activity card is shown in Figure 12-10 (page 268). Additional activity cards could be made for topics such as recycling materials, assessing the amount of litter on a vacant lot, learning how city sewage is treated, learning about local and state laws concerning air and water pollution, investigating ways to save energy in students' homes, studying the school cafeteria for ways to conserve materials used there, and developing ways to use some materials which would otherwise be thrown away.

Bulletin boards

ENVIRONMENTS WHICH WE ENJOY

Have students bring photographs, make drawings, or cut out magazine pictures showing places they enjoy. Be careful not to let them focus *only* on environments not greatly influenced by people; besides forests, deserts, and mountains, we also can enjoy clean homes, towns, and recreational environments, among others. Post the pictures and drawings on a bulletin board.

*James Kracht, *Folksongs*, Harris County Department of Education, 6208 Irvington Boulevard, Houston, TX 77022, 1977. This is one of six resource units from a project entitled *Environmental Education through Folklore.*

†Theodore Geiser (Dr. Seuss), *The Lorax*, Random House, New York, 1971.

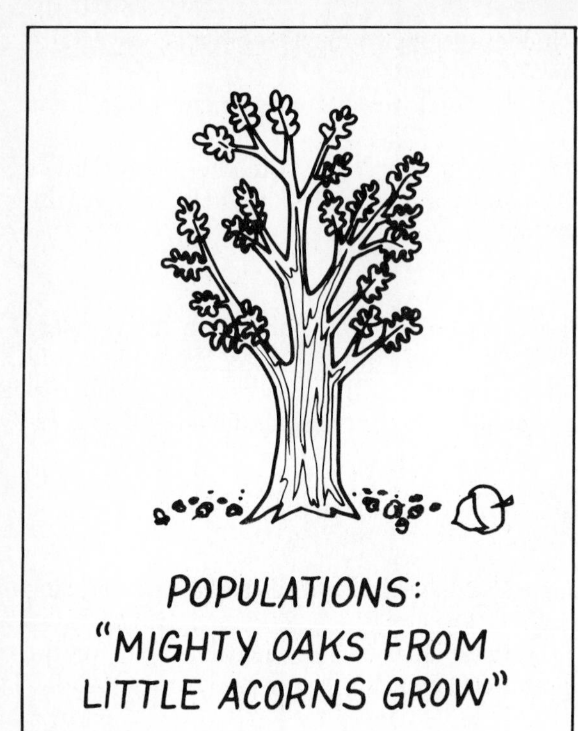

SOME TIME IN THE AUTUMN, LOOK UNDER AN OAK TREE AT ALL THE ACORNS. DO YOU THINK THERE ARE MORE THAN 1,000 ACORNS?

WHAT IF ALL THE ACORNS GREW INTO TREES? WHY DON'T THEY ALL GROW UP?

WHAT USES ACORNS FOR FOOD? HAS HUMANKIND EVER USED ACORNS FOR FOOD?

Figure 12-10. Activity card: Populations—"Mighty oaks from little acorns grow."

AIR QUALITY
Construct an informational bulletin board showing ways in which we as individuals can reduce air pollution and help to maintain air of high quality. Examples are using less energy (*turning off lights, walking or using bicycles*) and keeping home furnaces and other equipment in good operating condition. Have the students identify and illustrate other possibilities, and add them to the bulletin board.

Field trips

If possible, arrange to take a field trip to a local business or utility which is concerned about reducing pollution—perhaps a sewage treatment plant, a waste disposal company or utility, an electrical generating plant, a heavy-industry plant, or a mining operation.

An educational and productive field trip can be taken right on the school grounds: You can tell your students to find and clean up litter during a walk around the grounds. This will help them to become more aware of concrete things that they, as individuals, can do as part of their daily lives to improve their environment. Ask them if they can see air pollution from the school grounds, and if they can hear sound pollution. Then ask them what they think should be done about the air pollution and the sound pollution.

RESOURCES

Sources of assistance

FREE AND INEXPENSIVE MATERIALS

Here are some good sources of free and inexpensive materials for use in teaching about ecology and environmental quality:

1. *Department of Agriculture.** A large amount of material on several topics. Ask for a list of publications or indicate your grade level and topics of interest.
2. *Environmental Protection Agency.* A very large amount of material. Indicate your interests and ask to be placed on their permanent mailing list so you will receive new materials as they become available.
3. *Garden Clubs of America.* An environmental education packet.
4. *Soil Conservation Society of America.* Contact a local branch. Comic books on conservation are available.
5. *National Wildlife Federation.* Publishes an elementary education packet and *Ranger Rick* magazine.
6. *Department of the Interior.* Ask for "Conservation Notes"—short reports on specific animals and their habitats. Other materials are also available.
7. *Del Monte.* A booklet "Can Crafts" shows how throwaways (cans, bottles) can be used for art and craft projects.
8. *Department of Energy.* A set of science activities in energy folders. Each folder includes approximately 12 very good activities.
9. *American Petroleum Institute.* Several materials including "Conservation—A Picture Discussion Kit."

LOCAL RESOURCES

Local resources which can help in teaching about ecology and environmental quality include:

1. The county extension agent (even in large cities).
2. Resource people from local environmental organizations such as Audubon, Sierra Club, or a group organized for some local environmental concern.
3. Electrical and gas utility companies often offer teaching materials and resource persons.
4. Boy Scouts, Girl Scouts, and 4-H Clubs publish conservation and environmental education materials.

*The addresses of specific agencies, organizations, and businesses are listed in the Appendix.

ADDITIONAL RESOURCES

1. Magazines such as *Ranger Rick, World, National Geographic, National Wildlife,* and *International Wildlife.*
2. Almost all states have air and water quality boards, which often offer useful materials and may have a speaker's bureau.
3. Many states have an environmental education committee or board, which may publish teaching materials or plans.
4. The ERIC Clearinghouse for Science, Mathematics, and Environmental Education (The Ohio State University, 1200 Chambers Road, 3d Floor, Columbus, Ohio 43212) publishes a number of excellent books which describe environmental activities. Request a list of publications.

Strengthening your background

1. Select one animal on the endangered-species list and find out why it has become endangered. Is anything being done to save the animal? What could be done?
2. Read *A Sand County Almanac.**
3. Visit a large zoo and ask a member of the staff to explain how the natural food and environment of the animals are taken into consideration in caring for the animals.
4. Make a specific plan for driving your automobile less, and put the plan into effect in your everyday life.
5. Attend a Sierra Club meeting. If members are concerned about a controversial issue, ask them how they defend their position. Find people who take a different position on the same issue, and ask them the same question.
6. Go out to an undisturbed environment and observe what happens there. What kinds of plants and animals are present? (You don't have to know the organisms' names.) What are the animals doing? Focus on a single animal—perhaps an insect—and observe it closely for several minutes. What does it eat? Does it have an identifiable "home"?
7. Visit a local sewage treatment plant to find out how the sewage is treated.
8. Make a food chain for a local ecosystem.

Getting ready ahead of time

1. Write to the National Wildlife Federation and request an elementary education packet. When you receive it, check the teaching-materials list for other materials you can use. (Single copies of many materials are free.)
2. Obtain the folders of science activities available from the United States Department of Energy. (See "Free and Inexpensive Materials.")

*Aldo Leopold, *A Sand County Almanac*, Oxford University Press, New York, 1949.

3. Create 10 activity cards for use in teaching about ecology and environmental quality.
4. Make a food web mobile to hang in your classroom as an example for the students.
5. Create a guided fantasy about a mouse visiting your town to find out whether or not the environment is good enough for it to move to.

Building your own library

Alexander, Taylor, and George Fichter: *Ecology*, Western, Racine, Wis., 1973.

Benedick, Jeanne: *Science Experiences: Ecology*, F. Watts, New York, 1975.

Benson, Dennis: *Recycle Catalogue*, Abingdon, Nashville, Tenn., 1975

Cooper, Elizabeth: *Science on Shores and Banks*, Harcourt Brace Jovanovich, New York, 1960.

Cornell, Joseph: *Sharing Nature with Children*, Ananda Publications, Nevada City, Calif., 1979.

Darnell, Rezneat: *Ecology and Man*, Brown, Dubuque, Iowa, 1973.

Elementary Science Study (ESS): *Pond Water*, Webster, McGraw-Hill, Manchester, Mo., 1976.

Gardner, John F.: *Book of Nature Activities*, Interstate, Danville, Ill., 1967.

Gill, Don, and Penelope Bonnett: *Nature in the Urban Landscape: A Study of Urban Ecosystems*, York Press, Baltimore, Md., 1973.

Golden Nature Guide Series, New York, and Western, New York. The following titles are included: *Birds, Butterflies and Moths, Cacti, Fishes, Flowers, Insects, Mammals, Non-Flowering Plants, Pond Life, Reptiles and Amphibians, Seashores, Spiders, Trees, Weeds.*

Gross, Phyllis, and Esther Railton: *Teaching Science in an Outdoor Environment*, University of California Press, Berkeley, 1972.

Helfman, Elizabeth: *Our Fragile Earth*, Lothrop, New York, 1972.

Laun, H. Charles: *The Natural History Guide—A Study, Reference, and Activity Guide*, 2d ed., Alsace Books and Films, Alton, Ill., 1967.

Nickelsburg, Janet: *Nature Program for Early Childhood*, Addison-Wesley, Reading, Mass., 1976.

Peterson Field Guide Series, Houghton Mifflin, Boston. The following titles are included: *A Field Guide to Animal Tracks, A Field Guide to Birds' Nests, A Field Guide to Reptiles and Amphibians, A Field Guide to Birds, A Field Guide to the Butterflies, A Field Guide to the Ferns and Their Related Families . . . , A Field Guide to the Insects . . . , A Field Guide to the Mammals, A Field Guide to Trees and Shrubs, A Field Guide to Western Birds.*

Pringle, Laurence: *Natural Fire: Its Ecology in Forests*, Morrow, New York, 1979.

Roth, Charles E.: *The Most Dangerous Animal in the World*, Addison-Wesley, Reading, Mass., 1971.

Sale, Larry, and E. Lee: *Environmental Education in the Elementary School*, Holt, New York, 1972.
Simon, Seymour: *Science in a Vacant Lot*, Viking, New York, 1970.

Building a classroom library

Adler, Irving: *The Environment*, John Day, New York, 1976.
Allen, Dorothy: *The Story of Soil*, Putnam, New York, 1971.
Baylor, Byrd: *Desert Voices*, Scribner, New York, 1981.
Bendick, Jeanne: *A Place to Live: A Study of Ecology*, Parents Magazine, New York, 1970.
Berger, Gilda: *Coral Reef, What Lives There*, Coward-McCann, New York, 1977.
——— and Levin Berger: *"Fitting In": Animals in Their Habitats*, Coward-McCann, New York, 1976.
Buck, Lewis: *Wetlands: Bogs, Marshes, and Swamps*, Parents Magazine, New York, 1974.
Busch, Phyllis: *Dining on a Sunbeam: Food Chains and Food Webs*, Four Winds, New York, 1973.
———: *Exploring as You Walk in the City*, Lippincott, Philadelphia, 1972.
———: *Exploring as You Walk in the Meadow*, Lippincott, Philadelphia, 1972.
Cowing, Sheila: *Our Wild Wetlands*, Messner, New York, 1980.
David, Eugene: *Spiders and How They Live*, Prentice-Hall, Englewood Cliffs, N.J., 1964.
Edwards, Joan: *Caring for Trees on City Streets*, Scribner, New York, 1975.
Farb, Peter: *Ecology*, Time-Life, New York, 1963.
Fisher, Ronald: *A Day in the Woods*, National Geographic Society, Washington, 1975.
Geiser, Theodore (Dr. Suess): *The Lorax*, Random House, New York, 1971.
George, Jean: *All upon a Sidewalk*, Dutton, New York, 1974.
Gordon, Esther, and Bernard Gordon: *Once There Was a Passenger Pigeon*, Walck, New York, 1976.
Graham, Ada, and Frank Graham: *The Changing Desert*, Scribner, New York, 1981.
Grossman, Shelly, and Mary Grossman: *Ecology*, Grosset and Dunlap, New York, 1971.
Hahn, James, and Lynn Hahn: *Recycling: Reusing Our World's Social Waste*, F. Watts, New York, 1973.
Harris, John, and Aleta Pahl: *Endangered Predators*, Doubleday, Garden City, N.Y., 1976.
Harris, Lorle: *Biography of a Whooping Crane*, Putnam, New York, 1977.
Hartman, Jane E: *Living Together in Nature*, Holiday, New York, 1977.
Hinshaw, Dorothy: *Animal and Plant Mimicry*, Holiday, New York, 1978.

Jaspersohn, William: *How the Forest Grew*, Greenwillow, 1980.
Kane, H.: *The Tale of a Meadow*, Knopf, New York, 1959.
Keifer, Irene: *Poisoned Land: The Problem of Hazardous Waste*, Atheneum, New York, 1981.
Lavine, Sigmund A.: *Wonders of Terrariums*, Dodd, Mead, New York, 1978.
Laycock, George: *Beyond the Arctic Circle*, Four Winds, New York, 1978.
Learner, Carol: *Seasons of the Tallgrass Prairie*, Morrow, New York, 1980.
List, Albert, and Ilka List: *A Walk in the Forest: The Woodlands of North America*, Thomas Y. Crowell, New York, 1977.
McClung, Robert M.: *Vanishing Wildlife of Latin America*, Morrow, New York, 1981.
McCombs, Lawrence, and Nicholas Rosa: *What's Ecology?*, Addison-Wesley, Reading, Mass., 1978.
McCoy, J.: *Saving Our Wildlife*, Crowell-Collier, New York, 1970.
Millard, Reed (ed.): *Clean Air—Clean Water for Tomorrow's World*, Messner, New York, 1977.
Norris, LouAnne, and Howard Smith: *An Oak Tree Dies and a Journey Begins*, Crown, New York, 1979.
Patent, Dorothy: *Sizes and Shapes in Nature—What They Mean*, Holiday, New York, 1979.
Pitt, Valerie, and David Cook: *A Closer Look at Deserts*, F. Watts, New York, 1975.
Podendorf, Illa: *Everyday Is Earth Day*, Children's Press, Chicago, Ill., 1971.
Pringle, Laurence: *Animals and Their Niches*, Morrow, New York, 1977.
———: *City and Suburb: Exploring an Ecosystem*, Macmillan, New York, 1975.
———: *The Controversial Coyote*, Harcourt Brace Jovanovich, New York, 1977.
———: *Death is Natural*, Four Winds, New York, 1977.
———: *Estuaries: Where Rivers Meet the Sea*, Macmillan, New York, 1973.
———: *The Gentle Desert*, Macmillan, New York, 1977.
———: *The Hidden World*, Macmillan, New York, 1977.
———: *Lives at Stake: The Science and Politics of Environmental Health*, Macmillan, New York, 1980.
———: *Natural Fire: Its Ecology in Forests*, Morrow, New York, 1979.
Quinn, John R.: *Nature's World Records*, Walker, New York, 1977.
Rosier, Lydia: *Biography of a Bald Eagle*, Putnam, New York, 1973.
Ross, Wilda: *Who Lives in This Log?* Coward-McCann, New York, 1971.
Selsam, Millicent: *How Animals Live Together*, Morrow, New York, 1979.
———: *Land of the Giant Tortoise: The Story of the Galapagos*, Four Winds, New York, 1977.
———: *See through the Forest*, Harper and Row, New York, 1971.

Silverstein, A., and V. Silverstein: *Unusual Partners: Symbiosis in the Living World*, McGraw-Hill, New York, 1968.

Simon, Seymour: *Exploring Fields and Lots*, Garrard, Champaign, Ill., 1978.

———: *A Tree on Your Street*, Holiday, New York, 1973.

———: *Science Projects in Ecology*, Holiday, New York, 1972.

Stone, Harris, and Steven Collins: *Populations: Experiments in Ecology*, F. Watts, New York, 1973.

Stuart, Gene S.: *Wildlife Alert! The Struggle to Survive*, National Geographic Society, Washington, 1980.

Watson, Jane W.: *Deserts of the World: Future Threat or Promise?* Philomel, New York, 1981.

Wise, William: *Animal Rescue*, Putnam, New York, 1978.

PART 3

TEACHING THE EARTH SCIENCES

Survivors of volcanic eruptions, tornadoes, or earthquakes remember these events for a lifetime; and indeed all of us are awed by dramatic events like these. But we are surrounded by daily experiences which, because they are less dramatic, go virtually unnoticed and unappreciated. We tend to notice and remember a few days of terrible weather while ignoring the many days of good weather. We notice the earth's crust as such when it cracks and erupts; otherwise, we are not likely to think about it at all.

Learning more about the physical world, then, is an interesting challenge for all of us. The more we understand about it, the better we realize how much more there is to be understood. We hope that as elementary science teachers you will help your students discover the excitement of the world around them and whet their appetites for knowledge about it.

As you prepare yourself to become a teacher, consider the following passage by John Steinbeck, and the opportunities that you will have to introduce your students to the world around them.

"My eleven-year-old son came to me recently and, in a tone of patient suffering, asked, 'How much longer do I have to go to school?'

'About fifteen years,' I said.

'Oh! Lord,' he said despondently. Do I have to?'

'I'm afraid so. It's terrible and I'm not going to try to tell you it isn't. But I can tell you this—if you are very lucky, you may find a teacher and that is a wonderful thing.'

'Did you find one?'

'I found three,' I said. . . .

My three had these in common—they all loved what they were doing. They did not tell—they catalyzed a burning desire to know. . . .

I shall speak only of my first teacher because, in addition to other things, she was very precious.

She aroused us to shouting, bookwaving discussion. She had the noisiest class in school and didn't even seem to know it. We could never stick to the subject, geometry or the chanted recitation of memorized phyla. Our speculation ranged the world. She breathed curiosity into us so that we brought in facts or truths shielded in our hands like captured fireflies.

She was fired and perhaps rightly so . . . for failing to teach the fundamentals. . . . She left her signature on us, the literature of the teacher who writes on minds. I suppose that, to a large extent, I am the unsigned manuscript of that high school teacher. What deathless power lies in the hands of such a person.

I can tell my son who looks forward with horror to fifteen years of drudgery that somewhere in the dusty dark a magic may happen that will light up the years . . . if he is very lucky. . . .

I have come to believe that a great teacher is a great artist and there are as few as there are any other great artists. It might even be the greatest of the arts since the medium is the human mind and spirit."*

We agree with Steinbeck that a good teacher is one who leads students to discover the world. As you prepare to teach children about the earth and its properties, we challenge you to become a stimulating teacher who provides more questions than answers.

*Quoted in *The Education of Teachers—Curriculum Programs,* Report of the Kansas TEPS Conference, National Education Association, Washington, D.C., 1959, p. 71. See also Louise Sharp, ed., *Why Teach?* Holt, New York, 1957, p. 260.

CHAPTER 13

THE UNIVERSE

IMPORTANT CONCEPTS

The solar system is made up of the sun, the planets, their satellites, and a few less well known objects.

The sun is a star.

From our position on earth, the sun appears to rise in the east and set in the west.

The rotation of the earth is what causes the sun to appear to move across the sky.

Day and night on earth are the result of the rotation of the earth.

The sun supplies almost all the energy available on the surface of the earth.

The moon reflects the sun's light.

The movement of the moon relative to the earth and sun causes its lighted half to appear as a series of different shapes.

Tides are the result of the changing relative positions of the sun, moon, and earth.

Stars are great masses of incandescent gases—enormous powerhouses of energy.

There are millions of stars and star systems in the universe.

Analysis of light from stars allows us to obtain information about those stars.

Legends have been written about some groups of stars identified as constellations.

On a clear night, more than 2000 stars are visible to the unaided eye.

The paths of planets and other bodies in the universe are determined by gravitational attraction and the body's inertia.

The size of the universe is so great that distances are usually measured in terms of the time it takes light to travel between the objects in space.

The universe is believed to be dynamic and evolving.

Manned and unmanned space missions have greatly increased our knowledge of the moon, the solar system, and the universe.

There are many questions about the universe for which we do not presently have answers.

Consideration of the force of gravity and the laws of motion is very important in planning space travel.

Rocket engines must produce enough force to overcome gravity and must be supplied with all necessary materials for flight in the nearly perfect vacuum of space.

Spacecraft come under the increasing influence of gravity of any celestial body which they approach.

The property of inertia causes objects either to remain at rest if they have been at rest or to continue in motion if they have been in motion.

The application of a force is necessary to change the motion of a body.

Aiming a spacecraft for some destination is a very complex process.

Astronauts must undergo special training and conditioning before space travel.

Colonization of space is both intriguing to students and of potentially great significance to humankind.

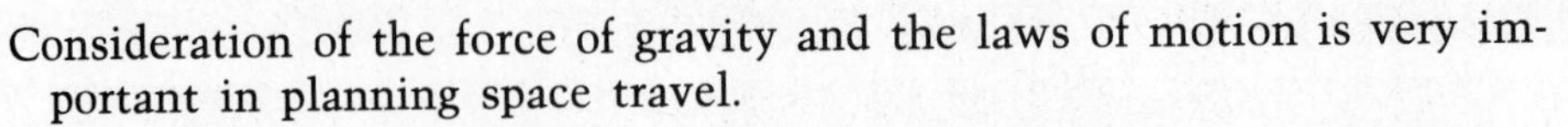

INTRODUCTION: STUDYING THE UNIVERSE

People have long been fascinated by the objects visible in the heavens. Skylore—folklore pertaining to celestial objects—has been generated by almost every group of ancient people for whom records have been found. A special new interest in astronomy has been kindled by humankind's efforts at exploring space. Perhaps one reason for the popularity of topics related to astronomy is the scope they give to our imagination.

Scientists have learned a great deal about space. One important reason for the advances has been the development of technology which facilitates the study of the universe. Telescopes, spectrometers, computers, space probes, and a host of other inventions have aided astronomers immensely. Much of our present knowledge about space is very abstract and beyond the intellectual capabilities of most elementary school students; however, there are many fascinating topics which are appropriate for them. This chapter includes a few activities on basic concepts for primary grades and several more complex activities for intermediate grades.

Space exploration captures the imagination of many people. Even though it has been going on for many years, it is still an adventure which leads to new frontiers. One early technological advance which helped in the exploration of space was the telescope, and both manned and unmanned spacecraft are later developments which enable scientists to explore portions of space more thoroughly.

Many aspects of space exploration are fascinating to elementary school students. The excitement, intrigue, and fascination of the scientific and technological exploration of space may be just the spark needed to set some of your students aglow. Both you and your students should fully realize that there are many unanswered (and even unasked) questions in astronomy and, further, that not all astronomers agree upon all information and hypotheses which are presently available.

When you teach space exploration, one or more of the films of space flights which are available from the National Aeronautics and Space Administration (NASA) will be especially useful (see the Appendix).

Remember to follow up students' inquiries and to encourage them to find answers to their own questions. Suggest activities whenever possible, but do not hesitate to suggest library research also. If a student has a real interest in a topic, reading books or magazine articles written at an appropriate level will be very meaningful.

THE UNIVERSE FOR PRIMARY GRADES

Activities

➤ DAY AND NIGHT

In a darkened room, have students demonstrate the cause of day and night. One student, representing the sun, directs a flashlight toward another, representing the earth. Or they can use the light from an overhead projector. (Whatever light source is used, make sure that the students understand that the real sun shines in all directions.) Ask the class these questions: "Does the light from the sun fall on all the earth? How about the side away from the sun? Which side of the earth is in daytime? In nighttime?"

And then ask the question that you most want answered: "How could the dark side of the earth become lighted?" (*Careful! There are at least two correct answers: The earth can rotate,* or the sun can move around to the other side of the earth.*) Try to get students to propose both possibilities; but if necessary, you can offer them. Have the students representing the sun and the earth demonstrate each possibility in turn. Then ask if any students know which is the correct explanation. Inform the students that scientists now have strong evidence indicating that it is indeed the rotation of the earth that causes night and day.

As a reinforcing activity, have the students again demonstrate the cause of day and night, this time using the flashlight and a globe. Ask the students how often the earth turns around.

➤ SOLAR SYSTEM

A simple model of the solar system will not only teach that there are other planets besides Earth but will also introduce the concept of the orbits of the planets and provide information about the locations of the planets in relation to the sun.

*Use this activity to introduce the terms "rotate" and "rotation" and make sure the students understand them.

Figure 13-1. Students assuming the positions of planets.

Prepare 10 labels: "Sun," "Mercury," "Venus," "Earth," "Mars," "Jupiter," "Saturn," "Uranus," "Neptune," "Pluto." Give a label to each of 10 students and have them stand up in front of the class in the order listed. Tell the students that they represent the order of the planets in the solar system. Have the planets circle around the sun, maintaining the same relative order. (See Figure 13-1.) Introduce the terms "revolve" and "revolution." If appropriate for your grade level, you may also indicate that the path of a planet around the sun is called its "orbit."

Ask these questions: "Is Earth the planet closest to the sun? Which planet is closest? Which planet is farthest from the sun? Which two planets are Earth's closest neighbors?" Have the students repeat the names of the planets in order of their increasing distance from the sun. Have pairs or small groups of students make drawings of this simple model of the solar system or make models of the solar system using styrofoam balls.

➤ PLANETARY MODELS FROM SPACECRAFT

After showing pictures of planets taken by spacecraft which flew close to or landed on the planets, have students construct models of either a

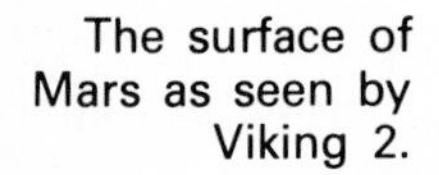
The surface of Mars as seen by Viking 2.

whole planet or the surface of a planet. Any medium may be used for models; one possibility would be to make a shadow box by gluing sand, rocks, and other such materials to the side of a shoebox, to show the surface of a planet.

THE SUN'S POSITIONS

In a little-used part of the school grounds which is in full sunlight all day, place a wooden stake in the ground, or use an existing pole, such as a basketball goalpost or a fence post, to enable the class to observe the positions of shadows early in the morning, about noontime, and late in the afternoon. If possible, mark the positions of the shadows. Where does each shadow fall relative to the sun?

Repeat the observations for several days. Are the shadows found in the same approximate locations on each day at the various times? Where would students expect to see the sun the next morning? Noon? Afternoon? Ask whether the students notice a pattern to the sun's location. (*The sun always rises in the same general–easterly–location and always sets in the same general–westerly–location.*) If they do not notice the pattern, point it out to them and reinforce it by having them make predictions and several more observations.

After making one of the observations, ask the students where they think the tip of the shadow will be in 5 minutes. Let them place a mark to indicate their prediction, and then check the actual location in 5 minutes. What will happen in 5 more minutes? Were you and they surprised at the amount of movement? Would the tip of the shadow from a tall stick (or a tall building) move farther than, the same amount as, or less than the shadow of a shorter stick?

THE MOON'S POSITIONS

Though we often think of the moon as a "night object," it is actually visible nearly as frequently during the day as during the night. This activity can therefore be done either by the class as a whole during the school day or by individual students at home at night (with their parents' supervision and cooperation). Tell the students to locate the moon and stand on a spot which makes the moon appear to be just at the top of a particular tree or at the corner of a building, or some other way definitely positioned. Because of children's varying heights, each child's "standing spot" will be different.

Make a record of the time and have each child mark his or her standing spot so that the children can return to exactly the same spots the next day. Each student should also make a sketch of the moon's location in relationship to the selected reference point. (See Figure 13-2.) Ask the students to predict whether or not the moon will be in the same location on the next day at the same time. If they predict "no," have them make another prediction—where they think the moon will be at that time. On the next day have them check their predictions and predict for the third day at the same time. Keep repeating the process for several days. Are their predictions accurate? Can the students make any generalizations

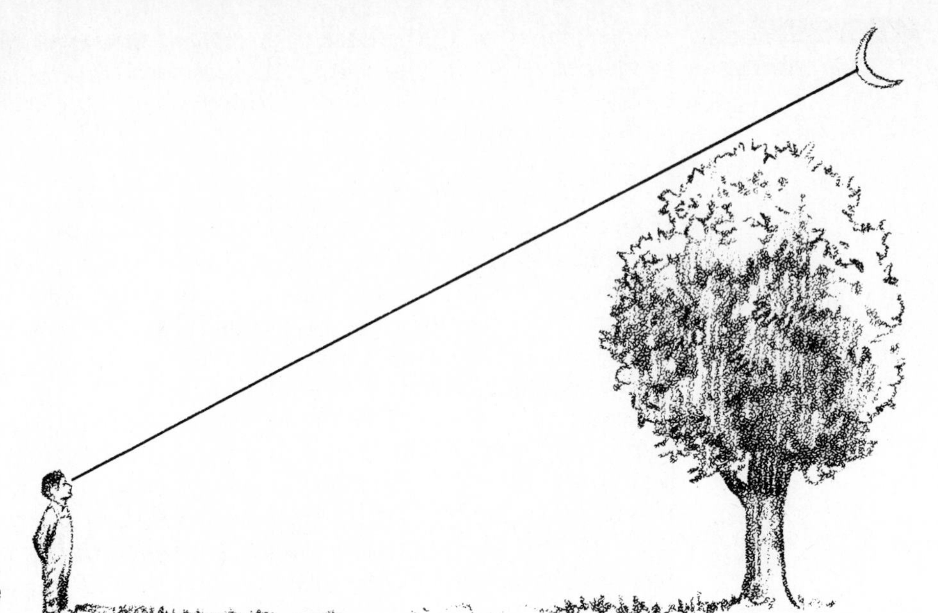

Figure 13-2. Locating the moon.

about the moon's motion? They may not be able to make generalizations, but if they attempt any, have them check the generalizations out.

▶ DESIGN A SPACECRAFT

Ask the students to name as many things as they can which must be taken into consideration when building a craft for space travel. Make a list of those requirements. You might begin the list with items such as:

The craft will have to be airtight.
Enough food will have to be taken.
A radio will be needed for communication

Throughout the study of space exploration, students can add items to the list. Have them make drawings of spacecrafts and later ask them to explain how they have taken some of the requirements into consideration in their drawings.

The purpose of this activity is both to demonstrate the complexity of space travel and to give students practice in planning the future. Thinking about the future causes people to pull together and apply all that they know about a topic and thus can be very fruitful in helping children put bits of information into a meaningful structure.

▶ ASTRONAUTS—FOOD, CLOTHING, EXERCISE

To familiarize students with some of the things which astronauts do, and which might also be healthy for students, have them read books (see "Building a Classroom Library") about the activities of astronauts, or request information from NASA.

Set aside a special day for study of astronauts. Prepare several samples of food as similar as possible to that which astronauts eat. Serve each student a sample. Also, have the class do some of the exercises which astronauts do. Ask the students if astronauts would be likely to lift weights while in space. (*No! Taking weights up into space would use unnecessary fuel; besides, in space the weights would be "weightless" and thus would offer little benefit in an exercise program.*) Show students pictures of astronauts' clothes for use both within the protected spacecraft and outside it. If possible, obtain samples of the material used in both. If some students—and their parents—become really interested, encourage them to make spacesuits. Ask students to explain why the special food, clothing, and exercises are used by astronauts.

If you live near a NASA center,* invite a NASA speaker, perhaps even an astronaut, to visit your class and demonstrate the food, clothing, and exercise used by astronauts.

➤ SPACECRAFT MODELS

Have students make models of several different spacecraft from pictures or drawings. The models can be made of clay, soap, paper, or any other materials which your students like to use.

Obtain the dimensions of a spacecraft and the rocket that launches it. Mark off the dimensions with chalk on a sidewalk near the school.

➤ SPACECRAFT WORD GAME

Beginning with the word "spacecraft" (see Figure 13-3, page 286), challenge students to add in "domino" fashion words which relate to space, space travel, and the universe. The words can run either vertically or horizontally, and any letter on the board can be used as part of the words they add.

➤ SPACE TRAVEL

Tell the students that they are to picture themselves aboard a spacecraft traveling through space and that they will be listening to music which may cause them to think about things that are happening on the trip. Then play the music *2001—A Space Odyssey.* The students can then discuss their trips. Ask what routine things they did on the trip. What special things happened?

Children's literature

Read either *The Day We Saw the Sun Come Up*† or *Let's Go on a Space Shuttle*‡ to your students or have them read one of these books, and then discuss with them the concepts presented.

Find other books on astronomy in the school library and identify them for the students or place them in an interest center.

*NASA centers are located in Moffett Field, California; Greenbelt, Maryland; Houston, Texas; Kennedy Space Center, Florida; Hampton, Virginia; Cleveland, Ohio; and Marshall Space Flight Center, Alabama.

†Alice Goudey, *The Day We Saw the Sun Come Up*, Scribner, New York, 1961.

‡Michael Chester, *Let's Go on a Space Shuttle*, Putnam, New York, 1975.

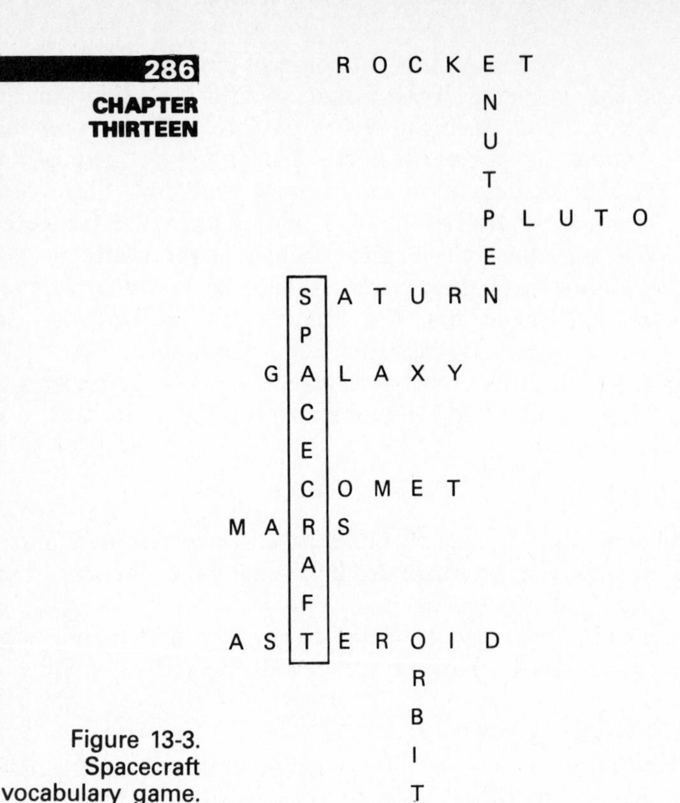

Figure 13-3. Spacecraft vocabulary game.

Activity cards

An example of an activity card is shown in Figure 13-4. Additional activity cards could be made for topics such as observing the moon; using a ball to duplicate moon phases; finding songs about the moon, sun, planets, or stars; writing stories about traveling in the solar system; making models of planets or the solar system; visiting a local planetarium; collecting space pictures; and making models of spacecrafts and space colonies.

Bulletin boards

MOON PHASES

Have the students observe and make sketches of the lighted part of the moon for at least 30 consecutive days. Post the sketches on a bulletin board as illustrated in Figure 13-5. Indicate cloudy days by posting appropriately labeled cards. If possible, take photographs of the moon each night, or find a student who can take them. (Exposure information can be obtained from a local photographer—perhaps a portrait photographer.) Does any pattern become evident?

- COLLECT PICTURES OF SOME SPACE EXPLORATION WHICH IS GOING ON OR ABOUT TO BEGIN.
- PRETEND YOU ARE GOING ON THE TRIP. WHAT ARE YOU GOING TO STUDY?

IF YOU COULD TAKE ONLY ONE ROLL OF FILM, WHAT WOULD YOU TAKE PICTURES OF?

HOW WILL YOU PREPARE FOR THE TRIP?

Figure 13-4. Activity card: What's going up, Doc?

Figure 13-5. Bulletin board: The moon as we saw it.

Figure 13-6. Bulletin board: Who am I?

ASTRONAUTS
Obtain from NASA photographs and biographical data on several astronauts. Post pictures and folded information cards (have students prepare the cards) for six astronauts on the bulletin board as illustrated in Figure 13-6. The visible side of each information card should give a short statement of some special bit of information about the astronaut pictured above it.

The students are to learn who the astronauts are. The astronauts' names and short biographical sketches should become visible when a student lifts folds of the card. The biographical sketches should include the astronauts' birthplaces, high schools, colleges, professional areas of preparation, and—if possible—hobbies.

Leave the first six astronauts on the board for a week or so and then post six others.

Field trips

During the study of astronomy, it is important to make a field trip to the school grounds, in the daytime, to observe the moon. The trip can be repeated, as discussed, in the activity "The Moon's Positions," or you can make a single visit simply to show the class that the moon is not only a nighttime phenomenon.*

*See the Elementary Science Study unit *Daytime Astronomy* (listed in "Building Your Own Library") for additional suggestions.

Another possible field trip might be made to a local observatory or planetarium. A nighttime stargazing session might be difficult to arrange but would be a great experience for you and your students. Invite a member of a local amateur astronomer's club to identify a few visible stars and planets and to bring along a telescope if possible—amazing and beautiful things can be seen through a telescope.

THE UNIVERSE FOR INTERMEDIATE GRADES

Activities

DISTANCES IN THE SOLAR SYSTEM

This exercise can help students imagine the vast distances in the universe, and specifically even within the solar system. Have the class make signs or models to represent the sun and the nine planets (they need not try to make the models to scale). The information given in Table 13-1 (page 290) is based upon letting 1 centimeter represent 1 million kilometers (or 1 inch represents 1,580,000 miles). Have students hold the signs or models and space themselves according to the scale indicated for distances between the planets.

Are the students surprised at the distances? Tell them that if the same scale were used to represent the sun and planets, the diameter of the sun would be 1.4 centimeters (about ½ inch); the diameter of Mercury, the smallest planet, would be 0.004 centimeter; the diameter of Jupiter, the largest planet, would be 0.143 centimeter; and the diameter of Earth would be 0.013 centimeter. How easy would it be to see model planets of these sizes if they were spaced appropriately far apart? How far away would the nearest star (other than the sun) be? (*The nearest star, Proxima Centauri, is 40,681,440,000,000 kilometers or 25,222,492,000,000 miles from Earth. The answer is 40,681,440 centimeters or 40.7 kilometers—16,016,314 inches or 25.3 miles.*)

SCALE MODELS OF THE PLANETS

Making scale models is a good way to learn to understand the relative sizes of the planets. Have the students use the information in Table 13-1 to calculate the sizes of models of the sun and planets, letting 1 centimeter (0.4 inch) represent 10,000 kilometers (33,000,000 feet). (*The diameters of the models would be: sun, 139.2 centimeters or about 55 inches; Mercury, 0.4 centimeters or about 0.2 inch; Venus, 1.2 centimeters or about 0.48 inch; Earth, 1.3 centimeters or about 0.50 inch; Mars, 0.7 centimeters or about 0.27 inch; Jupiter, 14.3 centimeters or about 5.6 inches; Saturn, 12 centimeters or about 4.7 inches; Uranus, 5.2 centimeters or about 2.0 inches; and Pluto, 0.6 centimeters or about 0.23 inch.*) Have the students make models of the planets by cutting paper disks or making clay balls; balloons might be used for Jupiter and Saturn. Draw a circle representing the sun on the chalkboard, and then compare the

TABLE 13-1 **DIAMETERS AND DISTANCES IN THE SOLAR SYSTEM**

OBJECT	DISTANCE FROM SUN: KILOMETERS	DISTANCE FROM SUN: MILES	DIAMETER OF OBJECT: KILOMETERS	DIAMETER OF OBJECT: MILES
Sun	—	—	1,392,000	804,000
Mercury	57,900,000	36,000,000	4,880	3,030
Venus	108,200,000	67,200,000	12,104	7,520
Earth	149,600,000	92,900,000	12,756	7,920
Mars	227,900,000	141,500,000	6,787	4,210
Jupiter	778,300,000	483,300,000	142,800	88,680
Saturn	1,427,000,000	886,200,000	120,000	74,520
Uranus	2,870,000,000	1,782,000,000	51,800	32,170
Neptune	4,497,000,000	2,793,000,000	49,000	30,430
Pluto	5,900,000,000	3,664,000,000	5,800*	3,600*

OBJECT	SCALE DISTANCE (1 CENTIMETER = 1 MILLION KILOMETERS)	SCALE DISTANCE (1 INCH = 1,580,000 MILES)	SCALE DIAMETER (1 CENTIMETER = 1 MILLION KILOMETERS)	SCALE DIAMETER (1 INCH = 1,580,000 MILES)
Sun	—	—	1.392	0.547
Mercury	58	23	0.004	0.002
Venus	108	43	0.012	0.005
Earth	150	59	0.013	0.005
Mars	228	90	0.007	0.003
Jupiter	778	307	0.143	0.056
Saturn	1427	562	0.120	0.047
Uranus	2870	1131	0.052	0.020
Neptune	4497	1772	0.049	0.019
Pluto	5900	2325	0.006*	0.002*

*These figures may not be entirely accurate.

models of the planets with the sun and with one another. Which is the largest planet? The smallest? Which other planet is nearly the same size as Earth? How large would a model of our moon be? (*Actual diameter = 3476 kilometers; scale model = 0.3 centimeter or about 0.13 inch.*)

Using the same scale (1 centimeter or 0.4 inch represents 10,000 kilometers or 6215 miles) how far would Earth be from the sun in a scale model? (*Actual distance, 149,600,000 kilometers; scale distance, 14,960 centimeters, or 5900 inches or 490 feet—about one city block.*) What would be the problems involved in attempting to view an object 1.3 centimeters (about 0.5 inch) in diameter at a distance of one city block?

➤ SUNDIAL

Although sundials are now used more as decorations than as indicators of time, they are fascinating timepieces and elementary school students can learn a great deal by constructing one. A very simple sundial can be made

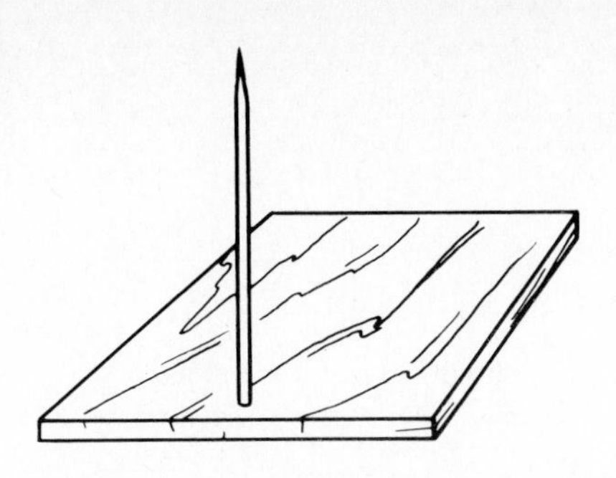

Figure 13-7.
Making a sundial.

by simply attaching a vertical "pole" (see Figure 13-7) to a piece of wood or heavy cardboard. The vertical pole might be a knitting needle or nail approximately 15 to 20 centimeters (6 to 8 inches) long. The sundial should either be permanently placed in a sunlit location or temporarily located in a sunny spot to which it can be *exactly* returned as needed.

Choose a sunny day to calibrate this simple sundial; mark the point where the shadow falls at each hour (e.g., 9 A.M., 10 A.M., . . . 3 P.M.). Then draw a line from the pole through that point and indicate the time. With the students, observe the sundial for several days. Does the shadow continue to fall on the lines at the times indicated? (*Yes; however, there will be some natural variations due to factors which will not be presented here. Check a book on sundials if you are curious.*) Ask the students to explain why the sundial works.

Some students may be interested in learning about different types of sundials and even attempting to construct some of them. Consult one of these references: (1) "Building a Sundial" by Robert Burnham,* (2) *The Great Sundial Cutout Book* by Robert Adzema and Mablen Jones.†

Challenge advanced or especially interested students by asking them questions: "Why does the length of the pole's shadow change?" (*Check it at ten o'clock each morning for 2 months and mark the position of the tip of the shadow.*) "Why is the shadow not shortest at noon? Or is it?" (*It should be.*) "How much time difference is there between real noon and twelve o'clock noon on our clocks? Why do we have time zones?"

▶ SUNRISE—SUNSET

A globe and a flashlight can be used to clarify for students the idea that the sun rises in the east and sets in the west. Ask individual students or small groups to demonstrate the concept with a globe. Which way must the globe turn for the sun to rise in the east? *(Counterclockwise, as viewed by looking down upon the North Pole.)* When the morning sun is just appearing in New York City, is it light or dark in San Francisco? When the people in San Francisco are eating supper, what time is it in Philadelphia? When it is noon in Honolulu, what time is it in Washington, D.C.? In your town?

*Robert Burnham, "Building a Sundial," *Astronomy*, vol. 8, no. 3, March 1980.
†Robert Adzema and Mablen Jones, *The Great Sundial Cutout Book*, Hawthorn, New York, 1978.

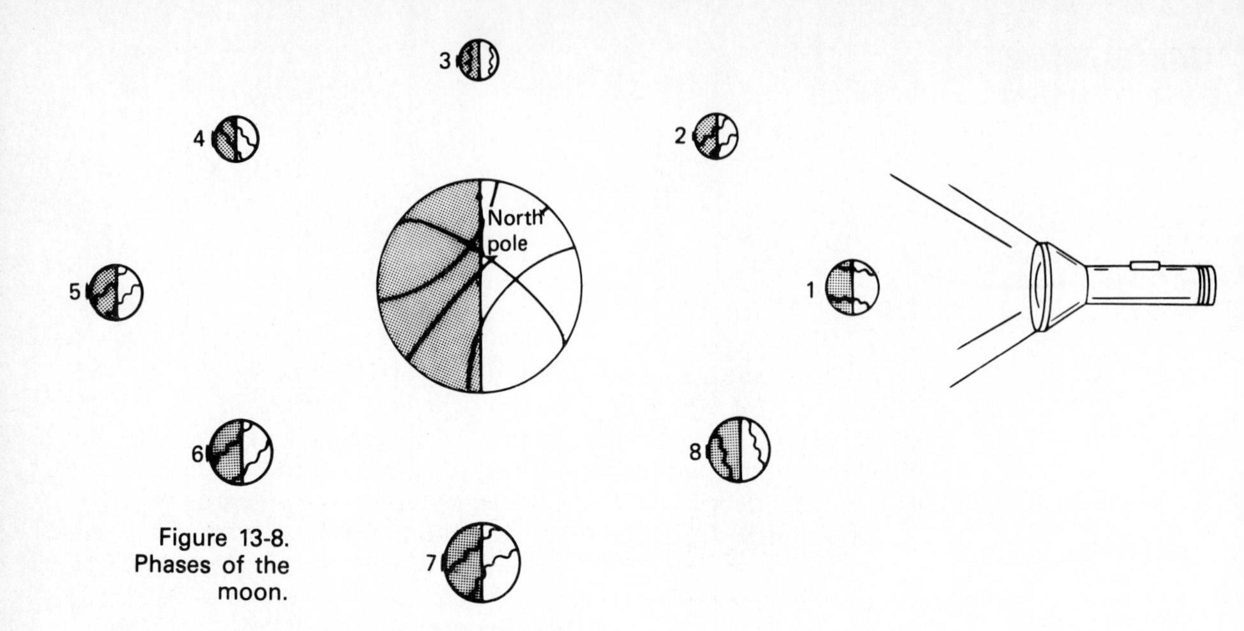

Figure 13-8. Phases of the moon.

➤ PHASES OF THE MOON

In a darkened room let students help you demonstrate why different amounts of the moon's visible surface are lighted at different times (see Figure 13-8). Use a basketball to represent the earth, a softball for the moon, and a flashlight for the sun. Have one student hold the earth stationery and another shine the flashlight on it. Have the moon revolve around the earth. Which side of the basketball represents the nighttime side of earth? (*The dark side, away from the sun.*) Where will the moon be when the whole visible surface is lighted? (*When it is directly opposite the earth from the sun—position 5 in Figure 13-8.*)

To help students see the various phases more easily, have each in turn place his or her head directly under the earth and look at the moon. When the moon is at right angles to the line between the earth and sun (positions 3 and 7) how much of the surface of the moon that is visible from earth is lighted? (*One-half.*) Is there any difference between the two positions? (*In position 3 we see the right half of the visible portion of the moon lighted, while in position 7 the left half is lighted.*) At what position is none of the visible surface lighted? (*Position 1.*)

Ask the students if they know the names of the phases represented by positions 1, 3, 5, and 7. (*New moon, half moon, full moon, half moon.*) How about positions 2, 4, 6, and 8? (*Crescent, gibbous, gibbous, crescent.*)

To be a little more challenging, ask some of these questions: "Can the moon be seen from the night side of the earth when it is in position 1?" (*Yes.*) "How can that be?" (*The earth's axis is tilted to the plane of its orbit around the sun, and furthermore, the moon's orbit is tilted to the plane of the equator. Thus, the moon can be quite high or low compared with the horizontal plane which includes our model earth and sun; at various times the moon's orbit is inclined from 0 to 50 degrees to the "horizontal" plane including our model sun and earth.*) "How can we see the moon in position 1 anyway, since its lighted surface is on the opposite side of the moon?" (*Sunlight is reflected from the earth, to the moon, and back to the earth; the "dark" side is lighted by earthlight.*) Have the students demonstrate their answers by using the models.

➢ THE MOVING SHADOW

Use masking tape to make a large X on a school window which faces the sun, placing the X in such a way that the shadow will fall on a wall if possible. What happens to the shadow of the X during the day? Mark the one o'clock position; does the shadow fall on the same spot the next day at one o'clock? How about in 2 weeks? Mark the one o'clock position each day for 2 or 3 weeks. What causes the shadow to move across the wall (or floor) on a given day? What causes the slight change in the one o'clock position during the 2 or 3 weeks? (*Between about December 21 and June 21 the sun is "higher" in the sky each day. The opposite is true between June 21 and December 21.*)

The ultimate challenge: Could you and the students somehow mark the locations of the shadows so that you could tell the time of year, especially the longest day, the shortest day, and days when there is sunlight and no sunlight for equal periods of time? (*The answer is "yes." But the real question is "How?"*)

➢ STAR TRACKS

Pictures of stars making "tracks" on a photograph are often seen in books about stars. Such a picture is relatively easy to take. What you need is a camera whose shutter can be kept open (such as one that has a "bulb" setting—most 35mm cameras will do the job very nicely); ASA 400 black-and-white film; and a tripod, a cloth bag containing sand, or some other device to hold the camera *very* steady. On most cameras a shutter-release cable will also be needed, to hold the shutter open during the exposure. The f-stop setting should be the lowest number present.

The picture must be taken where city or other lights are not visible.

From these "star tracks," can you tell where the North Star would be located? The white line is the trail of a meteor.

Aim the camera high in the sky at some fairly bright stars, and lock the shutter open for at least 60 minutes. Then try another photograph with the North Star (Polaris) centered.

After the photographs have been developed, ask students some of these questions: "What caused the light streaks?" (*The stars "moving" across the sky.*) "How could this movement have taken place?" (*Either the earth or the stars or both could have moved.*) "Are some light streaks longer than others? Why does this occur?"

➤ MAKE A CONSTELLATION

On a clear night hundreds of years ago people could gaze at the stars for long periods of time. (Remember, they had no television sets, no radios, no drive-in restaurants; about all they had to do was stargaze, socialize, and sleep!) The arrangements of some of the stars made them think of something—an animal, a person, an object. Creative people made up stories to tell to others. Thus were created what we refer to as "constellations" and their associated folklore, or skylore.

To help students understand constellations and their folklore, have the class create "constellations" and tell stories about them. They can do this in several ways: by using photographs of real stars (not including well-known constellations); by sprinkling sand, salt, or sugar on black paper; or by drawing random dots on paper. Drawing dots on a paper may be the simplest method. Have the students study their groups of "stars" and then draw pictures incorporating some of them; they have just created constellations. Now have them write or tell short stories about how the subjects of their pictures came to exist in the heavens.

For students who enjoyed this activity, point out books which record skylore (see "Building a Classroom Library"). They might especially enjoy American Indian skylore. If your school has a student newspaper, try to have a few of the "sky tales" published.

➤ IDENTIFYING CONSTELLATIONS

A classroom activity which can help students identify common constellations is to make and use a constellation box (see Figure 13-9). Place a small lamp in a medium- to large-sized box with a 15-centimeter-square (about 6-inch-square) hole cut in the top. Using pieces of dark paper or cardboard slightly larger than the hole, mark star patterns of constellations (taken from any star chart) on the pieces of paper. Then punch holes where the stars are indicated. Place the paper over the hole in the box, and the stars of the constellation will be lit up. Have students record the names of the constellations on the papers; then they can work individually or in pairs to learn to identify the constellations.

To locate the same constellations in the sky may be quite another matter. Suggest to the students that they go "stargazing" with a parent or an older friend or close acquaintance who knows the locations of at least a few of the constellations. Or they may use a simple star chart to identify a very recognizable constellation such as the Big Dipper or

Figure 13-9.
Constellation box.

Orion; then they can use the chart to find the positions of other constellations in relation to the known one. Using a star chart to find constellations sounds simple, but to a beginner it can be very frustrating. The assistance of someone who can identify a few constellations is very helpful.

➤ LIGHT-YEAR

The distances between stars are very great, and expressing those distances in miles would require use of extremely large numbers. Thus astronomers commonly use the "light-year" to measure distance. A light-year is the distance light travels in 1 year. So that students will understand how large a number this is, have them calculate the number of kilometers (or miles) that light does travel in 1 year. The speed of light is approximately 300,000 kilometers (or 186,000 miles) per second.

To help them do the calculating, give them this form:

Kilometers per second = 300,000
Kilometers per minute = 60 seconds × 300,000 = ______________
Kilometers per hour = 60 minutes × kilometers per minute = 60 × ______________ = ______________
Kilometers per day = 24 hours × kilometers per hour = 24 × ______________ = ______________
Kilometers per year = 365 × kilometers per day = 365 × ______________ = ______________

(Light travels 9,460,800,000,000 kilometers in 1 year. Using the speed of light as 186,000 miles per second, we can use a similar procedure to determine that it travels 5,865,696,000,000 miles in 1 year.)

Ask students to calculate the distance in kilometers (or miles) to our nearest star other than the sun. That star is Proxima Centauri, which is approximately 4.3 light-years from us. (*The distance from us is 40,681,440,000,000 kilometers or a little less than 41 trillion kilometers. In miles, the distance is 25,222,492,000,000 or a little over 25 trillion miles.*) The North Star is approximately 650 light-years from earth. How many kilometers (or miles) is that? How long would it take a spacecraft to get there if it traveled 1609 kilometers (1000 miles) per hour? Students will not have a real concept of such large numbers, but they should be impressed that the distances between objects in space are, indeed, very great.

▶ LIGHT SPECTRA: USING A PRISM

Have your students use a prism in sunlight to create a spectrum (a rainbow of colors) on a piece of white paper. Tell the students that if astronomers examine a spectrum very closely, they can learn much about the object which produced the spectrum. When astronomers study the spectrum of a star's light, for example, they can even find out what chemical elements are present in the star.

▶ SPACE EXPLORATION TIME LINE

On an adding machine tape, have students make a time line beginning in 1957, the date of the first artificial satellite—launched by the Soviet Union. Include all launchings or attempted launchings, if possible. Data for this activity can be obtained from NASA (see "Free and Inexpensive Materials").

Display the time line on bulletin boards, along with pictures which record information on several of the activities listed. Ask the students to identify the period of time when the most activity occurred. When did most launchings with astronauts occur? If the present time does not appear to be the time of most activity, might it be that space exploration has simply become so commonplace that we do not notice or record each effort?

▶ BALLOON ROCKET

To demonstrate the nature of the force used to propel a rocket, have the students inflate a balloon and release it. In what direction does it move? (*In general it moves in the direction opposite the opening.*) Then make a drawing on the chalkboard similar to Figure 13-10. Explain to the students that the arrows represent the air pressure within the balloon and that the air pressure pushes equally in all directions. If the balloon is tied shut, a force in any direction is equally balanced by a force in the opposite direction. What happens when the balloon is not tied shut? (*There is a force on the front side of the balloon, indicated by the arrow marked X; however, there is no part of the balloon upon which the air exerts its force in the back. Therefore, the balloon moves forward because there is an unbalanced force in that direction.*)

Ask the students to explain how this model is similar to a rocket which propels a spacecraft. What creates the force in the rocket? (*Burning fuel.*)

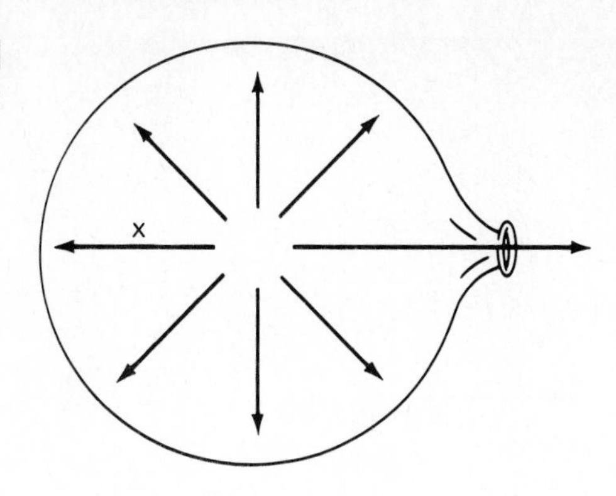

Figure 13-10. Air pushes outward equally in all directions.

ROCKET PROPULSION

Another demonstration of propulsion can be done with a skateboard. Attach three C clamps to the skateboard as indicated in Figure 13-11. Attach a rubber band between the two side clamps; then stretch it and attach it to the back clamp with a string. Put a rock in the rubber band near the string, and place the skateboard on a smooth surface. Be sure that the rock is not aimed at anyone or anything which it might damage (you may want to place a pillow where the rock will land). Ask the students what they think will happen when the string is cut. Try it.

Ask the students to explain why the skateboard moved. (*To cause motion there had to be an unbalanced force. The rubber band pushed on the rock and caused it to move. The rock also pushed on the rubber band and the skateboard and thus caused it to move.*)

Would there be any difference if you used a rock with twice the mass? If so, what would the difference be? If not, why not? Try it. How about if you used a tighter rubber band? Try it.

A similar demonstration can be done by having a child on a skateboard or roller skates throw a heavy object. Does the size of the person affect how far he or she rolls? What about how hard the person throws the weight?

Figure 13-11. Rockets move like skateboards.

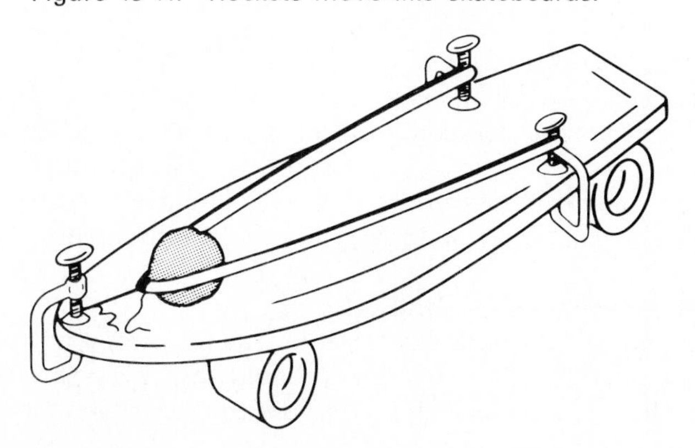

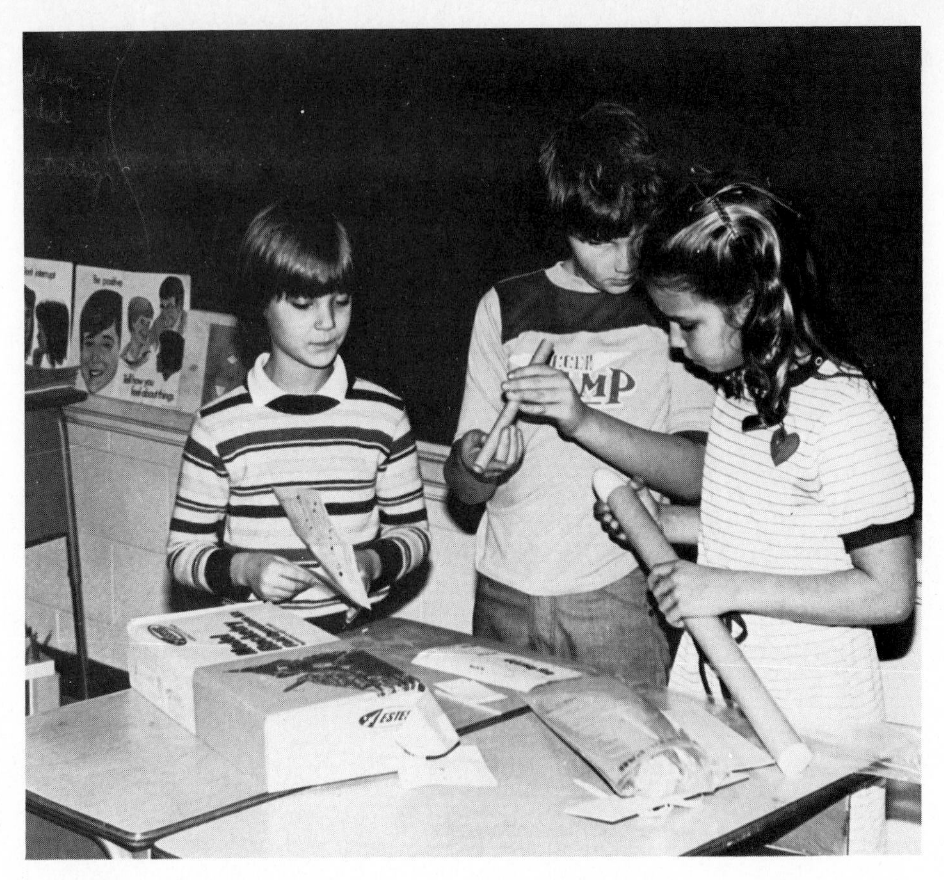

Building model rockets can add real excitement to your science class.

MODEL ROCKETRY

Think of how excited your class would be if they could actually fire a real rocket. It can be done—and fairly easily. Some science supply houses do make simple rocket kits for schools, and they also will provide the teacher with sample lesson plans and activities. At least one such rocket carries a camera and takes a picture while in flight.

To prepare for a rocket-launching class, contact a science supply house (see the Appendix for addresses).

LARGE SPACECRAFT MODEL

Give the class the measurements of some spacecrafts which have successfully completed space missions and have the students make ¼-size models of the one or ones they choose. Invite other classes to see the models and to hear your students explain the features of the spacecrafts as well as the types of investigations in which they were involved.

SIGHTING A SATELLITE

Observing a clear dark sky on any evening should bring into view at least one or two satellites per hour. Satellites are as bright as most visible stars; they move more rapidly across the sky than stars but not so rapidly as meteors. Unlike high-altitude aircraft, satellites do not have red, green, or flashing lights.

Encourage teams of your students (along with their parents) to spend a couple of hours some warm night watching for satellites. Tell the teams to record the time of the sighting, the approximate direction of travel of the satellite, and how long it was visible.

Do satellites travel more often in one general direction than others? Is there some direction in which they are seldom or never seen traveling? Are satellites more frequently seen at some times of the night than others? The students may wish to do some library research or to write to NASA to find out approximately how many satellites are presently orbiting the earth.

➤ MOON WRECK

This activity will help your students realize the value of cooperative work in small groups, as well as helping them understand some of the factors which go into planning a trip to the moon. It is modified slightly from one which has been used by NASA astronauts. To begin, read this story to your students:

You are a member of a moon exploration team which was scheduled to land near the mother craft for the expedition on the lighted surface of the moon. Owing to problems, your spacecraft was forced to land 300 kilometers from the mother craft. Upon landing, much of your equipment was damaged. Since your survival depends on reaching the mother craft, you survey the items which are still usable. Your task is to decide upon which items are most valuable in helping you complete the 300-kilometer trip to the mother craft. You have a sheet which lists 15 items left undamaged after your landing. Your task is to rank the items in terms of their importance for your trip to the mother craft. Place the number 1 by the most important item through 15, the least important item of those listed.

The sheet referred to is shown in Table 13-2. Give the students copies of the list of undamaged items, but do not tell them the astronauts' rankings at this time. Each student is to rank the items independently. After the individual rankings are completed, students should work in groups of three or four to arrive at a consensus ranking of the items. To discuss each of the items and then rank them may take about half an hour; do not rush the process.

After the group rankings have been completed, reveal the rankings and comments of the NASA astronauts. Tell the students to score their individual responses by calculating the numerical difference (absolute value) between each item as they ranked it and as the NASA group ranked it. The groups can follow the same procedure to find the group scores. Have the students write both the individual and the group scores on small pieces of paper (their names need not be written on the papers). List the individual scores and the group scores on the chalkboard, and have the students compute the average individual score and the average group score. Which average was lowest (and, thus, most in agreement with the NASA scores)? (*Almost without exception, group results will be*

TABLE 13-2 MOON WRECK

UNDAMAGED ITEMS*	NASA ASTRONAUTS' CHOICES AND REASONS†	
___ Box of matches	15	Little or no use on moon
___ Food concentrates	4	Supply daily food requirement
___ 15 meters (about 50 feet) of nylon rope	6	Useful in tying injured together; helpful in climbing
___ Parachute silk	8	Shelter against sun's rays
___ Portable heating unit	13	Useful only if party landed on dark side
___ Two 45-caliber pistols	11	Could be used to make self-propulsion devices
___ One case of dehydrated milk	12	Supplies nutrition when mixed with water for drinking
___ Two 45-kilogram (100-pound) tanks of oxygen	1	Fills respiration requirement
___ Stellar map (showing the moon's constellations)	3	One of principal means of finding directions
___ Life raft	9	Includes CO_2 bottles which could be used for self-propulsion across chasms, etc.
___ Magnetic compass	14	Probably no magnetized poles on moon; thus, useless
___ 20 liters (about 20 gallons) of water	2	Replenishes loss by sweating, etc.
___ Signal flares	10	Could be used for distress call when line of sight is possible
___ First-aid kit containing injection needles	7	Oral pills or injection medicine valuable
___ Solar-powered FM receiver-transmitter	5	Distress signal transmitter; possible communication with mother craft

*Give copies of this list to the students.
†Do not give this list to the students until *after* they have completed their individual and group rankings.

more closely in agreement with the experts. A group of people working together possess more knowledge than an individual does.)

➤ ASTRONAUTS' REQUIREMENTS AND TRAINING

Write to NASA for information on the present requirements and suggestions for qualities of astronauts, as well as the training program for astronauts. Also ask for photographs showing astronauts in their training activities.

Have students prepare a "want ad" to be used for hiring new astronauts—a half- or full-page advertisement which outlines requirements for becoming an astronaut as well as listing some of the benefits.

Students can also prepare a curriculum outline showing the subjects which astronauts study, the physical fitness program, and other major aspects of the training program.

➤ CONGRESSIONAL HEARING

Write to NASA for briefing documents on the benefits of the space program. Tell your students to use the documents to prepare information for a congressional committee, showing the many benefits of the space program for all humankind. Congress is considering cutting back on the space exploration effort, and your students' job is to convince Congress not to cut back (whether or not that would be what they would really want). Their position is that the costs will be minimal in view of the probable benefits, on the basis of past experience.

Encourage students to think about benefits which are often overlooked. Have they considered weather satellites? Communication? Astronomical observatories in space? Medicine? Food? Metals technology? Space blankets?

➤ MOON ROCKS

Participate in the NASA training program for teachers on the use of a moon rock in the educational program of their school district. Once the training program has been completed, your school district can have a moon rock on a short-term loan for use in teaching students. The opportunity to study a moon rock is a real attention getter for both teachers and students.

➤ SPACE COLONIES

One of the new frontiers of humankind will surely be the colonizing of space. Some things could be done better in space than on earth because of the low gravity and cleanliness of space. Another benefit of space is that solar energy can be uninterrupted if the earth's shadow is avoided.

Share with your students some basic information about space colonies (see "Building a Classroom Library" or recent articles in *Current Science* or an astronomy magazine). Then tell the class to plan a colony for 100,000 inhabitants. Have them consider the following concerns and others which they themselves think of: What safety precautions must be built into the colony? (The inhabitants should not have to wear spacesuits all the time.) How will energy be produced? How will daily necessities—oxygen, food, etc.—be provided? (*They will not be shipped from earth.*) Could or should the colony include a green forest?

This activity could develop into a year-long major project, incorporating many topics in science and social studies. What would have to be done about ecological balance? What type of government would be desirable? If your class does make this activity into a major project, show off the results to the whole school and invite local news reporters from television, radio, and newspapers to visit your space colony exposition.

➢ WEATHER SATELLITES

One benefit of the space program is 24-hour surveillance of weather by spacecraft. Viewing satellite "pictures" of weather conditions on television is now rather commonplace. Obtain (or have your students obtain) a sample weather photograph showing your area for a specific recent date. Have the students identify the phenomena shown in the picture.

You could also get from NASA some classic photographs—e.g., the first hurricane which was ever identified from space before it was spotted on earth. What are the benefits of weather satellites? How do they help weather researchers?

➢ STARS GROW OLD

Read aloud a brief account of the evolution of stars. Then have the students close their eyes and imagine they are stars. Describe the life of a star from birth to death. Emphasize the ideas that even stars are constantly changing and that huge amounts of time are hypothesized for the life cycle of a star.

Children's literature

The folklore of the sky is an especially interesting type of literature. One good collection of American Indian skylore is *The People.** Read aloud, or have students read, some or all of the stories in the book. Ask the students to identify concepts in the skylore which are now held to be correct (e.g., seasonal appearance of certain stars, the sun rising in the east and setting in the west). Some of humankind's first efforts to explain the astronomical phenomena which they observed led to the development of folklore which is still fascinating.

Read *Colonies in Space*† to your students or have them read it, and then discuss with them the concepts presented.

Find books on astronomy and space exploration in the school library and identify them for the students or place them in a classroom interest center.

*Mark Littmann, *The People: Sky Lore of the American Indian*, Hansen Planetarium, Salt Lake City, Utah, 1976.

†Frederic Golden, *Colonies in Space*, Harcourt Brace Jovanovich, New York, 1977.

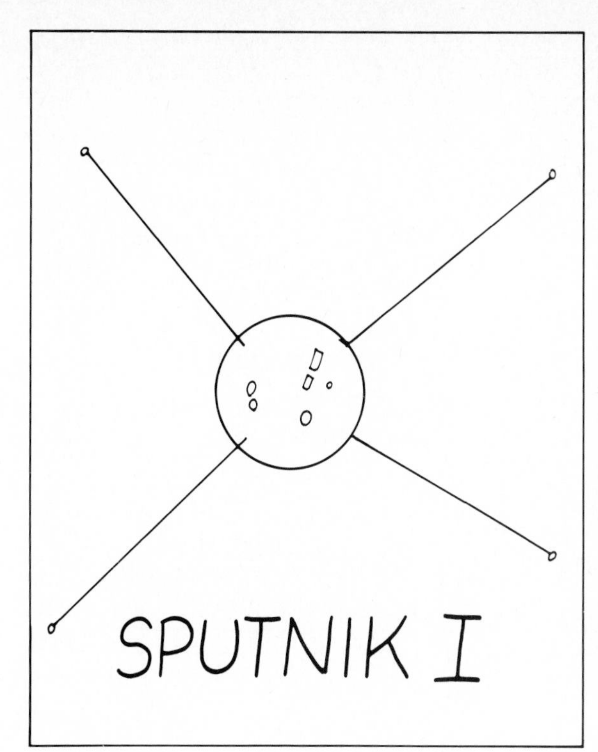

Figure 13-12.
Activity card:
Sputnik I.

Activity cards

An example of an activity card is shown in Figure 13-12. Additional activity cards could be made for topics such as looking at constellations, studying a particular planet in the solar system, studying the life cycle of stars, learning how the sun works, exploring the moon, learning what causes tides, finding "heavenly music" (songs about stars or planets), observing planets which are presently visible, studying meteors, colonizing space, understanding the value of satellites to farmers (and to fishers, pilots, energy researchers, etc.), demonstrating the laws of motion, and learning about Pioneer I (or any other space venture).

Bulletin boards

HISTORY OF ROCKETS
Have students research the history of rockets (at least as far back as ancient China) and depict that history on a bulletin board. Include as many of the students' pictures of ancient rockets as possible.

THE TALKING BOARD
An interesting way to get students to respond to a question or situation is to write a statement in a cartoon "balloon" and provide an empty balloon, labeled "your response," in which they can write their responses. Figure 13-13 is an example. After students have responded to one or two of your statements, have them create their own statement balloons.

Figure 13-13. Bulletin board: The talking board.

Field trips

The field trips described for primary grades are also appropriate for intermediate grades. In addition, a trip to a local museum which is showing meteorite samples or moon rocks, or which has a display on any other astronomy topic, would be beneficial. Also, consider taking one or more evening field trips on which students could both watch for satellites and identify constellations. You might want to take only one-fourth of your class on each trip—one trip per week for 4 weeks.

RESOURCES

Sources of assistance

FREE AND INEXPENSIVE MATERIALS

Here are some good sources of free and inexpensive materials for use in teaching astronomy and space exploration:

1. *General Electric.** Posters of the solar system and United States space efforts.
2. *National Aeronautics and Space Administration.* Photographs, charts, and information booklets on several topics in astronomy and space exploration.

*The addresses of specific agencies, organizations, and businesses are listed in the Appendix.

3. *Astrophysical Observatory.* Booklets on comets and satellites and a list of children's books on astronomy.
4. *Local observatories, such as McDonald Observatory.* Monthly publications on astronomical happenings during the coming month.
5. *Estes Industries.* Information on model rockets.
6. *National Air and Space Museum.* A periodical publication on flight and space exploration.
7. *Eros Data Center.* Photographs of much of the world taken from satellites.
8. *American Institute of Aeronautics and Astronautics.* Literature and films on space travel.

LOCAL RESOURCES

Local resources which can help in teaching astronomy and space exploration include:

1. Members of an amateur astronomy club.
2. An observatory or a planetarium.
3. Many high school teachers of earth and space science or physics have telescopes and may be willing to share them.
4. Colleges or universities often have at least one person who teaches a course in astronomy.
5. Museums of science and technology.
6. Rocket clubs (ask about them at a hobby shop).
7. Television stations and weather bureaus may be able to supply satellite weather photographs.
8. A county extension agent or a Soil Conservation Service may have Landsat map photographs available. (See the Appendix.)

ADDITIONAL RESOURCES

1. Magazines such as *Astronomy, Sky and Telescope, Odyssey, World, National Geographic, Natural History,* and *Popular Science.*
2. A few radio stations present a 5-minute "star-watch" announcement each evening which describes current celestial happenings.

Strengthening your background

1. Read an article on how ancient people built monuments which indicated special times of the year (e.g., the longest day). One such group of people were the Indians of Chaco Canyon, New Mexico.*
2. With the help of a friend, identify five constellations and check yourself for five consecutive nights by observing them. What are their positions relative to each other? Read an American Indian legend about at least one of your constellations.
3. Find out which are the two brightest stars visible from earth (not including the sun). Are they presently visible at night?

*See *Science 80,* November-December, 1979, p. 56–67.

4. Where are the zodiac constellations found? Can you find your sign in the sky?
5. Construct a sundial.
6. Each afternoon for a month find the moon if it is visible. Keep a record of your observations.
7. Photograph star tracks.
8. Describe at least two major theories of how the universe was formed.
9. Write three riddles about stars or planets.
10. Fly a model rocket.
11. Visit the NASA center (see the listing earlier in this chapter) which is nearest you. If possible, visit Washington, D.C., and include a half-day or longer visit to the National Air and Space Museum.
12. Read a book, written at a level suitable for your students, on colonizing space.
13. Answer the question "Why do we not aim spacecraft directly at the moon and fire?"
14. Find out what information may be obtained from a Landsat program. (See the Appendix.)
15. Buy and sample some food products which were developed for the space program.
16. Read a brief account of the contributions of the early rocketry inventor Robert Goddard to space exploration.
17. Spend 2 hours on a clear evening watching for satellites with someone who knows what they look like.

Getting ready ahead of time

1. Make a collection of photographs and drawings of the planets.
2. Prepare topics and sample response cards for a bulletin board on astronomy.
3. Develop a shoebox science kit on constellations.
4. Find out the locations of the nearest planetarium and the nearest observatory, and visit one or both of them to learn about programs and materials which might help you with your teaching.
5. Develop 20 game-board cards for your students to use in learning about astronomy.
6. Find and read four children's books on astronomy which will be appropriate for your students to read.
7. Since astronomy is a popular topic, design a learning center for astronomy and develop several activities for it.
8. Write to NASA to request their packet of teaching aids for elementary school teachers.
9. Prepare for your students a guided fantasy on taking a walk on the moon.
10. Make a collection of newspaper and magazine articles and photographs on space exploration.
11. Obtain a satellite photograph of your state or local area from the Eros Data Center.

12. Write to NASA to ask for autographed photographs of several astronauts.
13. Prepare a bulletin board on a space exploration mission scheduled for the near future.

Building your own library

Asimov, Isaac: *The Solar System*, Follett, Chicago, 1975.
Bergaust, Erik: *Colonizing Space*, Putnam, New York, 1978.
Berger, Melvin: *Planets, Stars, and Galaxies*, Putnam, New York, 1978.
Branley, Franklyn: *Comets, Meteors, and Asteroids*, F. Watts, New York, 1974.
———: *Eclipse: Darkness in Daytime*, Thomas Y. Crowell, New York, 1973.
Elementary Science Study (ESS): *Daytime Astronomy*, Webster, McGraw-Hill, Manchester, Mo., 1971.
———: *Where Is the Moon?* Webster, McGraw-Hill, Manchester, Mo., 1968.
Friedman, Herbert: *The Amazing Universe*, National Geographic Society, Washington, 1975.
Gurney, Gene: *Americans to the Moon*, Random House, New York, 1970.
Hartman, William K.: *Astronomy: The Cosmic Journey*, Wadsworth, Belmont, Calif., 1978.
Limburg, Peter: *What's in the Names of Stars and Constellations?* Coward-McCann, New York, 1976.
Reed, Maxwell: *Patterns in the Sky: The Story of Constellations*, Morrow, New York, 1951.
Simon, Seymour: *Look at the Night Sky: The Story of Constellations*, Morrow, New York, 1951.
Taylor, L. B., Jr.: *Space Shuttle*, Thomas Y. Crowell, New York, 1979.
Zim, Herbert, and Robert Baker: *Golden Nature Guide to Stars*, Golden Press, New York, 1951.

Building a classroom library

Adams, Florence: *Catch a Sunbeam: A Book of Solar Study and Experiments*, Harcourt Brace Jovanovich, New York, 1978.
Adler, Irving: *The Stars: Decoding Their Messages*, Thomas Y. Crowell, New York, 1980.
Anderson, Norman D., and Walter R. Brown: *Halley's Comet*, Dodd, Mead, New York, 1981.
Asimov, Isaac: *Comets and Meteorites*, Follett, Chicago, Ill., 1972.
———: *How Did We Find Out about Black Holes?* Walker, New York, 1978.
———: *How Did We Find Out about Outer Space?* Walker, New York, 1977.
———: *The Kingdom of the Sun*, Abelard-Schuman, New York, 1962.
———: *Space Dictionary*, Scholastic Book Services, New York, 1971.
———: *What Makes the Sun Shine?* Little, Brown, Boston, 1971.

Bendick, J.: *Space Travel*, F. Watts, New York, 1953.

Bergaust, Erik: *The Next 50 Years on the Moon*, Putnam, New York, 1974.

Blumberg, Rhoda: *The First Travel Guide to the Moon: What to Pack, How to Go and What to See When You Get There*, Four Winds, New York, 1980.

Branley, Franklin: *The Big Dipper*, Thomas Y. Crowell, New York, 1962.

———: *Columbia and Beyond*, Putnam, New York, 1979.

———: *Experiments in Sky Watching*, Thomas Y. Crowell, New York, 1967.

———: *The Moon Seems to Change*, Thomas Y. Crowell, New York, 1960.

———: *The Nine Planets*, rev. ed., Thomas Y. Crowell, New York, 1978.

———: *What Makes Day and Night?* Thomas Y. Crowell, New York, 1961.

Cipriano, Anthony J.: *America's Journeys into Space: The Astronauts of the United States*, Messner, New York, 1979.

Colby, C.: *Astronauts in Training*, Coward-McCann, New York, 1969.

Coombs, Charles: *Project Apollo, Mission to the Moon*, Morrow, New York, 1965.

Couper, Heather: *Space Frontiers*, Viking, New York, 1979.

Cromie, William: *The Story of Man's First Station in Space*, McKay, New York, 1976.

Crosby, Phoebe: *Junior Science Book of Stars*, Garrard, Champaign, Ill., 1960.

Dean, Annabel: *Up, Up and Away! The Story of Ballooning*, Westminster, Philadelphia, 1980.

Dietz, D.: *Stars and the Universe*, Random House, New York, 1968.

Fields, Alice: *Satellites*, F. Watts, New York, 1981.

Gallant, Roy A.: *Exploring the Planets*, Doubleday, Garden City, N.Y., 1967.

Golden, Frederic: *Colonies in Space*, Harcourt Brace Jovanovich, New York, 1977.

Goudey, Alice: *The Day We Saw the Sun Come Up*, Scribner, New York, 1961.

Harris, Susan: *Space*, F. Watts, New York, 1979.

Hendrickson, Walter: *Manned Spacecraft to Mars and Venus: How They Work*, Putnam, New York, 1975.

Kerrod, Robin: *The Challenge of Space*, Lerner, Minneapolis, 1977.

Knight, David: *American Astronauts and Spacecraft: A Pictorial History from Project Mercury through the Skylab Manned Missions*, rev. ed., F. Watts, New York, 1975.

———: *Colonies in Orbit*, Morrow, New York, 1977.

———: *Galaxies, Islands in Space*, Morrow, New York, 1979.

———: *The Moons of Our Solar System*, Morrow, New York, 1980.

Lambert, David: *The Earth and Space*, Warwick, Baltimore, and F. Watts, New York, 1979.

Moche, Dinah L.: *The Astronauts*, Random House, New York, 1979.

———: *The Star Wars Question and Answer Book about Space*, Random House, New York, 1979.
Morgan, Helen: *Maria Mitchell: First Lady of American Astronomy*, Westminster, Philadelphia, 1977.
Neely, Henry: *The Stars by Clock and Fist*, Viking, New York, 1972.
Nourse, Alan: *The Asteroids*, F. Watts, New York, 1975.
———: *The Giant Planets*, F. Watts, New York, 1974.
Patterson, Lillie: *Benjamin Bannider: Genius of Early America*, Abingdon, Nashville, Tenn., 1978.
Polgreen, John and Cathleen Polgreen: *The Earth in Space*, Random House, New York, 1963.
Poynter, Margaret, and Arthur L. Lane: *Voyager: The Story of a Space Mission*, Atheneum, New York, 1981.
Quackenbush, Robert: *The Boy Who Dreamed of Rockets*, Parents Magazine, New York, 1977.
Raviellie, Anthony: *The World Is Round*, Viking, New York, 1976.
Rey, H. A.: *Find the Constellations*, rev. ed., Houghton Mifflin, Boston, 1976.
———: *The Stars: A New Way to See Them*, Houghton Mifflin, Boston, 1976.
Ross, Frank, Jr.: *Space Shuttle: Its Story and How to Make a Flying Paper Model*, Lothrop, New York, 1979.
Ruchlis, Hy: *Orbit: A Picture Story of Force in Motion*, Harper and Row, New York, 1958.
Shapp, Martha, and Charles Shapp: *Let's Find Out about the Moon*, F. Watts, New York, 1975.
Shurkin, Joel N.: *Jupiter—The Star That Failed*, Westminster, Philadelphia, 1979.
Shuttlesworth, Dorothy E., and LeeAnn Williams: *The Moon, The Steppingstone to Outer Space*, Doubleday, Garden City, N.Y., 1977.
Simon, Seymour: *The Long View into Space*, Crown, New York, 1979.
———: *Look to the Night Sky*, Viking, New York, 1977.
Stine, Harry: *Handbook of Model Rocketry*, 4th ed., Follett, Chicago, 1976.
Taylor, G. Jeffrey: *A Close Look at the Moon*, Dodd, Mead, New York, 1980.
Thompson, Brenda: *Rockets and Astronauts*, Lerner, Minneapolis, 1977.
Valens, E. G., and B. Abbott: *The Attractive Universe: Gravity and the Shape of Space*, World, Tarrytown-on-Hudson, N.Y., 1969.
Veglahn, Nancy: *Dance of the Planets: The Universe of Nicolaus Copernicus*, Coward-McCann, New York, 1979.
Wicks, Keith: *Stars and Planets*, F. Watts, New York, 1977.
Wise, William: *Monsters from Outer Space?* Putnam, New York, 1979.
Wolfe, Louis: *Let's Go to a Planetarium*, Putnam, New York, 1958.
Worden, Alfred: *I Want to Know about a Flight to the Moon*, Doubleday, Garden City, N.Y., 1974.
Zim, Herbert: *The Sun*, rev. ed., Morrow, New York, 1975.

CHAPTER 14

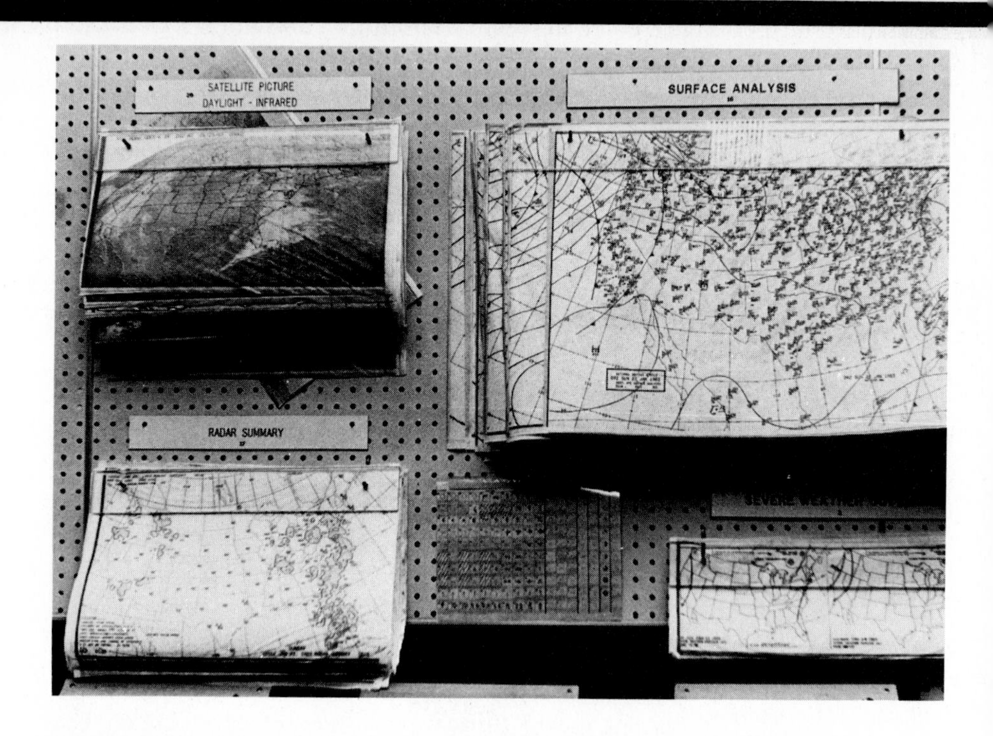

WEATHER AND CLIMATE

IMPORTANT CONCEPTS

Weather is a phenomenon of the atmosphere.

Weather is the condition of the atmosphere.

The atmosphere can be divided into zones or layers, each of which has distinguishing physical properties.

Water, an important part of the atmosphere, can exist in three forms—solid, liquid, and gas.

Weather can be related to the water cycle.

The condition of the atmosphere is greatly influenced by the nature of the earth's surface beneath it.

Available solar energy is a principal factor in the evolution and movement of weather systems.

The seasons on earth are a result of the revolution of the earth about the sun and the inclination of the earth's axis.

The earth may be divided into climatic regions.

Local and global measurements of atmospheric conditions—temperature, humidity, air pressure, wind velocity, precipitation—are used in predicting weather.

Weather and climate have had a great influence on people and their cultures.

Scientists attempt to modify weather but so far have had limited success because of the complexity of weather.

INTRODUCTION: STUDYING WEATHER AND CLIMATE

Weather and climate are usually taught beginning in first grade. Students are aware of the changing requirements for clothing through the school year in almost all localities. Study of weather and climate can help students understand what factors influence weather and how people can better understand and use weather forecasts.

Folklore of weather is a fascinating subject in itself, and it can also be "scientifically" tested. Folklore helps us realize some of the early ways in which people understood weather and its importance. While weather can be a complex topic, it can also be easily observed. Thus it offers many ways to challenge students to investigate problems by observation or experiment.

Scientists are now learning much about weather and are even making modest efforts to control it. Many occupations and industries are greatly influenced by weather, and people who are knowledgeable about weather will make greater and greater contributions. In addition, some of your students may even become meteorologists. Challenge all of them to investigate the weather and learn to understand it.

WEATHER AND CLIMATE FOR PRIMARY GRADES

Activities

➤ EVAPORATION AND CONDENSATION OF WATER

The teacher* places a pan of water on a hot plate. As the water begins to boil, point out to the class the rising water vapor. Ask the students to describe and explain, if they can, what is happening. Place a cool object (a large metal spoon works nicely) in the vapor. Ask the students to describe what is happening on the spoon. Explain the terms "evaporation" and "condensation."

Wipe a damp cloth across the chalkboard. When the water has evaporated, ask the children to explain what happened. Have each child place a few drops of water in a container and check the water frequently during the day. What happens? How is this related to the chalkboard experience with evaporation? After the water has completely evaporated, have the children relate all these experiences to the water cycle (evaporation—condensation—precipitation). The students can refer to an illustration of the water cycle, such as Figure 14-4 later in this chapter.

**Caution:* Do *not* allow the students to help with this demonstration; *do* warn them about the dangers of hot water.

➤ CLOUD WATCH

Clouds are fascinating to observe. They come in many shapes and sizes. Have the students observe clouds for several days and write descriptions of the different types they see, in their own words, and also describe weather on those days. After several days' descriptions have been recorded, ask the students to identify the types of clouds that are related to different types of weather.

Obtain a cloud chart (see "Sources of Assistance") and post it where students can consult it to learn the names of the cloud types they have been observing.

Have the students make up a story about a cloud or tell how they feel while watching clouds.

➤ WEATHER MUSIC

Ask students to bring to class records which include songs about weather—wind, rain, snow, temperature, other weather phenomena, climate, or seasons. Ask what kind of weather they like best. Which season? Does weather affect the way people feel?

➤ MEASURING TEMPERATURE

This activity will help students develop skill in measuring temperatures with a thermometer. Have students measure the temperature of water in three large containers of hot, lukewarm, and cold water. Make sure they notice that the level of the liquid in the thermometer changes as the temperature changes.

The students can also measure the air temperature at various heights in the classroom, or in a stairwell if your school has more than one floor. What can they conclude about the temperature at various levels in the room? *(The higher up in the room or stairwell, the higher the temperature. If air-circulation fans are operating, the differences will be less or may be unnoticeable. Take the readings when the fans are not operating; if necessary, ask the custodian to turn the fans off for several minutes. The students might observe or conclude that warm air rises.)* Which is heavier, hot or cold air?

Allow the students to take temperature readings in other places that they would like to explore. If they have thermometers at home (not clinical thermometers, however, for their range is much too small) or if you can allow them to take the classroom thermometers home, have them take readings in several locations (attic, garage, refrigerator, north side of home, south side of home) and report on those readings.

To find out how accurate the students have become in reading the thermometer, have them check room temperature, outside temperature, and a few containers of water at different temperatures while you are watching.

➤ WIND DIRECTION

In this activity, students can make a "fancy" wind vane as an art project. First, let them watch you make a simple arrow wind vane (as shown in

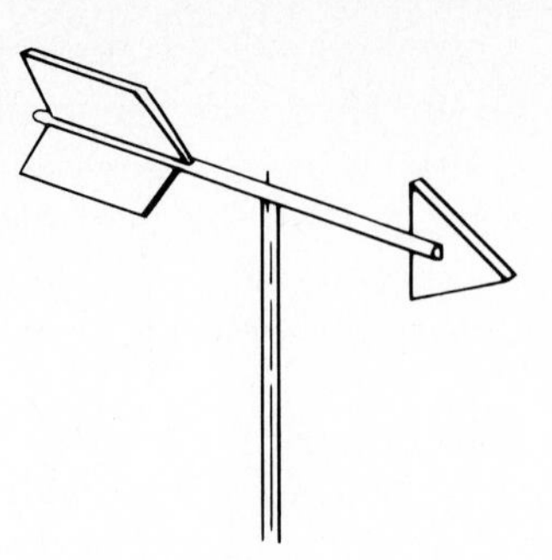

Figure 14-1. Making a wind vane. (Source: Adapted from G. C. Mallinson et al., *Science: Understanding Your Environment,* Silver Burdett, Morristown, N. J., 1975, grade 2, page 161.)

Figure 14-1), using a pencil with an eraser, a straight pin, a drinking straw, and some tagboard. On a windy day demonstrate how it works. Show the students pictures of ornate wind vanes from an encyclopedia or other sources. Then have each student design and construct an ornate wind vane. On a windy day let the students take their vanes outside to see how well they work.

Which way do the wind vanes point? *(The direction the wind blows from.)* Tell the students that is how we name the winds: a south wind blows from the south. Name north, south, east, and west for them if they don't already know the directions. Then ask them what direction the wind is blowing from while they are trying out their wind vanes.

RAIN GAUGE

Have your students make rain gauges to take home and give to their parents. Natural curiosity about just how much it did rain can be satisfied with this simple project.

For each student, obtain a small glass or—preferably—plastic container which has straight sides and an opening the same size as the bottom of the container. Prescription medicine vials will do nicely, but the containers do not have to be circular—they can be square or rectangular. A depth of 5 to 8 centimeters (2 to 3 inches) is satisfactory for most locations. Have the students either paint depth scales on the containers (using non-water-soluble paint) or construct *small* rulers which can be held next to the container or inserted into the water to determine the amount of rainfall.

Have students report and record the amount of rainfall for a month. As a longer-term project, they can graph the amounts of rainfall for several months.

WEATHER CHART

To focus students' attention on weather, have them keep a daily weather chart for 2 or 3 weeks during the time weather is being studied. They may record overall conditions (cloudy, sunny, rainy) with appropriate sketches or words, or they may record more specfic information (high and low temperatures, wind direction and speed, amount of cloud cover, and precipitation).

If possible, obtain a 24-hour weather radio (check with a radio shop or a branch of the National Weather Service)—or have the class listen to a regular radio or television weather forecast each day. The students can gather the weather data from the reports and then check the accuracy of the forecasts.

➢ CLIMATIC REGIONS

Have students collect magazine pictures which illustrate weather and climatic conditions. After all students have collected one or more pictures, ask one student to show a picture and tell what the climate may be like in the area pictured. Then ask others who have pictures from similar areas to show theirs and add any additional information they can. Would they like to live in these areas? Move on to other students with pictures of different regions. On a world map or globe, point out areas that may be like the ones pictured.

Which areas of the world seem to be coolest? *(Areas farthest from the equator at all seasons; areas in the interior of large continents in the winter.)* Which are warmest? What are islands like? Discuss specific islands, such as Hawaii, which the class may have heard about.

➢ WEATHER AFFECTS ANIMALS

Have students collect pictures or make drawings of animals which migrate to different locations for summer and winter. Tell them to make up stories about one of these animals, and to write their stories in first person, as though they were the animals and were telling their experiences.

In what other ways does weather affect animals, especially local animals? How can your students help the animals?

➢ SEASONS

In drawing pictures of seasons, students often draw what they have heard about seasons (e.g., snow for winter). A more meaningful activity would be to draw pictures of the seasons as they occur in the students' own location throughout the year. How do we dress throughout the year? What do plants look like at different times of year?

If you wish to expand the students' knowledge of other areas, have them draw pictures of the seasons in other places—farther north or south or on an island such as Hawaii. How is their home area different from the others pictured? Why?

➢ THROUGH THE WATER CYCLE

Have students close their eyes and pretend that they are small particles of water. Take them through the water cycle several times. On one trip they might end up in the city water supply and travel through pipes, through a water heater, into a clothes washer, and finally into the city sewage treatment plant where they evaporate back into the air. Ask students to help you think of other water trips.

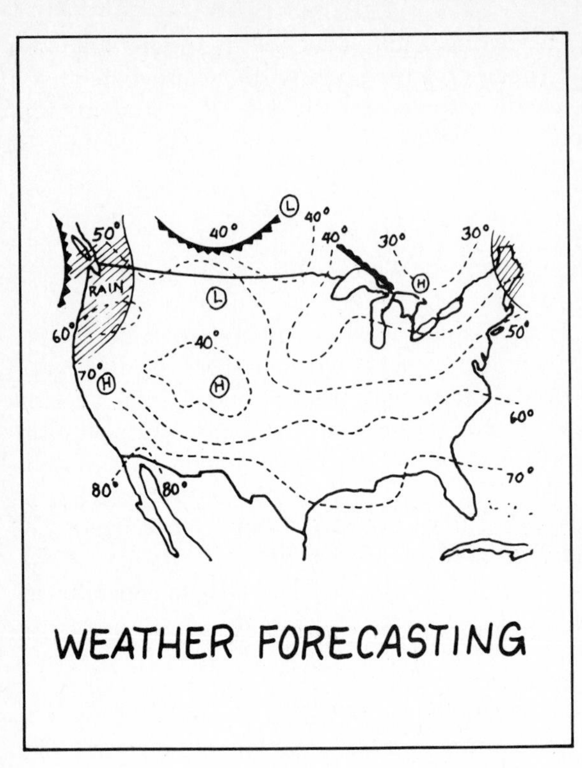

Figure 14-2. Activity card: Weather forecasting.

Children's literature

Read *I Like Weather** to your students or have them read it, and then discuss with them the concepts presented.

Find other books on weather or climate in the school library, and identify them for the students or place them in an interest center.

Activity cards

An example of an activity card is shown in Figure 14-2, above. Additional activity cards should be made for topics such as finding out what animals do in various seasons, listening to weather forecasts and checking their accuracy, checking temperatures at several locations in the local environment, and finding out what birds do when it is raining.

Bulletin boards

CLOTHING AND SEASONS

Divide a bulletin board into areas for the four seasons, and post a photograph or drawing of what each season is like in your locality. In a pocket

*Aileen Fisher, *I Like Weather,* Thomas Y. Crowell, New York, 1963.

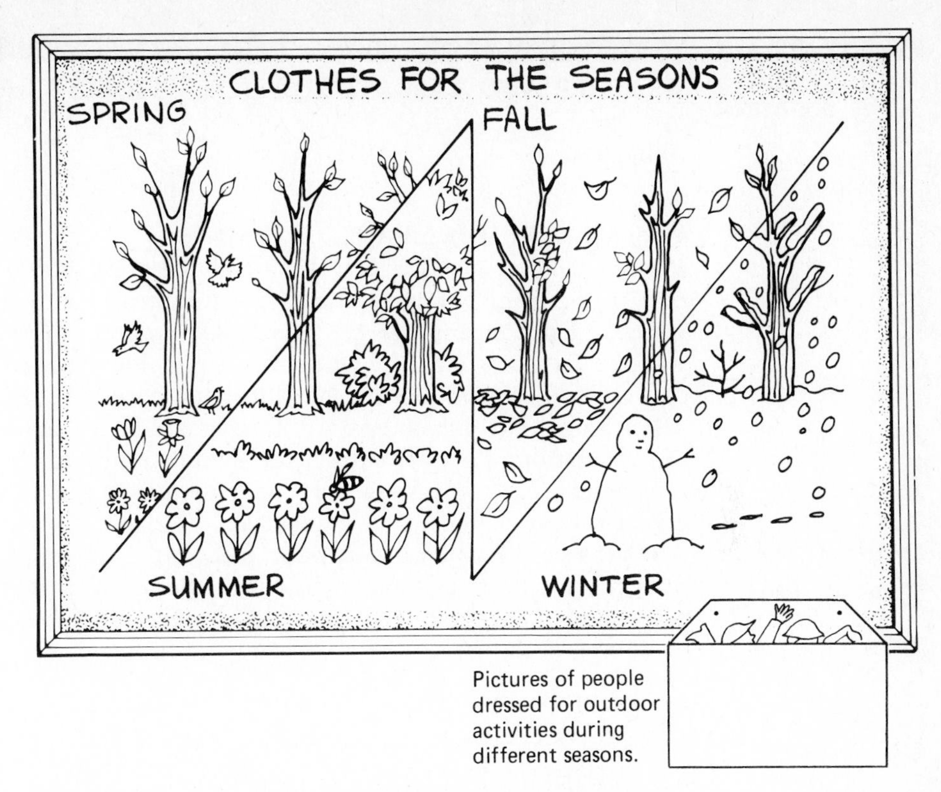

Figure 14-3. Bulletin board: Clothes for the seasons.

place pictures cut from magazines or catalogs which show people dressed in clothes appropriate for the various seasons. The students are to place the pictures in the appropriate seasons. (See Figure 14-3.)

WEATHER INSTRUMENTS
Make a bulletin board which names, shows pictures of, and indicates the use of several weather instruments, such as a thermometer, a weather vane, an anemometer, a hygrometer, and a barometer. Ask students to report on any which they have personally used.

Field trips

On several days take the students outside to observe clouds and compare the clouds with those pictured on a cloud chart (see "Sources of Assistance"). On a day when there are many clouds, have the students lie on their backs, look straight up at the clouds, and describe what they notice.

Other possibilities for field trips include going to a local television station to visit with the weather reporter, to the weather bureau, and to a museum which has a display on climate or weather phenomena.

Activities

➤ AIR WHICH IS HEATED EXPANDS

Place a balloon over an empty carbonated-drink bottle. Place the bottle first in a pan of hot water for a few minutes and then in a pan of cold water. What happens? *(In hot water the balloon inflates slightly; in cold water it deflates.)* Ask the students for possible explanations. *(Heating air causes it to expand; cooling it causes it to contract.)*

Point out that this activity illustrates one of the major factors in global air circulation. Air over equatorial regions is heated the most; and when it expands, it weighs less per cubic meter. It then rises, and air from cooler regions comes in to replace it.

What would happen to a balloon placed in a refrigerator? Have the children make predictions and then check them out. If it were placed in hot water after being removed from the refrigerator, what would happen?

➤ LAYERS OF THE ATMOSPHERE

Have the students work together to make a model or drawing—perhaps mural size—of the layers of the atmosphere. If the model or drawing is large enough, they can place various types of clouds in the troposphere at levels where these clouds are commonly found. Have them attach labels to both the layers and the cloud types.

The students can become "atmosphere estate" (real estate) agents and try to "sell" homes in selected layers of the atmosphere to prospective buyers. The agents should point out both desirable and undesirable features of their layers.

➤ AIR TEMPERATURES

This activity will illustrate the concept that the condition of the atmosphere (in this case the air temperature) is greatly influenced by the nature of the earth's surface beneath it. Have students measure the air temperature above several different types of surfaces—say, over a blacktop parking lot, a cement parking lot, a grassy area, open soil, and other surfaces available in the neighborhood. Do students note any temperature differences? What might have caused the differences? Where would the children rather be on a hot day? On a cold day?

➤ WATER CYCLE

Have students prepare an audio tape to explain the water cycle to second-grade students, as illustrated in Figure 14-4. How many ways can the students list in which water gets into the atmosphere? How many ways it gets out?

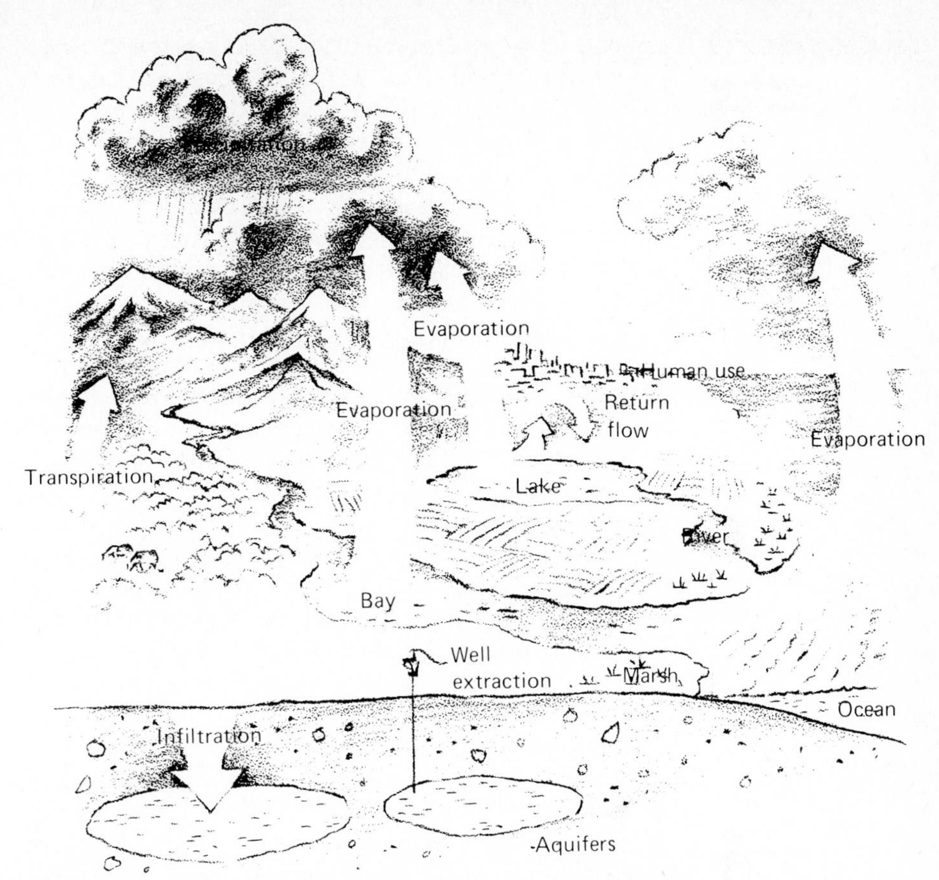

Figure 14-4. Water cycle. (Source: V. F. Lien, *Investigating the Marine Environment and Its Resources,* Texas A & M University, Sea Grant Program, 1979.)

➤ DAILY TEMPERATURE CHANGES

Have students take temperature readings every half hour during several school days. Plot each day's temperature on a graph. On the basis of several days' data, what is the hottest time of day? Noon? *(Generally 1 to 3 hours after noon.)* Why not exactly at noon? *(Air temperature rises when the atmosphere receives more heat from the sun than it loses, and drops when the atmosphere loses more heat than it receives. The most heat is received when the sun is highest overhead, but for some time after noon the amount received is still greater than the amount lost. The reverse is true for the coldest temperature in the mornings, and thus the lowest temperature generally occurs after sunrise.)* Do the daily patterns remain about the same throughout the seasons? When is "siesta" time in some countries? Could it be related to the daily temperature pattern?

➤ CLOUDS AS WEATHER PREDICTORS

Tell students that clouds alone can often be used to predict weather. Have a copy of *Weather,* a Golden Nature Guide book by Paul Lehr et al., on hand as a reference.* The students should record cloud types along with weather conditions for 1 and 2 days after each type of cloud is seen. Records should be kept for several days to a few weeks. As soon as they begin to notice patterns, the students should begin predicting what they think the weather will be like for the next 2 days, on the basis of the types of clouds present. Ask them to give reasons for their predictions. Divide the class into four to six predicting teams and have them compete to see which team can predict most accurately.

▶ CLOUD MURALS

Have students design and construct a mural which pictures and identifies major cloud types. If your hallways have painted walls, ask your principal for permission to make a permanent mural. Assure the principal that the job will be tastefully and beautifully done, and that you will use permanent wall paint. You may be able to convince a local or national paint company to donate the paint if you can make sure the company will be identified as donor in a newspaper article about the work your students and you have done.

Ask your students: "At what levels are particular clouds usually found? How thick might they be? What kind of cloud might have lightning under it?"

▶ CHANGING DAY LENGTH

Have students obtain information from an almanac on sunrise and sunset times for three cities at various distances from the equator—for example, Houston, St. Louis, and Duluth—on, say, the fifteenth day of each month. For each city, make a graph like the one shown in Figure 14-5. Which city has the greatest amount of daylight in summer? *(In our example, Duluth.)* Which city has the least amount of daylight in winter? *(In our example, Duluth.)* Which city has the least amount of difference between day length in the winter and in the summer? *(In our example, Houston.)*

Make sure the students know that the earth's axis is tilted, and then have them use a globe and try to explain the differences in amount of daylight in the various cities. This is a complex concept. The whole circle of latitude represents 24 hours. The portion of it that is lighted indicates the proportion of 24 hours that is lighted. *(In our example, they should note that in summer a greater portion of the circle of latitude which runs through Duluth is lighted by the sun than the circle of latitude which runs through the other cities. The opposite is true for winter.)* Does any point on the earth ever have 24 hours of daylight in a day?

▶ EFFECT OF TILT ON THE EARTH'S AXIS

In a darkened room have students hold a flashlight with a sharply defined beam of light directly over a piece of graph paper. Draw a circle around

*Paul R. Lehr, Will Burnett, and Herbert Zim, *Weather,* Golden Press, New York, 1975.

Daylight — Washington

	Jan. 21	Feb. 21	Mar. 21	Apr. 21	May 21	June 21	July 21	Aug. 21	Sept. 21	Oct. 21	Nov. 21	Dec. 21
Sunrise:	7:17 A.M.	6:46 A.M.	6:03 A.M.	5:15 A.M.	4:40 A.M.	4:31 A.M.	4:48 A.M.	5:17 A.M.	5:46 A.M.	6:16 A.M.	6:50 A.M.	7:18 A.M.
Sunset:	5:05 P.M.	5:42 P.M.	6:13 P.M.	6:44 P.M.	7:13 P.M.	7:32 P.M.	7:24 P.M.	6:49 P.M.	6:00 P.M.	5:12 P.M.	4:41 P.M.	4:38 P.M.

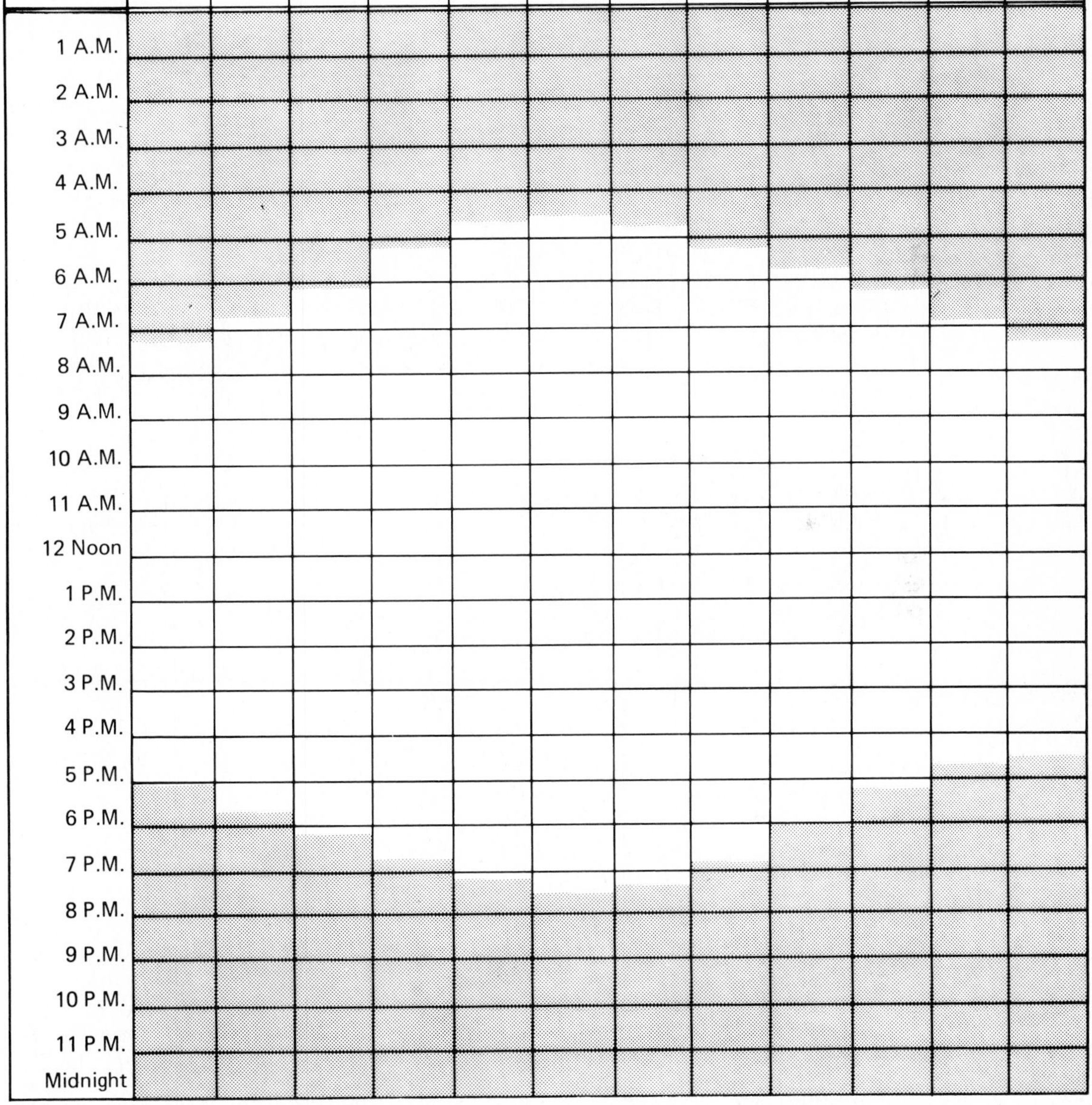

1 A.M.
2 A.M.
3 A.M.
4 A.M.
5 A.M.
6 A.M.
7 A.M.
8 A.M.
9 A.M.
10 A.M.
11 A.M.
12 Noon
1 P.M.
2 P.M.
3 P.M.
4 P.M.
5 P.M.
6 P.M.
7 P.M.
8 P.M.
9 P.M.
10 P.M.
11 P.M.
Midnight

Figure 14-5. Almanacs show the lengths of days. (Source: H. Tannenbaum et al., *Experiences in Science: Earth, Sun, and Seasons,* Webster Division, McGraw-Hill, New York, 1967, p. 22.)

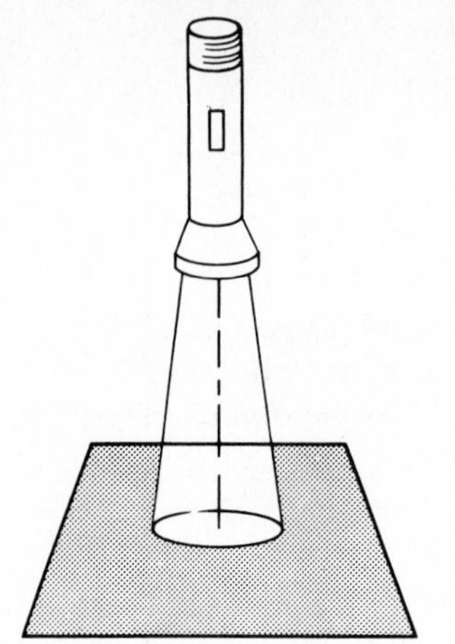

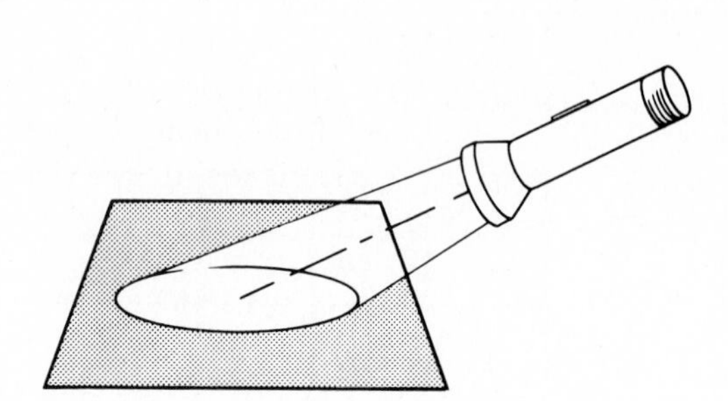

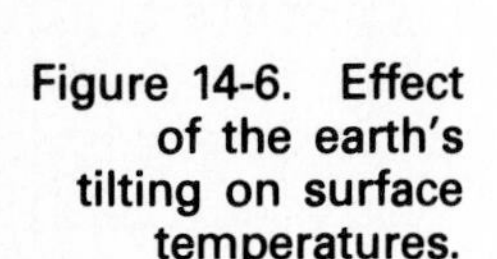

Figure 14-6. Effect of the earth's tilting on surface temperatures.

the beam. Then hold the flashlight at an angle (as in Figure 14-6) but still at the same distance above the paper, and again draw a line around the beam of light. Count the number of squares covered by the light in each instance. Have the students illustrate the same effect by using the flashlight and a globe.

Explain to the students that this experiment is analogous to the effect of sunlight striking the earth's surface at different angles. At noon in the summer, the sun is most directly overhead, and light striking the surface is most concentrated. At other times of the day and year, the light (and heat) is less concentrated, and thus less heat is received per square meter of the earth's surface at times other than noon or summer.

▶ MOVEMENT OF HEATED AIR

Here are two activities using water to represent air which will help students to understand the movement of heated air.

First place a small Pyrex container of water over a small heat source (a candle or small burner) in such a way that only one edge of the container is over the heat source.* Using an eye dropper, carefully place a drop of food coloring in the water near the bottom of the container just over the heat source. Have the students describe what happens. What could explain this occurrence?

Next, very slowly and carefully pour some warm colored (red) water into a container of clear, cool water. What happens? *(If you pour it slowly and carefully enough, the red water should stay near the top of the container.)* Similarly, very slowly and carefully pour cold colored (blue) water into a container of room-temperature water. What happens this time? *(The blue water should form a layer at the bottom of the container.)*

**Note:* Do this *yourself;* never let students work with open flames in the classroom.

Ask the students to draw conclusions about the observations which they have just made. *(As water—or air—is heated, it becomes less dense and begins to rise. It is replaced by cooler air. This phenomenon, the cause of wind, is one of the major influences on the movement of weather systems.)* Where are the hot and cold spots on earth?

EFFECT OF HUMIDITY ON EVAPORATION

We often hear people say, "It's not the temperature, it's the humidity that makes it seem so hot." If the students put some alcohol, which evaporates rapidly, on one arm, and some water, which evaporates more slowly, on the other, they will find out that the faster a liquid evaporates, the more they are cooled. People perspire constantly, and perspiration is like the alcohol—the faster it evaporates, the more we are cooled.

Have your students wrap a small piece of dry cloth around the bulb of a thermometer and record the temperature. Wet the bulb with room-temperature water. After several minutes record the temperature again.

One factor which must be controlled in taking the readings is differences in wind speed. Have students read the wet bulb and then fan the air around it. What happens? *(The temperature drops.)* Why? *(Moving air evaporates water faster. Anything which causes more rapid evaporation will cause greater cooling.)* Why do we fan ourselves on hot days?

Each day for several days, telephone a weather station or listen to a radio report to find out the relative humidity, and repeat the wet-bulb and dry-bulb temperature readings. Record the data for as many days as needed to get readings on days with differing relative humidity. Is there a relationship between the dry and wet thermometer readings and the relative humidity? *(Yes. When the relative humidity is high, there is less difference between the two readings than when the relative humidity is low.)* How do these findings relate to how comfortable we feel on days of differing humidity? *(When the relative humidity is high, perspiration evaporates more slowly and we are not cooled as much as on days of low relative humidity.)*

Taking dry-bulb and wet-bulb readings on the thermometer is actually a method of determining relative humidity. Relative humidity can also be determined by using Figure 14-7 (see page 324).

DEW POINT

The dew point is the temperature at which the air can no longer hold all the moisture in it. If the temperature reaches the dew point or goes lower, the moisture in the air begins to condense. The dew point can be relatively easily determined by placing some room-temperature water and a few ice cubes in a shiny metal can and stirring the water and ice cubes with a thermometer. When water first begins to collect, or show as a slight "frosting," on the surface of the can, take a reading from the thermometer. This reading is the dew point.

Challenge your students by having them note the dew point and the predicted low temperature on several nights. Then have them try to

RELATIVE HUMIDITY CHART

Difference (°C) Between Dry-Bulb and Wet-Bulb Temperatures

(°C) Dry-Bulb Temperature	1	2	3	4	5	6	7	8	9	10
-4	77	55	33	12						
-2	79	60	40	22						
0	81	64	46	29	13					
2	84	68	52	37	22	7				
4	85	71	57	43	29	16				
6	86	73	60	48	35	24	11			
8	87	75	63	51	40	29	19	8		
10	88	77	66	55	44	34	24	15	6	
12	89	78	68	58	48	39	29	21	12	
14	90	79	70	60	51	42	34	26	18	10
16	90	81	71	63	54	46	38	30	23	15
18	91	82	73	65	57	49	41	34	27	20
20	91	83	74	66	59	51	44	37	31	24
22	92	83	76	68	61	54	47	40	34	28
24	92	84	77	69	62	56	49	43	37	31
26	92	85	78	71	64	58	51	46	40	34
28	93	85	78	72	65	59	53	48	42	37
30	93	86	79	73	67	61	55	50	44	39
32	93	86	80	74	68	62	57	51	46	41
34	93	87	81	75	69	63	58	53	48	43
36	94	87	81	75	70	64	59	54	50	45

Figure 14-7. Relative humidity chart. (Source: *Science Activities in Energy: Conservation,* U.S. Department of Energy, HCP/U0033-02, 1977.)

predict what the low temperature will be on the basis of first hearing the dew point. They will receive immediate reinforcement when the weather reporter indicates the predicted low. *(In general, the low temperature will be a few degrees warmer than the dew point. Air cools very slowly when it contains much water.)* This method of predicting low temperatures is more successful in humid areas than in very dry areas, but students in dry areas also should be able to detect patterns.

▶ STORMS

Several types of severe weather—tornadoes, hurricanes, and electrical storms, to name just a few—can potentially cause great harm to humans. Obtain information on these phenomena from the National Weather Service (see "Sources of Assistance"), and have groups of your students prepare posters that will inform other students about the best ways to react to severe weather which may occur in your area. Place the posters in a display space in the school.

▶ SURVIVAL

Travelers and hikers are occasionally faced with weather- or climate-related conditions—such as traveling through deserts, getting caught in a snowstorm, and experiencing hypothermia (the lowering of body temperature)—that could prove fatal but can be handled relatively easily with appropriate knowledge.

With a little preparation, your students can easily learn to save their own lives or the lives of others. They do not have to fear deserts, snow, or conditions which may cause hypothermia—we fear what we do not understand. The crises which these three phenomena could generate can be avoided with advance preparation based on a little sound information.

Desert. Obtain desert survival information from DARES (see "Sources of Assistance"). Have students inspect an automobile to determine what survival equipment and supplies are built in. Have them design a low-cost desert survival kit which fits into a moderate-sized cardboard box and which would be useful in traveling though a desert area. If you are located in an arid area, first obtain the cooperation of your students' parents and then have your students make survival kits to be placed permanently in the trunks of their family cars.

Snowstorm. Obtain survival information from the Survival Education Association (see "Free and Inexpensive Materials"). Have students design a survival kit for use when traveling in areas where snowstorms can occur. If you are located in such an area, again first obtain the cooperation of your students' parents and then have your students construct snowstorm survival kits to be placed permanently in the trunks of their family automobiles.

Hypothermia. Hypothermia can occur in many environments. We will use mountainous areas as an example, because many people like to visit and hike in mountains. A hike into the mountains on a warm, clear summer day is very enjoyable—but weather can change rapidly. Quick rain showers and slightly lower temperatures can create conditions which could lead to hypothermia; the temperature does *not* have to be near freezing for hypothermia to occur. Obtain information on hypothermia from the U.S. Forest Service (see "Free and Inexpensive Materials"), and have your students describe what they can do to avoid hypothermia or to help someone who is experiencing it.

➤ COLD WEATHER AFFECTS INSECTS

What happens to insects when the temperature drops? Place some insects in a glass or plastic container* and cool it by partially immersing it in cold water. As the temperature drops, what happens to the rate of movement of the insects? *(It slows down.)* What happens to insects in winter? *(Some migrate. Many go into dormant stages in the ground, in logs, or in other protected places.)* Try to observe some insects on cool days.

➤ ADAPTATIONS TO WEATHER AND CLIMATE

Have students collect pictures of animals that show the characteristics of each animal which enable it to survive in the weather or climate condi-

**Note:* Punch holes in the lid of the container, if necessary, so that they can get the air they must have to survive.

tions in which it lives. Also have them obtain pictures which show how humans live in the same environments. Can we learn about adapting to an environment by studying the animals that naturally live there? *(We sure can!)* In what ways might such a study help us design homes? Make clothing? Form desirable eating habits?

PREDICTING WEATHER

Have the students cut out weather maps from newspapers for several days. What happens to a selected weather system (a high- or low-pressure area) during those days? After reviewing the maps for the past few days, have students predict where the various weather systems will be tomorrow. Have them predict the weather for the next 3 days and explain their predictions. Each day check on the accuracy of their predictions.

How accurate are weather predictions? Have students keep track of the successful and unsuccessful predictions made by the National Weather Service for several days. What proportion of the time are the predictions correct?

Have students note the direction in which the wind blows around the weather system. Ask them to predict what changes may take place in wind directions in the next day or two. Check their accuracy.

WEATHER MODIFICATION

Have a small group of interested students study what attempts have been made to modify the weather and what successes have been achieved. Have them identify both positive and negative aspects of weather modification. Information can be obtained from the National Weather Service (see "Free and Inexpensive Materials").

Is any research presently being done? What have the results been so far? What problems could arise? What kind of weather is being modified?

Efforts are presently being made to "seed" clouds.

▶ MIGRATING
Have the students relax and close their eyes. Tell them that they have become Canada geese. Take them through the seasons and migration route of the Canada goose. Stress the various activities of the goose in different weather and seasons, and also emphasize its food requirements. You can prepare yourself for this activity by looking up migration in an encyclopedia or by reading the appropriate sections of James Michener's book *Chesapeake.**

Children's literature

Read *Hide and Seek Fog*† to your students or have them read it, and then discuss with them the concepts presented.

Find other books on weather and climate in the school library, and identify them for the students or place them in an interest center.

Activity cards

An example of an activity card is shown in Figure 14-8 (page 328). Additional activity cards could be made for topics such as the different types of recreation in differing climates or seasons, animals' adaptation to different climates, microclimates (e.g., the temperature in a shoeprint in the snow or sand), music about weather or climate, and animal migration.

Bulletin boards

WEATHER CONDITIONS IN VARIOUS CITIES
Construct a bulletin board similar to the one shown in Figure 14-9 (page 328). Each day, have five different students post appropriate data.

WEATHER SATELLITES
Obtain information from NASA on weather satellites (see "Free and Inexpensive Materials"). Post some of the weather satellite pictures and some information pamphlets on a bulletin board. Can any storm system be seen? What happened in your locale on the day pictured?

Field trips

Skill in measuring weather phenomena (e.g., wind speed, temperature, wind direction, relative humidity, dew point) can be acquired on the school campus. A trip to a weather station can also be very informative. Ask the manager to show your students the facility and instruments, and also to describe how weather forecasts are made and what subjects the children could study if they wanted to become weather forecasters or researchers. Be sure to obtain samples of weather maps during the visit.

*James A. Michener, *Chesapeake,* Random House, New York, 1978.
†Alvin Tresselt, *Hide and Seek Fog,* Lothrop, New York, 1965.

WEATHER FOLKLORE

"WHEN THE MOON WEARS A HALO AROUND HER HEAD, SHE WILL CRY BEFORE MORNING AND THE TEARS (RAIN) WILL REACH YOU TOMORROW."
—AMERICAN INDIAN

FIND SOME WEATHER SAYINGS. ARE THEY TRUE? HOW CAN YOU FIND OUT?

"EVENING RED AND MORNING GRAY SETS THE TRAVELER ON HIS WAY; EVENING GRAY AND MORNING RED BRINGS DOWN RAIN UPON HIS HEAD." TRUE OR FALSE?

"WHEN THE BEES STAY CLOSE TO THE HIVE, RAIN IS CLOSE BY." TRUE OR FALSE?

Figure 14-8. Activity card: Weather folklore.

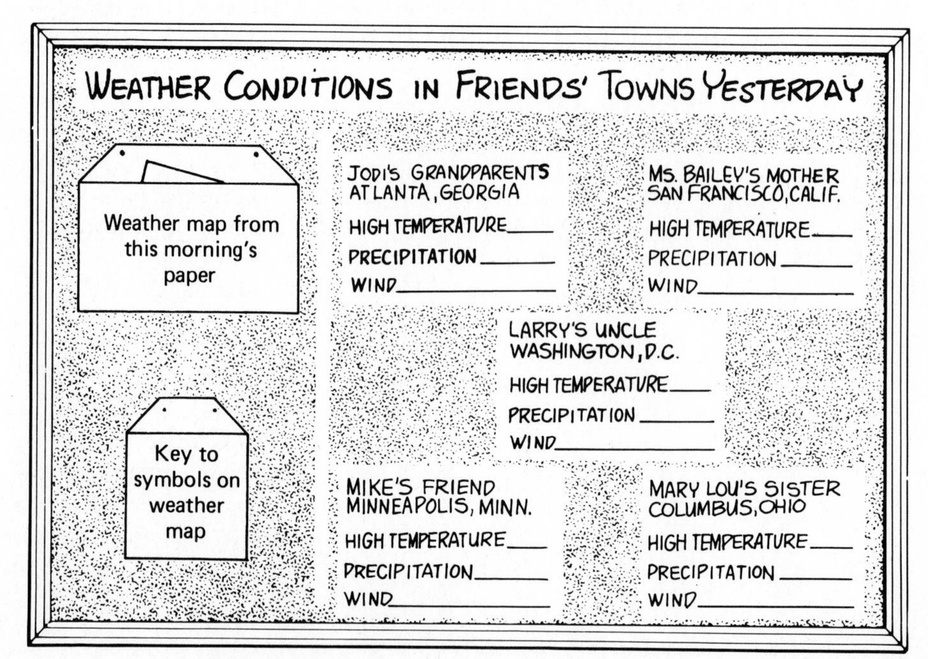

Figure 14-9. Bulletin board: Weather conditions in friends' towns yesterday.

RESOURCES

Sources of assistance

FREE AND INEXPENSIVE MATERIALS
Here are some of the best sources of free and inexpensive materials for use in teaching weather and climate:

1. *Taylor Instrument Company.** Pamphlets on weather and how to measure and record it. Classroom supplies of weather charts.
2. *U.S. Department of Commerce, National Weather Service.* A packet for teachers containing information on several weather phenomena.
3. *National Aeronautics and Space Administration.* Information on weather satellites.
4. *U.S. Department of Agriculture, Forest Service.* Woodsy Owl pamphlets on hypothermia.
5. *DARES.* Information on survival in deserts.
6. *Survival Education Association.* Information and programs on survival in mountains and snow.

LOCAL RESOURCES
Local resources which can help in teaching weather and climate include:

1. Weather maps in local newspapers.
2. National Weather Service or Flight Service at a local airport.
3. Weather reporters from local television stations.
4. Meteorology professors from local college or university.

ADDITIONAL RESOURCES
1. Magazines such as *Weatherwise, National Geographic World,* and *Ranger Rick.*
2. Publications available from the county extension agent.
3. Many teacher journals and general publications carry articles on the popular topics of weather and climate.
4. Some television stations include a short educational section in each weathercast.

Strengthening your background

1. Take slide photographs of five different varieties of clouds.
2. Watch a television weather report, and base a quick prediction of the next days' weather upon the weather map. Then see if your prediction is similar to the one given on television. The next day, note how

*The addresses of specific agencies, organizations, and businesses are listed in the Appendix.

accurate both your prediction and the reporter's prediction are. Continue this process for a few weeks and see whether your predictions become more accurate.

3. Talk to the pilot of a small aircraft to find out what weather conditions she or he checks on before a flight.
4. Read *Weather* by Paul Lehr, R. Will Burnett, and Herbert Zim (see "Building Your Own Library").
5. Learn some of the folklore of weather by talking to older persons about weather sayings they know and by reading *Folklore of American Weather* (see "Building Your Own Library").

Getting ready ahead of time

1. Prepare the materials for a bulletin board which identifies weather instruments and explains their use.
2. Identify six songs (popular or classical) which deal with weather or climate and could be used for your class.
3. Collect magazine pictures of weather or climate phenomena.
4. Produce some overhead transparencies or slides which illustrate the zones or layers of the atmosphere.
5. Make a plan for naturally air conditioning the school building by planting trees, bushes, and other vegetation.
6. Prepare a short class presentation on the strong influences of weather and climate on how people live in arid and semiarid regions such as Arizona.
7. Write for the free list of weather publications available from the Superintendent of Documents (see the Appendix).

Building your own library

Allison, Linda: *The Reasons for Seasons,* Little, Brown, Boston, 1975.
Harvey, Fred: *Why Does It Rain?* Harvey House, New York, 1969.
Lehr, Paul, R., Will Burnett, and Herbert Zim: *Weather,* Golden Press, New York, 1975.
Pine, Tillie: *Water All Around,* McGraw-Hill, New York, 1959.
Sloane, Eric: *Folklore of American Weather,* Hawthorn, Prentice-Hall, Englewood Cliffs, N.J., 1963.

Building a classroom library

Adler, Irving: *Weather in Your Life,* John Day, New York, 1959.
——— and Ruth Adler: *Air,* John Day, New York, 1962.
Alth, Charlotte, and Max Alth: *Disastrous Hurricanes and Tornadoes,* F. Watts, New York, 1981.
Aylesworth, Thomas G.: *Storm Alert: Understanding Weather Disasters,* Messner, New York, 1980.

Beer, Kathleen C.: *What Happens in the Spring,* National Geographic Society, Washington, 1977.

Bixby, W.: *Sky Watchers: The U.S. Weather Bureau in Action,* McKay, New York, 1962.

Branley, Franklyn: *Rain and Hail,* Thomas Y. Crowell, New York, 1963.

Buff, Mary, and Conrad Buff: *Dash and Dart: A Story of Two Fawns,* Viking, New York, 1942.

Davis, Hubert: *A January Fog Will Freeze a Hog, and Other Weather Folklore,* Crown, New York, 1977.

Ets, Marie Hall: *Gilberto and the Wind,* Viking, New York, 1963.

Farrington, Margaret: *Where Does All the Rain Go?* Coward-McCann and Geohegan, New York, 1973.

Fisher, Aileen: *I Like Weather,* Thomas Y. Crowell, New York, 1963.

Freeman, Mae Blacker: *Do You Know about Water?* Random House, New York, 1970.

Gallant, Roy: *Exploring the Weather,* Doubleday, Garden City, N.Y., 1969.

Helm, T.: *Hurricanes: Weather at Its Worst,* Dodd, Mead, New York, 1967.

Hitte, Kathryn: *Hurricanes, Tornadoes, and Blizzards,* Random House, New York, 1960.

Keen, M. L.: *Lightning and Thunder,* Messner, New York, 1969.

Kwitz, Mary DeBall: *When It Rains,* Follett, Chicago, 1974.

Mizumura, Kazue: *I See the Wind,* Thomas Y. Crowell, New York, 1966.

Podendorf, Illa: *The True Book of Weather Experiments,* Children's Press, Chicago, Ill., 1961.

Ryder, Joanne: *Fog in the Meadow,* Harper and Row, New York, 1979.

Schneider, Herman: *Everyday Weather and How It Works,* rev. ed., McGraw-Hill, New York, 1961.

Spilhaus, Athelstan: *Weathercraft,* Viking, New York, 1951.

Tresselt, Alvin: *Hide and Seek Fog,* Lothrop, New York, 1965.

Wyler, Rose: *The First Book of Weather,* F. Watts, New York, 1956.

Zim, Herbert S.: *Weather,* Golden Press, New York, 1957.

CHAPTER 15

THE EARTH

IMPORTANT CONCEPTS

The earth was formed billions of years ago and has been changing ever since.

The study of present geologic processes and products is a basis for the interpretation of the history of the earth.

The earth has a core, a mantle, and a crust.

Earthquakes are caused by the sudden movements along zones of weakness (faults).

Rocks may be classified on the basis of how they were formed.

Wind, water, ice, temperature variation, and living things are agents which affect the surface of the earth.

Land wears down in some places and is raised in others.

Soil is a mixture of broken-down rock particles and organic material such as plant and animal remains.

Scientists are attempting to predict changes in the earth's crust (e.g., earthquakes and volcanic activity).

Coal, oil, and other fossil fuels were formed by geologic processes acting upon the remains of organisms over long periods of time.

Geologic features—rivers, flood plains, deserts, mountains—influence the population distribution of the world.

Evidence in rocks provides information about the earth's history.

Rocks and other features and materials in an area give clues to that area's geologic history.

At times in the earth's history, great masses of ice have covered large portions of the continents.

Fossils provide evidence about plants and animals that once lived on earth.

Fossils are formed in several ways.

Some modern animals and plants are similar to ones that lived long ago.

A small fossil part may provide much information about the whole organism and its environment.

Ancient animals, like modern ones, were adapted to their environments.

As environments change, the organisms which live in them also change.

Many species of animals and plants which lived long ago are no longer found on earth.

INTRODUCTION: STUDYING THE EARTH

Students can make direct observations of many of the materials and processes studied in physical geology. Rocks and minerals, water and wind moving rock particles, geologic features (e.g., rivers, deserts, mountains), and soil are parts of the child's real and immediate environment. In addition, events—such as earthquakes, volcanoes, severe storms which change land, erosion problems, and flooding—which are studied in physical geology are frequently reported in newspapers and on television and radio.

Knowledge of physical geology has many practical applications. Is the nice flat land upon which a new house is being constructed a flood plain? If another house is on a hillside, what is the likelihood that a landslide may damage it? What natural building materials are most durable? Where are fossil fuels likely to be found?

People are naturally curious about many phenomena in physical geology, and other phenomena sometimes receive publicity and thus become topics of curiosity. Earthquakes, volcanic activity, continental drift, and the structure of the earth are interesting to most students. When people visit mountains, or deserts, or large expanses of plains, they frequently wonder how the areas became the way they are. When students learn about geologic processes and the phenomena which result from them, they will not only come to understand how mountains, for example, are produced but will also acquire the ability to see more of what mountains are like.

Historical geology is a prime example of scientific "detective" work. Evidence is gathered and an explanation is finally generated. As new evidence is discovered, it either supports or changes the explanation.

Detailed study of historical geology is probably too complex to be appropriate for elementary students; but some of the basic geologic concepts and processes are very interesting to young students. As an example, just think of their interest in fossils: almost every elementary school classroom has its expert or experts on dinosaurs.

At an introductory level, much attention in historical geology is focused upon fossils; however, historical geologists are just as concerned with the physical environment as they are with life forms. Lava flows indicate volcanic activity, limestone deposits indicate a marine environment, and fossils give clues to the types of life which were once present.

In recent years historical geologists have been very active in locating energy resources. Finding certain kinds of ancient environments often leads to finding energy resources.

Geologists identify sites for drilling oil wells.

Many complex geologic phenomena are not as yet completely understood. Perhaps some of your students will one day develop explanations for these phenomena. Challenge them with important questions: "Will we ever be able to predict when an earthquake will occur? When a volcano will erupt? Mount Saint Helens erupted—are other volcanoes in the contiguous 48 states likely to erupt?"

Many geological problems can be resolved only after acquisition of a great deal of background information, but children's curiosity about them can be developed. Elementary school students can answer many questions in the realm of physical geology, incluing the ones suggested by activities in this chapter. Remember to reply to your students' inquiries by asking them questions: "What kind of information would you need to answer that question? How could that information be obtained? What kind of an experiment would help you find the answer? I know a geologist: Dr. Jones. What questions could you ask if I made an appointment for us to talk with Dr. Jones?"

THE EARTH FOR PRIMARY GRADES

Activities

➤ CLASSIFYING ROCKS BY OBSERVABLE CHARACTERISTICS

Ask students to collect three local rocks and bring them to class. The three should be different in some observable way and should be approximately hand-sized. If rocks are scarce in your area, send a note home to ask parents to help their children get samples; rocks may have been collected on family trips, for instance. If necessary, you could contact a local rock shop or rock club member, or you could obtain samples from a science supply house (see the Appendix).

Select a dozen different samples and ask the students to divide them into two groups on the basis of a visible difference. Let them choose whatever characteristic they wish to use in grouping—the differences noted will probably include color (mostly dark or light); grains, that is, minerals (visible or not visible); texture (smooth or rough). Let the class as a whole group the rocks in several different ways. Then ask students in small groups to classify the rocks on another characteristic, and have another group of students guess what characteristic they used. This activity is useful in developing skill in observation and classification.

▶ SOIL

To help students understand soil, have them closely examine a sample. If possible, obtain a soil sample from some area where many plants grow. Give each student a sample of the soil and a hand lens or magnifying glass. Tell the students to carefully break the soil apart and examine the particles with the hand lens. Then ask them to describe what they observe and to attempt to name the kinds of materials they see. *(They may be able to identify small pieces of rock—sand—and bits of plant parts.)*

Ask the students to try to explain how soil is formed. *(Rocks break down and organic material is added.)* Of what value is soil? Can soil ever be destroyed or lost? *(Yes—e.g., by erosion.)* How long does it take for new soil to form if the original is removed? *(Hundreds or thousands of years.)*

Try to collect samples of several different types of soils. Ask students to describe any differences which they notice. Color? Size of rock particles? Amount of identifiable plant material? Describe the productivity of each soil type and ask students to think of ways the materials in or characteristics of the soil samples are related to the productivity of the soil. *(Obviously, your students will not have the background of a soil scientist, but they may be able to identify some relationships.)*

▶ EROSION BY WATER

To demonstrate water as an agent which changes the earth's surface, have the students help you construct a sand or soil pile near an outside water faucet to which a hose can be attached. Ideally, the pile should be at least 50 centimeters (about 20 inches) high and 1 meter (a little over 1 yard) wide. Place a meter stick (yardstick) or another stick of similar length into the pile. Note or mark how high the soil is in relation to the stick. Place small stakes in the ground around the perimeter of the pile. (See Figure 15-1.)

Attach the hose, turn on the water, and spray water onto the pile for several minutes. Hold the hose in one position—do not try to distribute

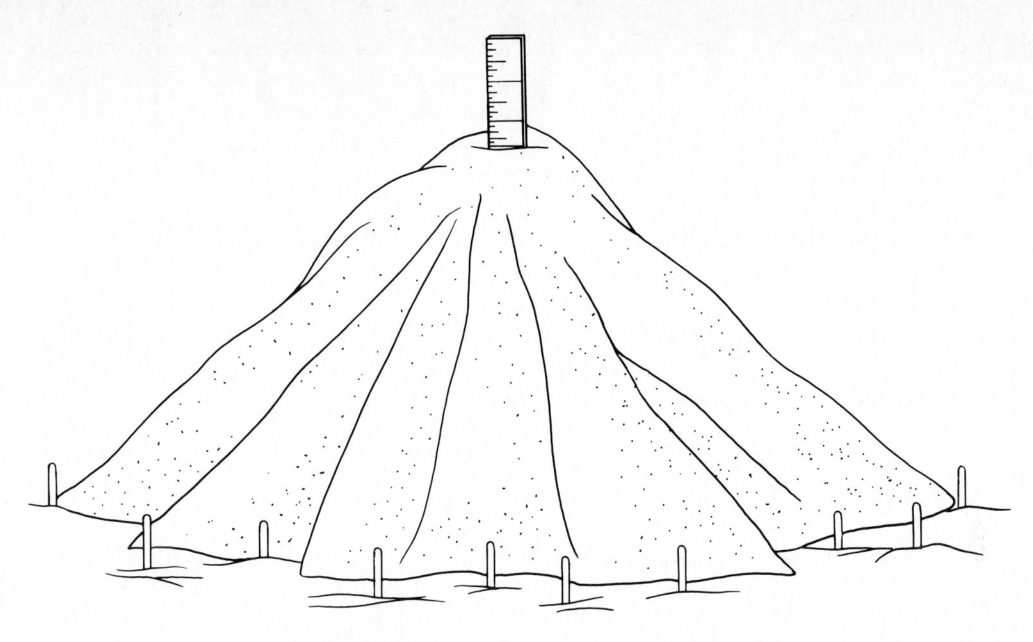

Figure 15-1. Demonstration of erosion.

the "erosion" evenly. While the water is running, ask the children to explain what is happening. Does the water flow straight down the sides of the pile? Is there more than just water running off? How do you know? After it is clear to the students that erosion is occurring, turn off the water and ask them where the eroded soil went.

➤ EROSION BY WIND

The role of wind in changing the earth's surface also can be rather easily demonstrated. Take the class outside and, either in a very large cake or pizza pan or on a flat surface, spread out a layer of fine sand. Then use a variable-speed fan on slow speed to simulate wind.* Ask the students to notice whether any movement of the sand is taking place.

Ask the students to predict what will happen if the wind speed is increased. Then turn up the fan and show them. What happens if a small barrier such as a pebble or a stick is placed on the sand? *(Try it and find out.)* Present the students with a little information on sand dunes and ask if they can suggest ways to form sand dunes on their sandy area. *(Try several different ways and ask the students to help you evaluate their effectiveness.)*

➤ LAYERS OF SEDIMENT

At least one possible means of formation of sedimentary rocks is easy to demonstrate. Fill a 4-liter (1-gallon) clear glass jar (or a similar container)

**Note:* All students and the teacher must stand *behind* the fan, to avoid eye damage.

approximately three-quarters full of water. Give students a soil mixture which includes a range of particle sizes from small pebbles through very fine powdery material. Have a student dump about 200 to 250 milliliters (¾ to 1 cup) of the soil into the water. Ask the class to notice whether or not all the material falls to the bottom at the same rate. What kind of material falls fastest?

Give the material at least 2 hours to settle, and then have another student dump another batch of the soil into the container. Repeat the procedure several times. Have the students examine the material after it has settled. Are there any noticeable features? *(Some evidence of layering should be visible.)* Point out that layering occurs in some sedimentary rocks. The materials—sediments—which make up some sedimentary rocks may have been laid down in bodies of water in a manner similar to what was done in this activity. After years of compaction and consolidation, rocks may have been formed from the loose material.

▶ OLD AND NEW ANIMALS—DIFFERENCES

Have students collect pictures or make drawings of modern animals and of some extinct animals. Then they can group the animals into sets of similar animals. This activity will teach them that some animals which formerly lived on earth no longer do so, that some extinct animals are very similar to some modern animals, and that some extinct animals are not closely related to any animals presently living. Students are generally interested in dinosaurs and hairy mammoths, and a book such as *Prehistoric Monsters Did the Strangest Things** can stimulate their interest in other ancient animals.

Students should be made aware that people did not live on earth at the same time that dinosaurs did, because popular cartoons have led many children to believe wrongly that people did coexist with dinosaurs. Students might be interested to determine what extinct animals really did exist after people began to inhabit the earth. The mammoth is one, and the students can use library books to find others. After several animals have been identified, encourage the students to use their imaginations to describe how the animals lived. Each students can choose one animal and write or tell the story of his or her life as that animal.

▶ OLD AND NEW ANIMALS—THE SAME?

Students may be surprised to learn that a number of animals, especially insects, which lived in the distant geologic past still exist in substantially the same form today. Have the class use reference materials to identify prehistoric animals and then collect pictures or make drawings of modern animals which are similar.

Ask the students to find out what types of environments the modern animals live in and then to describe the environments the prehistoric

*Leonora Hornblow and Arthur Hornblow, *Prehistoric Monsters Did the Strangest Things,* Random House, New York, 1974.

animals probably lived in. Try to find out how accurate the students' ideas are.

➤ MAKING MUD FOSSILS

Making mud "fossils" will familiarize students with how plant and animal remains are preserved. Have them mix a large batch of mud (soil and water) and pour a little of it into the bottom of a large container such as a plastic dish basin. Push bones, feathers, leaves, and other materials from living things down into the mud, and pour in more mud. Continue to add materials and mud until the total depth is at least 15 centimeters (6 inches).

Allow the mud to dry and harden for several days until it is thoroughly dry. Then tell the students that their task now is to recover the materials from the mud block by breaking it apart. After they have recovered a few materials and prints, ask them to describe any difficulties they may have had. Have them make suggestions about how they can remove as many fossils as possible without breaking them into small pieces. Ask the class to compare what they have done with what scientists who study ancient animals must do to collect animal remains.

If some students become especially fascinated by this activity, tell them that animal footprints are occasionally found as fossils. Challenge them to make fossil footprints in mud.

➤ DINOSAUR CARD GAME

Make or have students make cards with pictures or drawings of dinosaurs on one side and the names of dinosaurs (with pronunciation guides) on the other side. Individually or in small groups the students can then try to name the dinosaurs and check their answers against the names printed on the reverse side of the cards.

The cards can also be used to play a game like "go fish" if there are enough cards for each type of dinosaur. Another use would be to have students group the cards on the basis of the types of food they think each dinosaur ate. Students can do some library research to check out the accuracy of their ideas.

➤ PREHISTORIC ANIMAL SKIT

Determine the identities of a number of prehistoric animals, trees, and plants which lived in a given environment. Share the information with your students, and have each student draw or paint a picture of a different animal, tree, or plant on a piece of cloth or a T shirt.* Only one animal or plant is to be painted on each cloth or T shirt. Then have the students wear their T shirts or tie the cloths about their chests while they develop a skit in which each animal, tree, or plant demonstrates its behavior or function and its relationship to the other living things in the environment.

**Note:* Be sure to choose paints that are safe for use on clothing.

IMAGINARY TRIP TO A MUSEUM

Using *Frozen Snakes and Dinosaur Bones** as a reference, have your students take an imaginary trip through a natural history museum. Identify the work that people do in a museum, and place some emphasis on how scientists interpret ancient environments from rock and fossil evidence.

Children's literature

Read *Rocks and Their Stories*† or *Is This a Baby Dinosaur?*‡ to your students or have them read one of these books, and then discuss with them the concepts presented.

Find other books on geology in the school library, and identify them for the students in an interest center.

Activity cards

An example of an activity card is shown in Figure 15-2. Additional activity cards should be made for topics such as making a rock collection, finding pictures of several different kinds of landforms, making a stream table, finding pictures of landscapes which students like, finding ways in which rocks are used in your town, studying effects of erosion, collecting rocks which indicate the nature of ancient environments (e.g., volcanic rocks), making drawings of ancient environments, making fossils, and collecting models of ancient plants and animals.

Bulletin boards

ROCK TYPES

Post drawings or photographs of several common rocks—granite, limestone, schist, shale, slate, etc.—on a bulletin board. In a nearby box place an example of each type or rock pictured. The students are to match the rock samples with the pictures. You may also wish to post the names of the rocks on the bulletin board. Pictures and rocks for this bulletin board can be obtained from a science supply house if they are not easily available otherwise (see the Appendix).

ANCIENT LIFE

Collect pictures of a group of prehistoric animals and plants which lived at a specific time in geologic history (e.g., the "age of the dinosaurs"). Sets of such pictures can be obtained from a science supply house (see the Appendix); or you can use a historical geology textbook to identify such a group of animals and plants and then either collect pictures from magazines or other sources or make sketches of the animals and plants. Post

*Margery Facklam, *Frozen Snakes and Dinosaur Bones,* Harcourt, Brace and World, New York, 1976.

†C. L. Fenton and M. A. Fenton, *Rocks and Their Stories,* Doubleday, Garden City, N.Y., 1951.

‡Millicent Selsam, *Is This a Baby Dinosaur?* Harper and Row, New York, 1972.

USING CLAY, MAKE A MODEL OF SOME PREHISTORIC ANIMAL.

IF YOU WANTED TO PAINT THE ANIMAL'S ENVIRONMENT ON A BACKGROUND PAPER, WHAT WOULD BE THERE?

WHAT DID YOUR ANIMAL EAT?

Figure 15-2. Activity card: Clay models.

the pictures or sketches on a bulletin board, as in Figure 15-3, page 342. In an envelope attached to the board place tags which name the animals and plants. Tell the students to place the tags on or near the corresponding organisms.

Field trips

Many local field trips may be made while studying physical geology. Just outside the school building, students may collect rocks or look for examples of erosion (for example, plants growing in cement cracks and thus contributing to the breakdown of the cement). A small gully, if available, not only will show the action of running water, but also may show soil layers and yield some rock samples which have been eroded out of the soil.

Two attractive sites for field trips are museums of natural history and rock exposures which contain easily identified rocks and fossils. Fossil sites are not abundant; but if you can find one, it can be a great motivator for students. An earth science teacher in junior or senior high school may be able to tell you about local fossil-collecting sites. Museums of natural history often present educational displays on ancient life and environments and also can provide staff members to speak to your class about such topics. Additional field trips may be made to local rock shops and to a nearby college or university which has a geology department or division.

Figure 15-3. Bulletin board: Prehistoric animals.

THE EARTH FOR INTERMEDIATE GRADES

Activities

➤ EARTH LAYERS

Models and illustrations are frequently used to represent phenomena on a scale which is not easy to observe directly (e.g., an atom or the solar system), which are by their nature difficult to observe directly (e.g., an animal's digestive system or ear), or which are theoretical and have never been observed directly (e.g., the layers in the earth or how light travels). Constructing models causes us to pull together all our information about the subject of the model. Students should be encouraged to make models or drawings of many things they study.

Making the model in this activity will show students how models are developed. First discuss with the class the internal structure of the earth, including the layers identified as inner core, outer core, mantle, and crust; then have the students either construct a model using clay or papier-mâché or draw an illustration of those layers.

After the model or illustration is completed, ask the students to try to explain how scientists first created a model of the layers of the earth. How did they obtain information which enabled them to develop the model?

➤ MAKE A ROCK

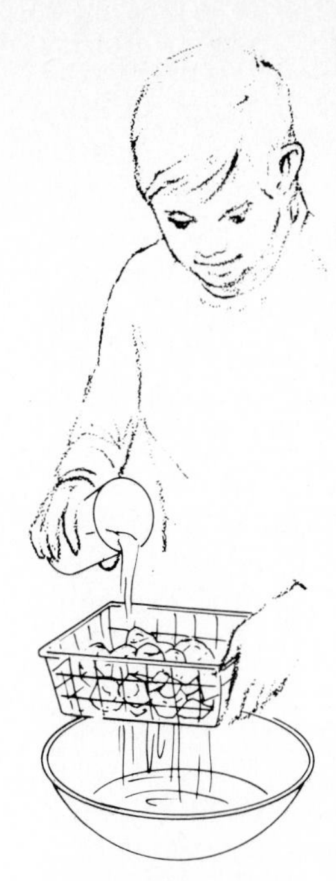

Figure 15-4. Constructing rocks. (Source: Adapted from V. N. Rockcastle et al., *Science: Level 5,* Addison Wesley, Reading, Mass., 1980, p. 52.)

Sedimentary rocks are formed in a process in which sediments are deposited, buried, and compacted, and then some "cementing" material bonds the sediments into rocks. To demonstrate the process, have each student place a few pebbles in a porous container (a styrofoam cup with several holes punched in the bottom and sides or a plastic strawberry box will do nicely) and then pour a small amount of thinned white glue (such as Elmer's) over the rocks. (Have the students work on old newspapers or other disposable material.) Allow the glue to dry and repeat the process three or four times. (See Figure 15-4.)

After the glue has dried, tell the students to break the containers off their "rocks." Ask the students what the glue has done. Has it filled all the spaces between the pebbles? *(It should not have.)* Do real sedimentary rocks have spaces in them? *(Yes, many do.)* Can you think of any evidence which supports the conclusion that these spaces exist in some sedimentary rocks? *(How about petroleum and water moving through them?)*

➤ GROWING CRYSTALS

Crystals are objects of beauty and fascination, and several kinds are very easy to grow. Each kind of crystal has a characteristic shape. If possible, obtain a quartz crystal to show the class; they are inexpensive and may be bought at almost any rock shop.

One kind of crystal which can be easily grown is alum. Powdered alum can be obtained from almost any drugstore. Heat a small amount of water (250 milliliters or 1 cup) and mix in as much alum as will dissolve. Then grow the crystals in one of two ways.

One way (see Figure 15-5) is to tie a string onto a pencil, place a small

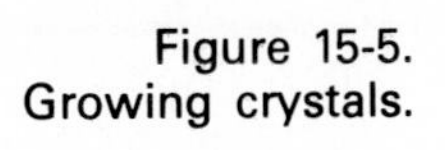

Figure 15-5. Growing crystals.

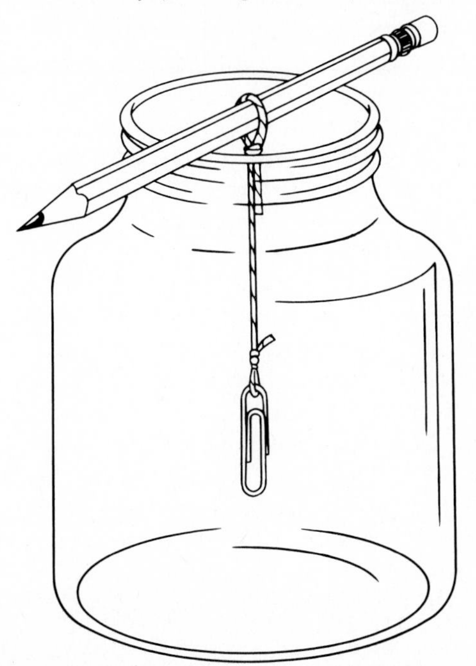

weight such as a paper clip on the string, and suspend the weight in the water. As the water cools and evaporates, crystals will begin forming around the weight.* A number of small crystals will then grow. If larger crystals are desired, remove all except two or three of the small crystals. Using a ruler with millimeter divisions, have the students measure the size of a specific crystal each day and graph the information. Does the crystal grow at a constant rate? Are there some days when the crystal becomes smaller? Why does this happen? *(How about temperature?)*

A second way to grow a crystal is to pour the alum solution into a wide, shallow dish such as a soup bowl. After several small crystals have begun to grow, remove all but one. As the water evaporates, the crystal will become larger. The crystal should be turned two or three times a day so that it will grow evenly. After much of the water has evaporated, more alum solution can be added. With a great deal of care and patience, beautiful crystals can be grown. After students have grown their crystals, set aside a special day when they can show off products of their work.

▶ WEATHERING ROCKS—MOVING WATER

One way in which rocks are broken down is by moving water. The water itself helps break the rocks down; but what is even more effective is that, as the water moves them, they bump into and grind on each other. This process can be relatively easily shown in the classroom.

For this activity, you will need a large plastic jar—1 liter (or 1 quart) to 4 liters (or 1 gallon)—with a lid that tightens securely and about 125 milliliters (½ cup) of stones with sharp edges. The kind of crushed rock that is used for driveways or in flower gardens will work well. Wash the stones and weigh them. Fill the jar half full of water and then add the stones. Tighten the lid and shake the jar, counting the number of shakes —for the initial try, shake the jar 1000 times.

Open the jar and notice the condition of the water. *(The cloudiness is due to fine pieces of rock which have broken off the stones.)* Are the stones smoother than they were? How many shakes would it take to make them really smooth? Try it. Did the mass of the stones decrease? *(What was the final mass in grams or ounces as compared with the original mass?)*

▶ WEATHERING ROCKS—ACTION OF ICE

Rocks are also broken down by ice in parts of the earth where freezing occurs. When water changes to ice, it increases in volume by approximately 10 percent. To demonstrate the force that this process can exert, *completely* fill a plastic container with water and fasten its lid securely. Set the container in a cake pan and place it in a freezer. If there is no freezer at school, have several students, with their parents' permission, do this activity at home and describe the results to the class.

What happens to the container when the water freezes? *(Either the lid*

*Consult reference books on growing crystals for more detailed directions and other materials from which crystals can be grown.

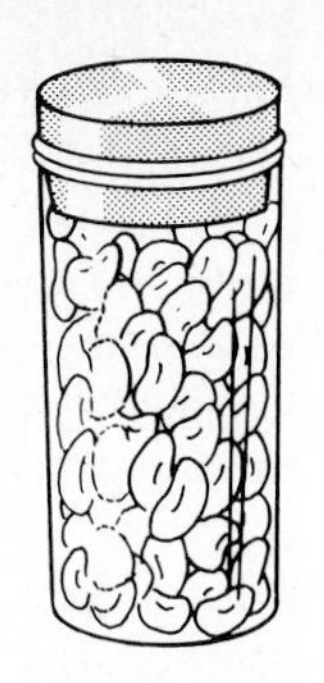

Figure 15-6.
Weathering rocks.

is forced off or the container ruptures.) What would happen if water were trapped in cracks in rocks and then froze? *(It would force the parts of the rock apart, thus breaking the rock.)*

➢ WEATHERING ROCKS—LIVING THINGS

Living things assist in weathering rocks in several ways. One way is the force exerted by seeds and roots of plants as they expand in small cracks in a rock. To demonstrate this phenomenon, fill a glass or plastic container as full as possible with bean seeds, leaving at the top just enough room to close the container with a stopper. Then pour just enough water into the container to fill the spaces between the seeds. Insert the stopper, place the container in a flat pan, and put the pan in a location in the room where it can be easily observed. (See Figure 15-6.)

Tell the students to look at the container two or three times a day. What happens? *(The seeds swell up and force the stopper out of the container. In time the seeds will sprout and occupy even more space.)* How does such swelling help break rocks apart? *(By enlarging the cracks.)* Living things also tend to produce acids—either in their life processes or when they decay after death—which help break rocks down.

➢ ROCK IDENTIFICATION

Obtain at least four hand-size examples each of sedimentary, metamorphic, and igneous rocks. You may buy them from a science supply house (see the Appendix) or rock shop, or you may be able to borrow them from an earth science teacher, or students and their parents may be able to help you find them.

Have the students develop a dichotomous key for the samples (see Chapter 9, Plants). Developing the key will cause students to focus on the observable properties of the rock samples. After they finish the key, tell them the names of the samples and describe how each was formed. Have each student write a brief description of each rock, based upon its observable characteristics. Then have them learn each rock's name, its most noticeable characteristics, and how it was formed.

Ask students to find examples of use of the rocks in local buildings, cemeteries, or other public places. Post a list of the places where they find the rocks used. Also have the class do some research to identify parts of the United States where the different rocks might be found in natural sites. Then the students can be on the lookout for the rocks if they take trips. By the way, students who travel often bring samples back to their teachers.

➢ EROSION STUDIES

To help students understand erosion, make a stream table out of a large plastic dishpan. Drill a hole in the bottom at one end, glue a drain hose (an air hose for aquariums works well) to the hole, and place the other end of the hose in a pail. (When you want to form a lake, the drain hose can be closed off.) Place a board under the other end of the pan, and pile sand on the high end in the shape of a mountain, a plateau, or some other natural land formation. Use a sprinkling can to "rain" water on the landform.

Students can use the stream table to investigate erosion in many ways. Here are questions to get them started: What happens to the sand after it is washed away from the high land? Does sand which is washed into a lake differ from that washed out onto other land? Does putting a few leaves on the upland area affect erosion? How about just one leaf? *(Careful: You may have to see this one to believe it.)* Is there a natural landform similar to what is produced in the stream table by one leaf? Can you make a model of a meandering stream? How about badlands?

▶ FOLDING ROCKS

Rock layers visible in mountains have often been folded and rippled. How this takes place can be demonstrated by stacking several sheets of colored construction paper on a table top and then pushing them together as shown in Figure 15-7. Point out to the class that rocks do not fold nearly so easily as paper but that strong pressure over many years accomplishes similar results.

To demonstrate the same process with a substance that is more resistant to folding, use several layers of modeling clay in different colors. Push the layers together with a block of wood at each end. Clay takes longer to fold than paper, and more force must be exerted.

As a research assignment, students can identify examples of folded mountains in the United States. How long do geologists estimate that it took to form the Rocky Mountains? Can examples of folded rocks be found near your school?

▶ EARTHQUAKES

Do earthquakes occur more frequently in some areas than others? Post a map of the world; and each time an earthquake is reported, have students mark the location of the earthquake on the map. This activity should be continued for several months, or you may even wish to make the record continuous for several years.

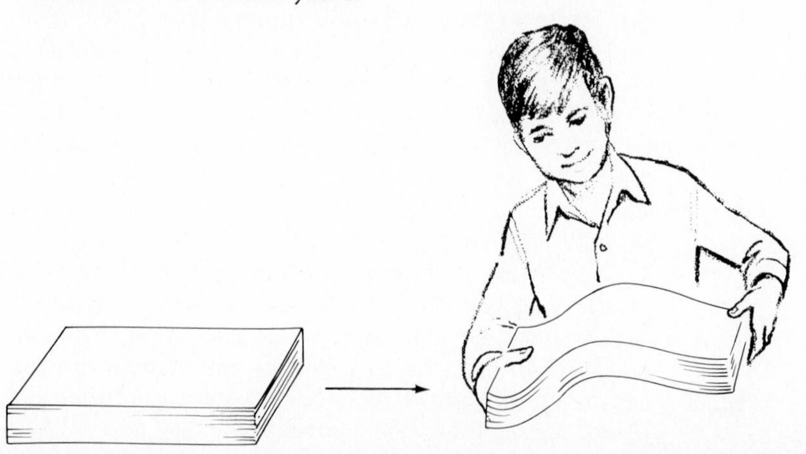

Figure 15-7. Demonstration of rock folding.

After the students become aware that the most active earthquake locations in the contiguous United States are along or near the Pacific Ocean, ask them why they think this is true. Encourage them to investigate and report on this topic. *(Give the class a hint: Part of the reason might be found by investigating the topic "plate tectonics.")*

To help students understand the relative severity of reported earthquakes, have them make a chart showing the kinds of damage which typically occur at levels 1 to 10 on the Richter scale. They can start the chart simply by using newspaper reports about destruction caused by particular earthquakes. Later a geology textbook can be used to complete the chart.

Here are some other questions which can be considered in studying earthquakes: Are earthquakes likely to occur in our area? What major earthquakes have affected populated areas? What should you do during an earthquake? How are scientists trying to predict when and where earthquakes will occur? Can earthquakes be controlled?

➤ VOLCANOES

Volcanoes, like earthquakes, occur more commonly in some parts of the world than in others—and in fact, the two phenomena often occur in the same areas.

Since the eruption of Mount Saint Helens in 1980, a great amount of information has become available to the general public on volcanic activity. Many magazines and journals, including *National Geographic, Science 80,* and *Natural History,* have published excellent articles on Mount Saint Helens and on volcanoes in general.

Appoint a "volcano experts committee" of students to prepare a world map recording volcanic activity. Where in the United States are there active or dormant volcanoes? Have there ever been active volcanoes in your area?

Also ask the class to investigate some of these questions: How do volcanoes differ from one another? What are the benefits of volcanic activity? After an active volcano becomes dormant, why do people move back into the area (e.g., consider Vesuvius)? Where in the United States are there areas of volcanic activity?

➤ NAME THAT ENVIRONMENT

Identify for students a small set of animals which all come from one environment—for example, a fish, a clam, and a snail—and ask them to describe the environment. If the students are not familiar with some of the animals, describe them briefly.

After you present two or three such sets, ask the students to work in groups, using reference materials, to form three sets per group—one set of modern animals and two sets of prehistoric animals. After all the groups have developed three sets, have each group present one set at a time to the whole class, and ask the class to describe the animals' environment.

Ask the students to explain why animals live in their particular environments. *(Animals' environments satisfy their life requirements.)*

What would happen if an environment changed? *(Possibilities include the following: First, the animals might exist much as they previously did, if their requirements were still being satisfied. Second, the animals might also change, in ways that would enable them to live in the new environment. Third, the animals might move to a different location. Fourth, the animals might all die.)*

PRESERVATION IN ICE

The woolly mammoth is one animal which has been preserved by being frozen into a mass of ice, and other animals and plants have been preserved similarly. If some students are interested in learning more about the mammoth's remains, have them do the necessary research and report their findings to the class.

To demonstrate the effectiveness of freezing, collect a few *dead* insects, put them into paper or styrofoam cups of water, and place them in a freezer. After the water is completely frozen, take the ice blocks out of the cups and note the condition of the insects. Place the insects, frozen in the ice blocks, into an airtight container and return them to the freezer. After several weeks or months, again check the condition of the "fossils." Near the end of the school year, let the ice blocks melt, and examine the insects. Were they well preserved? Ask the students to explain how prehistoric animal and plant remains could have been preserved for thousands and millions of years. How did they get into the ice in the first place? How have they remained there so long?

PLASTER FOSSILS

Students can make "fossils" by placing parts from living things in plaster. Mix a batch of plaster of paris and place it in a throwaway aluminum pan. When it has become somewhat firm, have students make imprints by pressing leaves and animals' footprints into the plaster. The could also let a small animal such as a pet mouse, gerbil, hamster, or a rat walk across the plaster.* After the plaster hardens, the students can inspect their fossils. Ask them how their fossils are similar to real leaf and footprint fossils. How are they different?

Another way the class can use plaster to learn about fossils is to embed chicken bones in blocks of plaster. Either you can prepare these "fossiliferous rocks" ahead of time or you can let the students help you prepare them. After the plaster hardens, tell the children that their job is to recover the fossils from the rocks. Encourage the class to plan ways of removing the fossils without destroying them. The children will discover that removing their fossils from plaster (a very soft "rock") not only is difficult but also leads to many broken chicken bones. They can do what geologists do under similar circumstances—piece the bones back together.

If any of the students are especially interested in fossils, arrange for them to talk to a geologist about the methods used in removing fossils from rocks. The students should have ready beforehand a short list of

**Note:* Be sure to clean the animal's feet carefully and immediately.

questions to ask the geologist. If there are no geologists in your immediate area, let students make a telephone call to a geologist at a museum or university. If possible, tape-record the interview.

➤ PUTTING A SKELETON TOGETHER

To help students understand the job of putting a skeleton together after the pieces have been collected, prepare a chicken skeleton. Buy from a butcher a cleaned chicken, complete with head and feet if possible. Boil the chicken, remove the meat from the bones, save all the bones, and spread them out to dry.

Every scientist who reconstructs a skeleton would like to have all or most of the bones. You will have a far greater percentage of the bones of your chicken than scientists usually have of a prehistoric animal.

Working individually or in small groups, the students should treat the bones as if they were a puzzle and put them together as best as they can. In a similar way, a scientist reconstructs the skeleton of an ancient animal. One difference is that the scientist would have carefully recorded the location of each bone relative to the others in the rock from which they were taken. Another difference is that your students are familiar with the appearance of live chickens—at least from pictures, if not from having seem them in a barnyard or zoo—whereas scientists have never seen live specimens of prehistoric animals whose bones they work with.

Ask the students: "How did you know that the leg bones belonged where you placed them? What about the order of the vertebra? What problems would you have had if some bones had been missing? Were there some bones that really did not seem to belong anywhere? How did you finally decide where to place those bones?"

Encourage the students to be on the lookout for skeletons of other animals that could be placed in the classroom collection. Wouldn't it be nice to have a collection of several skeletons—maybe even including a cow?

➤ HAS YOUR TOWN BEEN COVERED BY AN OCEAN?

Obtain the free publication "Our Changing Continent" from the U.S. Geological Survey (see "Free and Inexpensive Materials"). Using this publication as a reference and working on a set of outline maps of the United States, students can outline the extent of the oceans at various stages in geologic history. Was your locality covered by water? How many times? What about New Orleans? Chicago? Miami? New York City?

If your city was covered, can fossils of seashells be found near your location? If not, why not? *(They could have been covered over later by other materials—e.g., volcanic deposits.)*

Encourage interested students to make three-dimensional models to represent several different times in the geologic past. Suitable materials for such models include clay, layers of cardboard (cut out), layers of styrofoam, plaster, and balsa wood.

At one time glaciers covered much more of the earth than they presently do.

➢ THE GREAT ICE AGE

Have students draw on a map of North America the greatest extent to which ice masses covered the continent during the Pleistocene epoch. They can get the information they will need from historical geology books or from encyclopedias. Also, have them collect pictures or make drawings of animals and plants which lived during that time. They should find out what evidence geologists use to conclude which areas were covered with the great ice masses.

Did the ice cover your area? If not, was your area in any way different during the ice age from what it is now? Ask the students how they think it might have been different—and then have them check the correctness of their guesses.

Are we still in the ice age? Or is another one coming soon? Scientists do not unanimously agree on the answers to those questions. Have interested students investigate possible answers and reasons for the disagreement.

➢ POEMS ABOUT ANCIENT ANIMALS

Have students write poems about ancient animals, plants, and environments. Encourage them to make the poems describe the animal, plant, or environment. Here is an example of a simple poem:

Brontosaurus
A dinosaur it was
Who lived long before us.

Its size was so great
80 tons was its weight.

Each child in the class can compose one or more verses to add to a poem like this. If the students have studied or are studying special forms of poetry in reading or language arts classes, they can use those forms in writing poems about ancient phenomena. The limerick, for instance, is a form which fascinates children and which they can often use to good effect. Encourage students to make drawings or use magazine pictures to illustrate their poems.

➤ RECONSTRUCT AN ENVIRONMENT FROM A BONE?

To help students understand that scientists can know a great deal about animals and their environments without having complete skeletons, bring a few bones to class—say, leg bones from a cow, pig, sheep, chicken, and turkey, and a few teeth (or models of teeth from a cow, horse, dog, and cat). A butcher can help you to get some of the bones, and a science supply house can sell you others (see the Appendix).

Hold the cow's leg bone up and ask the students if they think it belonged to a mouse. The point is, of course, that an animal's size can be estimated from a single bone. A scientist can recognize certain kinds of bones as usually being leg bones, others as usually being toe bones, etc. Display all the leg bones and ask the students to make a list of animals which they think each bone might have come from.

Supervise the students closely while they use a hacksaw to cut each bone so that they can see its internal structure. They will note that some are more solid and heavy than others. What might that indicate? *(Large, heavy animals have solid, heavy bones.)* The chicken and turkey bones are relatively light. Why might that be true? *(Light bones are beneficial for animals that fly.)*

The students should notice some distinct differences when they examine the teeth of the different animals. In general, horse and cow teeth are large and have flat surfaces, whereas many dog and cat teeth are sharp and jagged. Ask the students what the shapes might indicate about each animal's food. *(Flat teeth may indicate extensive use in grinding food such as plants. Sharp, jagged teeth may indicate cutting and tearing at food such as meat.)* Have students examine their own teeth and note that their teeth have similar shapes to those of some of the animals and are also used similarly.

No single piece of evidence can tell the whole story about an animal and its environment, but a few bones can indicate a great deal. If we conclude that an animal eats chiefly plants and is very large, what might its environment look like? Could it be a desert? What kind of animal might have very small bones in relation to its body size? *(The animal may have lived in water, for water animals do not require a strong skeleton to support their body weight.)* If another animal had chiefly large grinding teeth, what might be indicated? *(The animal may have been a plant eater.)* Have students generate additional scenarios and ask their classmates to hypothesize the animal types and environments represented.

▶ COLLECTING FOSSILS

Fossils make fascinating collections. Students can have fun collecting them and can also learn a great deal.

In many states, the department of highways or the agency concerned with mineral resources (usually the state geologic survey) puts out publications for amateur fossil collectors. Other sources of information on local fossil-collecting sites include rock shops, gem and mineral societies, and geologists.

After the children visit a site and collect some fossils, they can identify the organisms with the help of reference books. Encourage the students to learn how the various organisms interacted with one another and what the environment was like at the time they lived. Have individual students or small groups make presentations to the class after they assemble sets of fossils.

Try to arrange for a display of the collections near the end of the school year.

For hints on how to collect and display the fossils, refer to *The Natural History Guide.** A visit to a local museum will show you and the children effective ways of displaying fossils.

▶ WATCH THE APPALACHIANS

Tell the class about James Hutton's theory that the geologic processes and products of modern times are the key to an understanding of the geologic processes and products of the past. (Do some library research yourself, if necessary, in order to make this an interesting and lively presentation.) Then have your students imagine that they have traveled back in time to when the Appalachian Mountains were very tall and rugged. Have them watch the Appalachians in speeded-up time. The mountains are wearing down and the rock particles are being washed out of the mountains. At the feet of the mountains, sediments are piling up deeper and deeper, and the height of the mountains is being reduced. The sediments are becoming rock—the same rock which can be found today.

While the above scenario is brief (you may wish to expand it or have students expand it), it can allow you to raise several questions to help students understand how geologists use knowledge about many present-day processes to explain past events. Here are a few questions to get you started: What kinds of processes caused the rock to break down? How were some of the rock fragments moved from where they broke off the large rock mass to their present sites? What present-day places in our country are similar to the way the Appalachians used to be? (Try, if possible, to include questions which relate this topic to your own area.)

*H. Charles Laun, *The Natural History Guide,* Alsace Books and Films, Alton, Ill., 1967.

Children's literature

Read *What Does a Geologist Do?** or *The Riddle of the Stegosaurs*† to your students or have them read one of these books, and then discuss with them the concepts presented.

Find other books on geology in the school library, and identify them for the students or place them in an interest center.

Activity cards

An example of an activity card is shown in Figure 15-8, page 354. Additional activity cards could be made for topics such as making "rocks" or "fossils"; collecting rocks; studying gemstones; making models of volcanoes, rock layers, rock folds, faults, mountain glaciers, features of volcanoes, rivers, etc.; studying soil; gathering information on and examples of fossil fuels; collecting fossils or rocks which give evidence of ancient environments; tracing the history of horses (or camels or cockroaches); visiting earth-history displays at local museums; making diagrams (or other representations) of ancient environments; and investigating careers in geology.

Bulletin boards

GEOLOGIC TIME

Post the names of major geologic time divisions (Precambrian times, Cenozoic Era, Mesozoic Era, Paleozoic Era) on a bulletin board (see Figure 15-9, page 354), with each division name on a different-colored background. In three pockets near the bulletin board place cards, with or without pictures or drawings, that give the following information:‡

1. The names of the geologic periods in the Paleozoic, Mesozoic, and Cenozoic eras and the names of the epochs of the Cenozoic Era.
2. Information on plants through geologic time. Examples are: time of first-known plants, time of first flowering plants, time of earliest known algae.
3. Information on animals through geologic time. Examples are: earliest fish, age of dinosaurs, first known mammals.

WHERE DO CITIES DEVELOP?

On a bulletin board post an outline map of the United States (including Alaska and Hawaii) which indicates only large bodies of water and major rivers. Have the students place thumbtacks where large cities are located.

*R. V. Fodor, *What Does a Geologist Do?* Dodd, Mead, New York, 1977.

†D. Ipsen, *The Riddle of the Stegosaurs,* Addison-Wesley, Reading, Mass., 1969.

‡This information may be easily found in any historical geology book and in many encyclopedias.

· DURING OR JUST AFTER A RAIN, OBSERVE WHAT HAS HAPPENED WHERE WATER HAS RUN OFF THE LAND IN SMALL STREAMS.

· FIND A PLACE WHERE SOIL WAS CARRIED AWAY. CAN YOU FIND PLACES WHERE IT WAS DEPOSITED?

IF YOU EVER FIND OTHER EXAMPLES OF EROSION AND SOIL DEPOSITION LIKE THE ONES YOU JUST OBSERVED, HOW WILL YOU GUESS THEY WERE CREATED?

ARE BIG RIVERS LIKE YOUR LITTLE STREAM?

Figure 15-8. Activity card: The changing earth.

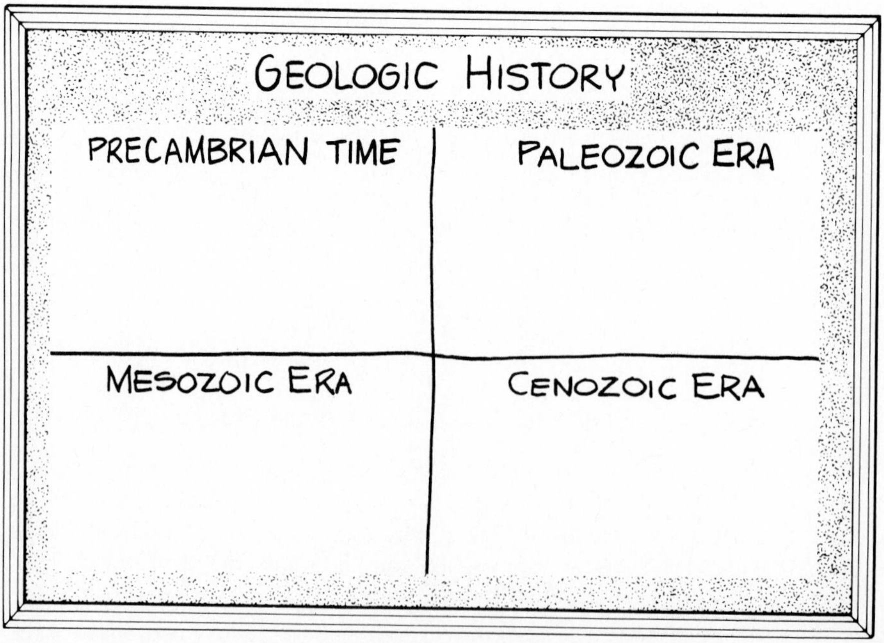

GEOLOGIC HISTORY

PRECAMBRIAN TIME	PALEOZOIC ERA
MESOZOIC ERA	CENOZOIC ERA

Figure 15-9. Bulletin board: Geologic history.

What can be noticed? *(Almost all the large cities are located on or near large bodies of water or major rivers.)* People settled near water chiefly for reasons of transportation. Thus, geological features certainly influenced peoples' lives in the past.

Do geological features influence people today? Where do people like to go for vacation trips? How are homes built in areas which have frequent earthquakes?

Field trips

Any place where rocks—and especially a variety of them—can be found is an excellent place for a field trip. Students can not only examine and collect rocks but also search for evidence of how the small pieces which they collect were weathered from the larger masses of rock and how they were moved to the locations where they were found.

Other field trips may be made to sites where erosion is occurring, museums, and rock shops.

RESOURCES

Sources of assistance

FREE AND INEXPENSIVE MATERIALS

Here are some good sources of free and inexpensive materials for use in teaching geology:

1. *American Petroleum Institute.** Information, posters, and teaching lessons and units on oil resources.
2. *U.S. Department of the Interior, Geological Survey.* A teacher packet which includes many pamphlets ("The Ice Age," "Volcanoes," "Our Changing Continent," "Gold") and additional teaching aids.
3. *National Coal Association.* Charts on coal usage and production, a map of United States coal areas, and several pamphlets.
4. *Rock of Ages Corporation.* A kit on granite, its composition, and quarrying processes.
5. *U.S. Department of the Interior, Bureau of Mines.* Two charts, "The Coal Products Tree" and "The Petroleum Tree."
6. *Shell Oil Company.* A booklet, "Let's Collect Rocks and Shells."
7. *Eros Data Center.* Information on how to order aerial photographs of most areas of the United States.
8. *National parks.* Booklets on the geology of the parks. Consult travel guides for addresses of specific parks.

*The addresses of specific agencies, organizations, and businesses are listed in the Appendix.

LOCAL RESOURCES
Local resources which can help in teaching geology include:

1. Museums and their staff persons.
2. Geologists, including those at any nearby college or universities.
3. Rock shops.
4. County extension agents can be especially helpful on soil topics.
5. Rock and mineral societies.

ADDITIONAL RESOURCES
1. Magazines such as *World, Lapidary Journal, National Geographic, Natural History, Science 80,* and *Arizona Highways.*
2. State geological survey (in some states the name is slightly different).
3. Many oil and coal companies produce teaching materials and have geologists as employees. They often have speakers' bureaus. (Let them know you are looking for information on earth history, *not* on their view of the energy controversy.)

Strengthening your background

1. Read at least the introductory chapters of a historical geology and a physical geology textbook.
2. Collect and identify at least five different local rocks.
3. Go outside on a rainy day and watch soil being eroded in a small gully.
4. Make up a funny riddle or cartoon about a rock breaking up (weathering).
5. Obtain or make a chart which lists the sequence of the eras, periods, and epochs of earth history and identifies a *few* major events, plants, and animals which lived in each time division.
6. Visit a museum which has an earth-history display.
7. Take a trip to the Grand Canyon, a well-known mountain range, or another well-known geologic feature. Learn about the history of that site, and take as many slides as you can.
8. Determine what kinds of rocks are formed when deep seas cover an area. How about shallow seas?
9. Find out about several ways in which fossils can be formed.
10. Collect samples of at least six fossils.
11. Study the geologic history of the area in which you teach. Ask a geologist to recommend books and other pertinent materials which you can read.

Getting ready ahead of time

1. Collect magazine articles on earthquakes and how they have affected people.
2. Make five activity cards on the topic "the changing earth."

3. Collect sedimentary, metamorphic, and igneous rock samples. (It will be easier to get them now, from a friendly university geologist than later, when you may have trouble locating a geologist.)
4. Identify the types of landforms present in the state in which you are teaching, and start collecting slides that show the landforms.
5. Buy a field guide to rocks and minerals, and use it to identify any rocks and minerals you see.
6. Obtain an aerial photograph of a large section of your state which includes the city or town in which you will be teaching.
7. Have your school order a set of sample fossils.
8. Obtain a set of study prints of prehistoric animals.
9. Make plaster copies of fossils which are in your classroom for study, and then write an instruction sheet for your students on how to make plaster casts.
10. Collect many pictures of ancient plants, animals, and environments from magazines and any other available source.
11. Prepare two shoebox science kits, one on collecting fossils and the other on collecting rock specimens which give information about ancient environments.
12. Make a list of books on earth history available in the school library.

Building your own library

Beiser, Arthur: *The Earth,* Young Readers Edition, Life Nature Library, Time-Life, New York, 1977.

Elementary Science Study (ESS): *Mapping,* Webster, McGraw-Hill, Manchester, Mo., 1971.

———: *Rocks and Charts,* Webster, McGraw-Hill, Manchester, Mo., 1974.

———: *Sand,* Webster, McGraw-Hill, Manchester, Mo., 1970.

———: *Stream Tables,* Webster, McGraw-Hill, Manchester, Mo., 1971.

Facklam, Margery: *Frozen Snakes and Dinosaur Bones,* Harcourt, Brace, and World, New York, 1976.

Foster, Robert J.: *General Geology,* 3d ed., Merrill, Columbus, Ohio, 1978.

Leopold, A. Starker: *The Desert,* Young Readers Edition, Life Nature Library, Time-Life, New York, 1977.

Levin, Harold L.: *The Earth through Time,* Saunders, Philadelphia, 1978.

Matthews, William H.: *The Story of the Earth,* Harvey House, New York, 1968.

Pringle, Laurence: *Dinosaurs and Their World,* Harcourt, Brace and World, New York, 1968.

Tarbuck, E. J., and F. K. Lutgens: *Earth Science,* 2d ed., Merrill, Columbus, Ohio, 1962.

White, Anne Terry: *All about Mountains and Mountaineering,* Random House, New York, 1962.

Zim, Herbert S., and Paul Shaffer: *Golden Nature Guide to Rocks and Minerals,* Golden Press, New York, 1957.

Adler, Irving, and Ruth Adler: *Rivers,* John Day, New York, 1961.
Asimov, Isaac: *How Did We Find Out about Earthquakes?* Walker, New York, 1978.
Atwood, Ann: *The Wild Young Desert,* Scribner, New York, 1970.
Bauer, Ernst: *Wonders of the Earth,* F. Watts, New York, 1973.
Baylor, Byrd: *If You Are a Hunter of Fossils,* Scribner, New York, 1980.
Bradenberg, Aliki: *Fossils Tell of Long Ago,* Thomas Y. Crowell, New York, 1972.
Branley, Franklyn: *Shakes, Quakes, and Shifts: Earth Tectonics,* Thomas Y. Crowell, New York, 1974.
Carrick, Carol: *The Crocodiles Still Wait,* Houghton, Mifflin, Boston, and Clarion, 1980.
Cassanova, Richard, and Ronald P. Ratkevich: *Illustrated Guide to Fossil Collecting,* 3d ed., Naturegraph, Healdsburg, Calif., 1981.
Clayton, K.: *The Crust of the Earth: The Story of Geology,* Natural History Press, Doubleday, Garden City, N.Y., n.d.
Cohen, Daniel: *The Age of Giant Mammals,* Dodd, Mead, New York, 1969.
Colbert, Edwin H.: *Millions of Years Ago: Prehistoric Life in North America,* Thomas Y. Crowell, New York, 1958.
Cole, Joanna: *Saber-Toothed Tiger and Other Ice Age Mammals,* Morrow, New York, 1977.
Cook, Brian: *Gas,* F. Watts, New York, 1981.
Creative Editors: *Geology of the Earth,* Creative Education, Minneapolis, Minn., 1971.
Dabcovitch, Lydia: *Follow the River,* Unicorn Press, London, and Dutton, New York, 1980.
Eiting, Mary, and Ann Goodman: *Dinosaur Mysteries,* Platt, New York, and Grosset and Dunlap, New York, 1980.
Farb, Peter: *The Story of Life: Plants and Animals through the Ages,* Harvey House, New York, 1962.
Fenton, C. L., and M. A. Fenton: *Rocks and Their Stories,* Doubleday, Garden City, N.Y., 1951.
Forbes, Duncan: *Life before Man: The Story of Fossils,* 2d ed., Transatlantic, Central Islip, N.Y., 1974.
Fodor, R. V.: *Earth Afire! Volcanoes and Their Activity,* Morrow, New York, 1981.
———: *What Does a Geologist Do?* Dodd, Mead, New York, 1977.
Gallant, R. A.: *Exploring under the Earth,* Doubleday, Garden City, N.Y., 1960.
Gallob, Edward: *City Rocks, City Blocks and the Moon,* Scribner, New York, 1973.
Gans, Roma: *The Wonders of Stones,* Thomas Y. Crowell, New York, 1963.
Gilbert, John: *Dinosaurs Discovered,* Larousse, Paris, 1981.
Goetz, Delia: *Mountains,* Morrow, New York, 1962.

———: *Valleys,* Morrow, New York, 1976.
Goldin, Augusta: *Geothermal Energy: A Hot Prospect,* Harcourt Brace Jovanovich, New York, 1981.
Goldreich, Gloria, and Esther Goldreich: *What Can She Be? A Geologist,* Lothrop, New York, 1976.
Greene, Carla: *After the Dinosaurs,* Bobbs-Merrill, Indianapolis, 1968.
Hamilton, E.: *The First Book of Caves,* F. Watts, New York, 1956.
Heady, Eleanor B.: *The Soil that Feeds Us,* Parents Magazine, New York, 1972.
Ipsen, D.: *The Riddle of the Stegosaurs,* Addison-Wesley, Reading, Mass., 1969.
Kerrod, Robin: *Rocks and Minerals,* Warwick and York, Baltimore, and F. Watts, New York, 1978.
Klaits, Barrie: *When You find a Rock: A Field Guide,* Macmillan, New York, 1976.
Knight, D.C.: *Let's Find Out about Earth,* F. Watts, New York, 1975.
Land, Barbara: *The New Explorers: Women in Antarctica,* Dodd, Mead, New York, 1981.
Lauber, Patricia: *All about the Ice Age,* Random House, New York, 1959.
———: *Junior Science Book of Icebergs and Glaciers,* Garrard, Champaign, Ill., 1961.
Leutscher, Alfred: *Dinosaurs and Other Prehistoric Animals,* Grosset and Dunlap, New York, 1975.
Markus, Rebecca: *First Book of Glaciers,* F. Watts, New York, 1962.
Matthews, W. H.: *Exploring the World of Fossils,* Children's Press, Chicago, Ill., 1964.
May, Julian: *Land Is Disappearing,* Creative Education, Minneapolis, Minn., 1971.
———: *They Turned to Stone (Prehistoric Animals),* Scholastic Book Services, New York, 1965.
Maynard, Christopher: *Prehistoric Life,* F. Watts, New York, 1971.
McGowan, Tom: *Album of Prehistoric Animals,* Rand McNally, Chicago, 1974.
Miklowitz, Gloria D.: *Earthquake,* Messner, New York, 1977.
Millard, Reed: *Careers in the Earth Sciences,* Messner, New York, 1975.
Morris, Dean: *Dinosaurs and Other First Animals,* Raintree, Milwaukee, Wis., 1977.
Navarra, John Gabriel: *Earthquake!* Doubleday, Garden City, N.Y., 1980.
Nixon, Hershell H., and Joan Lowery Nixon: *Earthquakes: Nature in Motion,* Dodd, Mead, New York, 1980.
——— and ———: *Glaciers: Nature's Frozen Rivers,* Dodd, Mead, New York, 1980.
——— and ———: *Volcanoes: Nature's Fireworks,* Dodd, Mead, New York, 1978.
Olney, Ross R.: *Offshore! Oil and Gas Platforms in the Ocean,* Dutton, New York, 1981.
Parish, Peggy: *Dinosaur Time,* Harper and Row, New York, 1974.

Penzler, Otto: *The Grand Canyon: Journey through Time,* new ed., Troll Associates, Mahwah, N.J., 1976.

Poole, Lynn, and Gray Poole: *Deep in Caves and Caverns,* Dodd, Mead, New York, 1974.

Poynter, Margaret: *Volcanoes: The Fiery Mountains,* Messner, New York, 1980.

Reuter, Margaret: *Earthquakes: Our Restless Planet,* Raintree, Milwaukee, Wis., 1977.

Ruchlis, Hyman: *Your Changing Earth,* Harvey House, New York, 1963.

Sattler, Helen Roney: *Dinosaurs of North America,* Lothrop, New York, 1981.

Schwartz, Julius: *Earthwatch: Space-Time Investigations with a Globe,* McGraw-Hill, New York, 1977.

Selsam, Millicent: *Is This a Baby Dinosaur?* Harper and Row, New York, 1972.

Shuttlesworth, Dorothy: *Story of Rocks,* rev. ed., Doubleday, Garden City, N.Y., 1966.

Silverberg, Robert: *Mammoths, Mastodons, and Man,* McGraw-Hill, New York, 1970.

Simon, Seymour: *Beneath Your Feet,* Walker, New York, 1977.

———: *Danger from Below: Earthquakes: Past, Present and Future,* Four Winds, New York, 1979.

———: *The Rock Hound's Book,* Viking, New York, 1973.

Stephans, William, M.: *Islands,* Holiday, New York, 1974.

CHAPTER 16

OCEANS AND THEIR RESOURCES

IMPORTANT CONCEPTS

Earth is predominantly a water planet—water covers more than 70 percent of the surface.

Water is a basic requirement of all living things.

Aquatic environments, like land environments, include a variety of components.

Plants, especially algae, are at the base of the food chain in aquatic environments.

The topography of ocean floors, like that of exposed land masses, is varied.

Physical characteristics of oceans—temperature, currents, tides—have a great influence upon adjacent land.

More than half the population of the United States lives within 1 hour of an ocean or the Great Lakes.

Resources from the marine environment have been and are important to humankind.

As land resources become less adequate, resources from the marine environment will become more and more important.

Aquatic environments are the least explored environments on earth.

INTRODUCTION: STUDYING OCEANS AND THEIR RESOURCES

Someday the aquatic environment, both freshwater and saltwater, will probably be studied as a part of almost all science curricula. Animals, plants, weather systems, and matter and energy relationships are just a few phenomena in the aquatic environment. At present, however, aquatic examples are seldom used in the study of living things—that is, land-based food chains are used as examples more often than water-based ones. When aquatic environments are included in an elementary science textbook, they are usually placed at or near the end of the book and seldom get the coverage they deserve.

Since the surface of earth is more than 70 percent water, it is necessary for everyone to understand more about aquatic environments. We must be educated, and we must educate ourselves, to make wise decisions about the use of aquatic environments. Humankind has always been fascinated by the mystique of the sea, as evidenced by the wealth of folklore about the sea—tales of sea monsters, pirates, storms, hardship, beauty, and wealth. A longing for adventure has always lured humankind to the large bodies of water, and continues to do so.

We still know relatively little about large bodies of water, and the challenge is strong and alluring. Perhaps some of your students will grow up to accept the challenge and answer some as yet unanswered questions.

Even if you do not plan to teach a special unit on the aquatic environment in a given year, this chapter can easily be incorporated into other units. A wealth of children's literature has been written on the aquatic environment; and many students are eager to read it, either as class assignments or, with the teacher's encouragement, on their own.

OCEANS AND THEIR RESOURCES FOR PRIMARY GRADES

Activities

➤ STUDY A GLOBE

Ask your class members to point out the most outstanding features of earth on a large globe. Someone will probably mention great amounts of water. A photograph of the earth from any space vehicle will show that the most prominent features are blue oceans and white clouds.

Ask the students whether they know more about water environments or land environments. (*Almost certainly, they know more about land environments—even if they live on a coast.*) Why is this? (*Because the land environment is where we live.*)

Read the book *Seashore Story** to your class. Ask the students to name

*Taro Yashima, *Seashore Story*, Viking, New York, 1967.

some real things which can be found in the water environment. Tell the students that to learn about the water environments we must do more studying in and around them. Emphasize that learning more is really very important, since water covers such a large part of our planet (over 70 percent).

STUFFING SEA ANIMALS

The goal of this activity is to make students more aware of the great diversity of living things in the sea. Collect pictures of a wide variety of sea animals from magazines and other sources, and place the pictures in a convenient location in the classroom. Have the class study the pictures, along with library books and other books which contain similar pictures.

After the students have become familiar with the pictures, have them draw outline pictures of the animals on sturdy 8½- by 11-inch (or larger) paper. Encourage them to draw a wide variety of animals, including sea cucumbers, sea anemones, clams, snails, several kinds of fish, turtles, and snakes. They can paint or color the drawings if they want to. Then have them cut out two copies of each animal and staple them together around the edges, leaving an opening on one side. They can use crumpled scraps of paper and any other handy fillers (paper towels, for instance) to stuff the animals, and then staple shut the remaining side. Hang the stuffed animals around the room and tell the class some interesting facts about each animal.

Some students might enjoy making stuffed sea animals out of cloth to give to their families for use as pillows. (Cloth pillows will, of course, have to be sewn or glued rather than stapled, and the stuffing should be scraps of cloth or other soft material.)

SAND PAINTING

Making sand paintings will also help to increase students' awareness of the diversity of life in aquatic environments. The materials needed for this activity are pencils, fine sand (preferably white or light-colored), colored chalk, sandpaper, white glue, small paintbrushes, and cardboard. You can use poster board or any scrap cardboard, cut from clean boxes or taken from shirt packages. A good size cardboard to begin with is approximately 23 by 30 centimeters (8½ by 11 inches).

Have the students use pencils to draw simple outline pictures of sea animals or plants on their cardboards. Then they should rub the colored chalk on a piece of sandpaper to make it into a powder and mix it with the sand to obtain the desired color. (Repeat for each color to be used in a given picture.) Dilute the glue with an equal amount of water, and use a paintbrush to spread a thin layer of glue on areas of the cardboard which are to be the same color. Sprinkle the appropriate color sand over the glue and allow it to dry for a few minutes. Gently shake off the excess sand, and repeat the process until all the colors have been applied.

Display the sand paintings in your classroom, and have the students learn something about their animals or plants and give reports to the

Sand paintings not only use one element found in several water environments—sand—but also can depict phenomena that take place in such environments.

class. Later they can give the pictures to their parents for use as wall decorations.

➤ FOOD FROM THE SEA

In the United States, food from the sea is not as common a part of our diet as it is in almost all other countries in the world. This activity will help your students become acquainted with seafoods which they may never have heard of before. It will also help them begin to think of the sea and other bodies of water as sources of materials which are important to humankind in addition to being environments for living things.

Buy a couple of varieties of crackers which contain sea plants (e.g., "Sea Crunchies") from a health food store or a large food market. Look in a seafood cookbook for a seafood dish which your students may not have eaten and which they would probably like the taste of—perhaps shark steaks, clams, or scallops. Purchase a small amount of the seafood, just enough to give each child a sample. If you send information about the seafood-tasting day to your students' parents ahead of time, some of them may want to help you prepare the food or even add a few extra tidbits.

Make a special occasion of the seafood-tasting day. Have the children help you decorate the room appropriately. While you prepare the food, play the record album *Music of the Sea** or some other sea music.

Have students do research at the school library and at a supermarket or fish market to find out about other types of marine and aquatic organisms that can be used for food.

➤ LIFE IN A LOCAL POND

Collecting pond water and examining it under a magnifying glass or microscope is an activity frequently performed in science classes, and a useful one—but much more is possible.

Set up a large (about 20-liter or 5-gallon) aquarium containing only water and an airstone through which a small aquarium pump is bubbling air. (If you do not have the equipment, try to borrow it for a week from a student or another teacher.) Let the water age for at least 2 days and then go with two students on a collecting trip to a nearby pond. (They may be your own students, high school students, or college students—you really should not have much trouble getting help for such an interesting trip.) Take along a plastic bucket (4 to 20 liters or 1 to 5 gallons), a collecting net (borrowed from a high school biology teacher or made out of an old Nylon stocking and a wire clothes hanger—the homemade one may not be overly efficient, but it will work), and a smaller net or a jar.

Find a shallow place in the pond with a great deal of detritus—decaying plant and animal matter—on the bottom. Dip up a few inches of pond water into the bucket. Then push the net down into the detritus; use the small net or jar to pick up the living creatures which you should see when you pull the net up and to place them in the bucket. If you don't see organisms at first, swish the net in the pond to clear the water in the net. You should find a large number of living things—mostly insects. If not, move to another location or scoop a little deeper into the muck. Also gather some detritus and plants.

Back in the schoolroom, place the collected organisms, detritus, and plants in the aquarium. The next day, your students will probably be amazed to see how many organisms came from a little pond. They will probably notice that some of the organisms are eating others—a valuable lesson in predator-prey relationships. The supply of organisms will probably dwindle during the week, so try to visit the pond again to replenish the organisms.

If some of the organisms die and you wish to preserve them in a collection, place them in alcohol. Your students may want to know what the organisms are; if you do not know, admit it and tell them that you would be happy if they could find out and tell you. You could have some identification guides available in the classroom—or better yet, let them go to a library and find out on their own. Individual research is the best way to get to know a library and become familiar with its resources.

*RCA Camden CAL 639.

Aquatic insects are much more numerous than we generally realize. Many are found in the soil beneath the water.

➤ SOUNDS OF THE SEA

Listening to the sounds of sea animals will be an interesting activity for your students; and after listening, they may be in a better position to evaluate stories about sea monsters. In preparation for the activity, ask your librarian to obtain a copy of the recording *Sounds of Sea Animals*,* which presents sea animal sounds and then identifies the organisms that make the sounds. Listen to the record by yourself, and practice until you can set the stylus down only on the parts of the record that you want to play for the class (see below).

To begin the activity, tell your students to imagine that they are sailors in the 1700s, making their first voyage (see Figure 16-1). They have just finished a hard shift at work, and now they go below to sleep for a few hours. When they lie down, however, they begin hearing noises. Darken the room by lowering the shades or turning off the lights. Tell the students to close their eyes and listen carefully. They should try to imagine what is making the noises they will hear. Then play parts of the record—some of the sounds only; at this time, do not play the parts that identify the sources of the noises.

Tell students that they can open their eyes, and turn on the lights or open the shades. Either have a class discussion or ask the students to write stories about what may have produced the sounds. Then play the parts of the record that identify the sources of the sounds.

*Folkways FX 6125.

Figure 16-1.
Sounds of the sea.

Finally, tell the students that many sea-monster stories have been created to explain the sounds and that, if they would like to, they can make up some sea-monster stories.

➤ OCEAN FOOD CHAIN AND MOBILE

To help your students understand the concept of food chains in an aquatic environment—as well as become familiar with more organisms, strengthening their concepts of predator-prey relationships and the complexity of aquatic environments—have them build mobiles to represent ocean food chains. Figure 16-2 indicates the appropriate order in the chain. The mobiles do not have to be large—when constructed, each one may be only 30 to 40 centimeters (12 to 16 inches) long.

To start the activity, give each student a copy of your sketches of the following organisms (see Figure 16-2): phytoplankton, zooplankton, insect, small fish, medium fish, large fish. The students are to cut out the organisms, color them if they wish, and tape or staple them to a piece of string in the order that they would eat each other, beginning with the phytoplankton at the bottom of the chain. While the class members are working on their chains and again after they finish, discuss with them the organisms in the chain and give specific examples of fish. Stress that water-based food chains, like land-based food chains, begin with green plants.

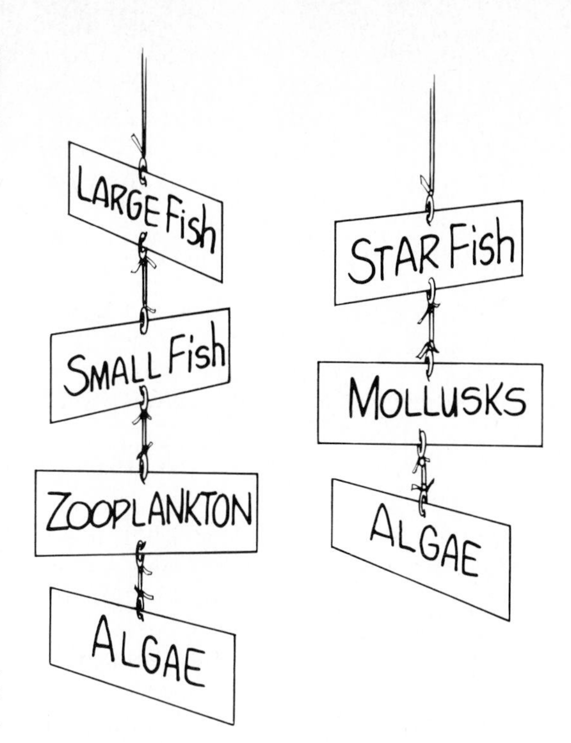

Figure 16-2. Ocean food chain for a mobile.

An excellent follow-up to this simple activity is to have students construct another mobile, with specific organisms in the same relative order given above. Have each student choose an aquatic organism and find out what it eats and what it is eaten by. If the students overlook some kinds of organisms, you can suggest them—examples are sea turtles, sea anemones, squid, coral, sponges, clams, and snails. The students should trace the food chain down to green plants and up as far as they choose. After doing their research, they can make drawings of their food chains and put together the new mobiles.

Ask the students what would happen if one organism were eliminated from the ecosystem. (*The organism that it ate might become overpopulated. And its predators would have to find other food in order not to starve—which would affect still more predators.*) What would happen if one kind of organism became overpopulated? (*That organism would consume all its food supply and then possibly starve.*) Talk about balance within an ecosystem. Ask the students what might change the balance.

➤ SEASHELL IDENTIFICATION

Elementary school students, like people of all ages, just naturally seem to be fascinated by seashells. Obtain at least two specimens each of a variety of shells, and number and identify one set. The students can identify the unmarked shells by matching them with the numbered ones. Ask the students to describe the characteristics which they used in matching the shells.

Tell the students that organisms created and lived in the shells, and that shells with their organisms in or near them may appear quite different from the shells in a collection. Suggest that each student learn something about one or more of the organisms from which the classroom shells came. Then have the students tell the class about their selected organisms.

➤ EXPLORING

Have the students close their eyes, relax, and pretend that they are explorers who are about to set out by sailboat across a large body of water. Have them think about what the journey will be like most of the time. What will be some of the dangers? What beautiful things will they see? What do they expect to find?

After the guided fantasy is over, ask the students to share their experiences with the class.

Tell your students that in many cases modern-day children know more about what explorers might find than did the real sailors who were explorers in earlier centuries. Ask your students how they think the sailors must have felt. Tell them that it is common and natural for people to fear things which they do not understand. As we learn more about things, we fear them less. Ask if they can name anything about aquatic environments which they may once have feared but which they no longer fear.

Children's literature

A great deal of children's literature on aquatic environments has been published, as a look at "Building a Classroom Library" will indicate. Both fiction and reference books are widely available. An excellent reference guide has been developed by Norma Bagnall which identifies children's literature of the sea and describes associated learning activities and learning centers.*

Read either *ABC's of the Ocean*† or *Seashore Noise Book*‡ to your students or have them read one of the books, and then discuss with them the information presented.

Find other books on aquatic environments in the school library, and identify them for the students or place them in an interest center.

Activity cards

An example of an activity card is shown in Figure 16-3. Additional activity cards could be made for such topics as foods from the water, water recreation, collecting seashells, boats and ships, and aquariums.

*Norma Bagnall, *Children's Literature—Passage to the Sea: A Guide for Teachers*, Sea Grant College Program, Texas A. & M. University, College Station, Tex., 1980. Available from publisher at minimal cost (see the Appendix).

†Isaac Asimov, *ABC's of the Ocean*, Walker, New York, 1970.

‡Margaret Wise Brown, *Seashore Noise Book*, Harper and Row, New York, 1941.

Figure 16-3. Activity card: Draw a water animal.

Bulletin boards

FOOD WEBS
Place the names and pictures of several aquatic organisms on a bulletin board (you can easily get the pictures from an inexpensive book on sea life or pond life; such books are often found in food or discount stores). Tell the students to tack a piece of yarn between each organism and another organism that eats it. After making all the possible links, the students should be able to see that the organisms in a body of water are interdependent.

BEAUTY UNDER WATER
Obtain pictures or drawings of aquatic organisms which have unusual shapes or colors or which attract attention in some other way. Place the pictures on a bulletin board, along with instructions for establishing and maintaining a simple aquarium. While the bulletin board is up, invite a representative from a local tropical fish shop to visit your class to tell and show students how to set up and care for an aquarium.

Field trips

A field trip to an aquarium or pet shop may enable students to see aquatic animals which they have not seen before. Ask the manager to tell your class about natural foods which the organisms eat.

A trip to a local pond or stream can also help students understand how complex aquatic ecosystems can be. If possible, take along someone who is knowledgeable about aquatic organisms—perhaps a high school or college student who has a special interest in aquatic life. The person you take does not have to know the names of all the organisms present—but the trip will be much more valuable to your students if someone can show them many varieties of aquatic organisms.

OCEANS AND THEIR RESOURCES FOR INTERMEDIATE GRADES

Activities

➤ WHO'S FOR DINNER?*

An excellent game which can help students become familiar with several saltwater organisms, predator-prey relationships, and the complexities of the marine environment is "Who's For Dinner?" A few of the cards for the game are shown in Figure 16-4. The rules of the game are as follows:

1. Use the deck of marine organism cards.
2. As life in the salt marshes and bays goes on and on, so does this game. 15 to 30 minutes should be adequate.
3. Two to six players are best. More may play.
4. Deal all cards out. Some students may get more than others.
5. Person to the left of dealer starts play by asking any individual for a *showdown.*
6. At the same time, each of the two players lays down one of his cards. If one card "eats" another, that player takes the "eaten" card. If neither eats the other, it is a *standoff,* and each returns his card to his hand. Play goes on to the next person. If both cards "eat" each other, it is also a *standoff.*
7. When a player's turn comes and he knows where a certain card is that he can take with one of his cards, then he can *challenge* rather than ask for a showdown.
8. In a *challenge,* the player demands a certain card and shows the card with which he can take it. (Sally, I want your GRASS card, and I'm taking it with my PLANT EATING INSECT card.) He then wins the card and is entitled to another turn. As long as he can win cards in a challenge, he is entitled to another turn. (This is not so in showdown, even if someone wins.)
9. If the challenger was wrong and the person he challenged did not have that card, he must give up his challenging card to the person wrongly challenged and his turn is over.

*Taken from *Investigating the Marine Environment and Its Resources* by Violetta Lien. Published by Sea Grant College Program, Texas A. & M. University, College Station, Tex., 1979. Used by permission.

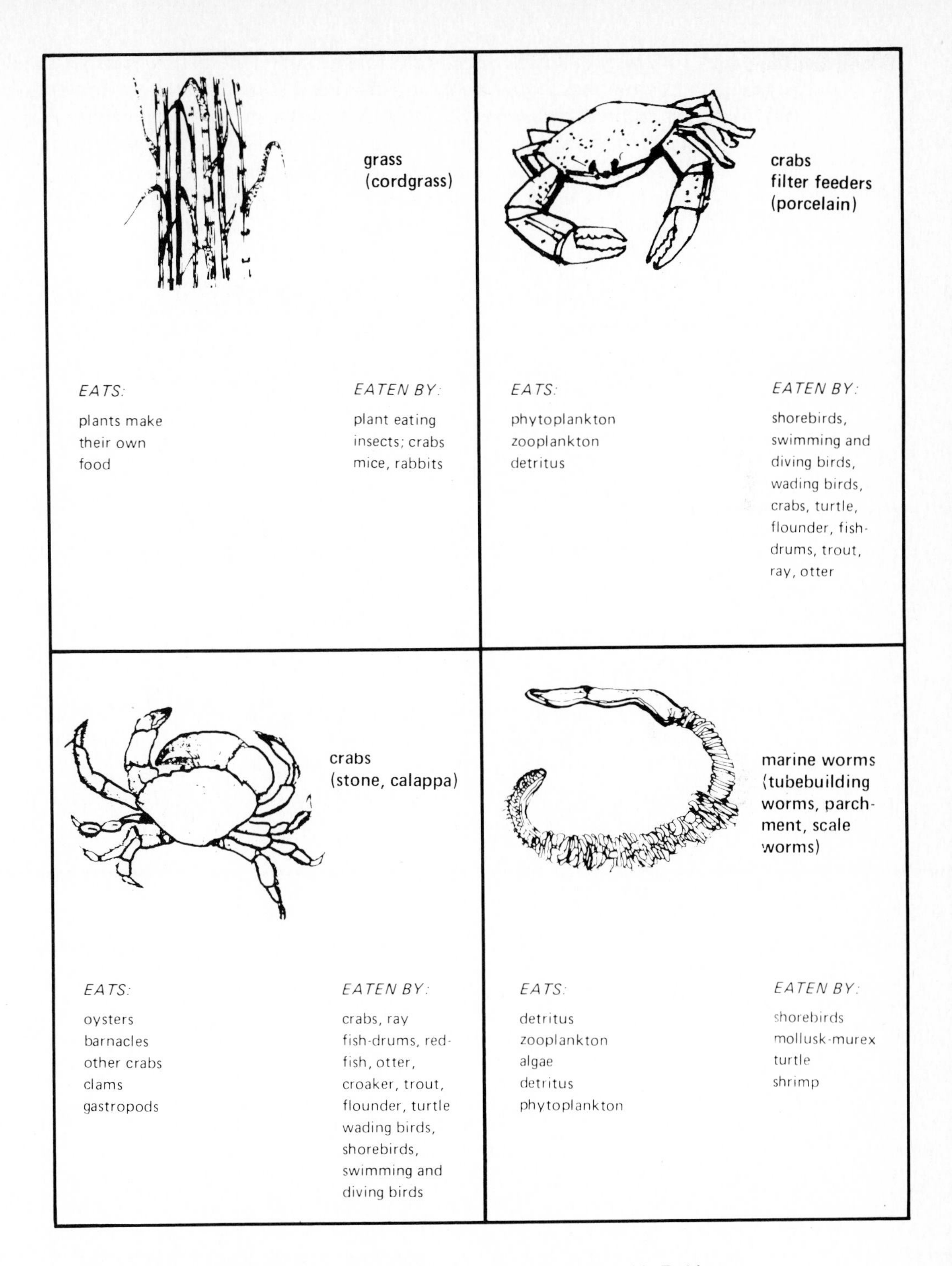

Figure 16-4. Sample "Who's for Dinner?" cards. (Source: V. F. Lien, *Investigating the Marine Environment and Its Resources,* Texas A & M University, Sea Grant Program, 1979.)

10. When a challenger is no longer sure where the cards are that he can take, he should ask someone for a showdown, and this ends his turn.
11. Sometimes, two kinds of animals can eat each other. For instance, fish eat each other. In a *showdown*, neither takes the other. But in a *challenge*, the challenger does take the other card.
12. No cards are laid down, they are all kept in the hand. No player may use the same card consecutively.
13. The two DEATH AND DECAY cards are very powerful. Their use is restricted: (*a*) DEATH AND DECAY may be used as a challenging card only once in a person's turn. (*b*) There are organisms that can take DEATH AND DECAY cards: anything that consumes decaying or decayed material. In a showdown, they provide a standoff with DEATH AND DECAY. In a challenge, the challenger wins. A person may capture only one DEATH AND DECAY card by challenging in any one turn.
14. As players become familiar with the cards and game, it may be necessary to restrict a player's time to one minute. That is, a player must challenge or ask for a showdown within one minute of his turn. Select a timer, someone who is not playing.
15. When the time set for the length of the game expires, the player with the most cards wins.

IS THE BOTTOM OF THE OCEAN FLAT?

Many people have the mistaken idea that the ocean floor is flat, except for a few interruptions by mountains which form islands. This activity will help dispel that notion. The activity is also a meaningful exercise in graphic representation of data. Make a copy of the graph grid in Figure 16-5*a* for each student, or have the students use pieces of graph paper which have been divided into the approximate scale shown.

The data illustrated in Figure 16-5*b* and listed in Table 16-1 are depth readings along a line from Galveston, Texas, to a point south of Key West, Florida. Have your students find the location of that line on a map of the United States.

Have the students make a profile of the sea floor using the information in Table 16-1. There is, of course, a vertical exaggeration of the profile, and thus the slopes are really much more gentle than they are shown on the graph. (Note that the vertical distance on the graph is less than 3 miles, while the horizontal distance is 800 nautical miles) Ask the students to describe the nature of the floor of the Gulf of Mexico along the line selected.

The Gulf of Mexico does not have as many outstanding bottom features as some other bodies of water. Try to find a map—perhaps one published by the National Geographic Society, Washington—which shows the nature of the ocean basins. Have students point out some of the more noticeable features of the ocean basins. Is there a difference in the sea floor just off the east and west coasts of the United States? How do those coasts compare with the Gulf of Mexico?

Ask your students whether or not they think the bottom of a local lake is flat and level in the middle. Does the lake have gently sloping sides all

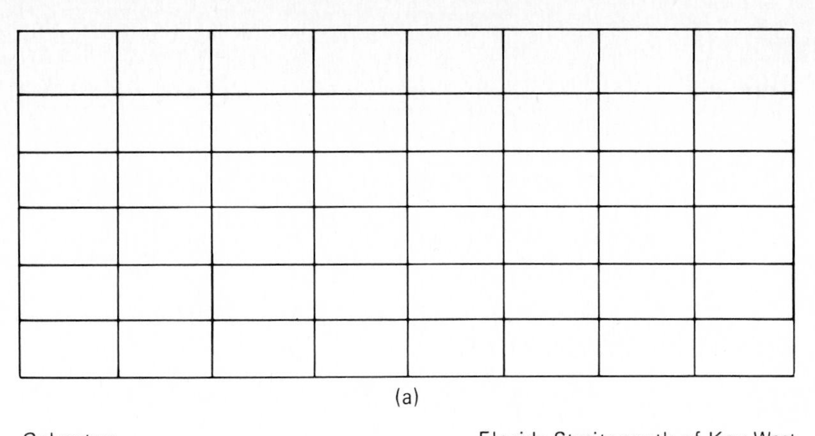

(a)

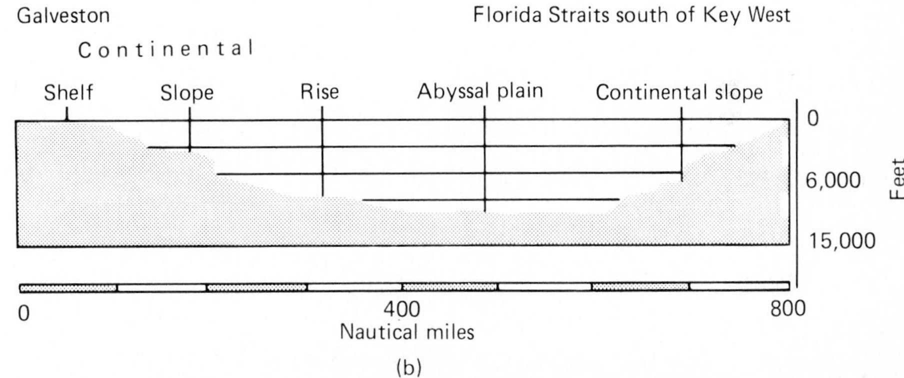

(b)

Figure 16-5. (*a*) Grid for mapping the ocean floor. (*b*) Cross section from Galveston, Texas, to the Florida Straits south of Key West. (Source for *b*: V. F. Lien, *Investigating the Marine Environment and Its Resources*, Texas A & M University, Sea Grant Program, 1979.)

TABLE 16-1

DISTANCE FROM GALVESTON, NAUTICAL MILES	DEPTH, FEET	DISTANCE FROM GALVESTON, NAUTICAL MILES	DEPTH, FEET
25	100	425	10,000
50	300	450	10,000
75	500	475	10,000
100	600	500	11,000
125	3,000	525	10,000
150	3,500	550	11,000
175	4,000	575	12,000
200	4,200	600	11,000
225	7,000	625	12,000
250	8,000	650	10,000
275	9,000	675	7,000
300	9,000	700	5,000
325	7,000	725	3,100
350	9,000	750	2,800
375	10,000	775	1,500
400	10,000	800	200

Note: Measurements can be converted to metrics. To convert depth to meters, multiply feet by 0.305. To convert distance from Galveston to kilometers, multiply nautical miles by 1.85.

around it? What do they think happens where streams enter and leave the lake? Have your students make a plan for finding out what the bottom of the lake is really like. (*One method is to drop a thin line with a heavy weight attached to it, to determine depth at preselected points. What would happen with this method in a swiftly moving current?*) Go to a local fishing supply shop and ask for a map of the lake. If you are able to get the map, show it to the students and ask them to pick out good fishing spots. Invite a member of a local fishing club to visit your class and explain how to use the map and why some locations are better for fishing than others.

▶ EXPLORATION OF AQUATIC ENVIRONMENTS

Since people cannot take oxygen directly from water and since air exerts great pressure even at shallow depths, scientists must use special equipment in order to make detailed studies of an aquatic environment. Some studies are made by gathering information with collecting equipment and sounding devices which can be operated without personally getting into the water. In other studies, scientists dive, sometimes to great depths, using special equipment such as scuba gear, diving suits, and small scientific submarines.

Have your students do library research on the types of equipment used in aquatic studies. There are many books available (especially by such authors as Jacques Cousteau, who has also made a popular television series which your students may enjoy and learn from), and a large number of pertinent articles have been published in *National Geographic*. The students can either make oral reports on their findings or prepare drawings and information to be used on a bulletin board. Encourage those who are interested to make models of the equipment, or even to make equipment which they can actually use—with appropriate supervision and permission from their parents—to learn about a local body of water.

▶ OCEAN CURRENTS

It is not especially important that your students learn the names and positions of the various currents in the ocean, but they should know that those currents have profound influences on both climate and oceanic transportation. In considering the influences, students will learn to understand important concepts.

Using an ocean current map (ideally in the form of an overhead transparency or a wall map; see Figure 16-6), point out the major ocean currents. Ask the students what effects the current along the Alaskan coast might have. Have the students use an almanac and compare the average winter temperature of Anchorage, Alaska, with those of several other cities in the northern United States—good choices would be Duluth, Minnesota, Helena, Montana, and Fargo, North Dakota. (*The temperature of Anchorage is relatively mild because of the warm ocean current.*) Also, have the students use the almanac to determine the average summer temperatures of several cities on both the east coast and the

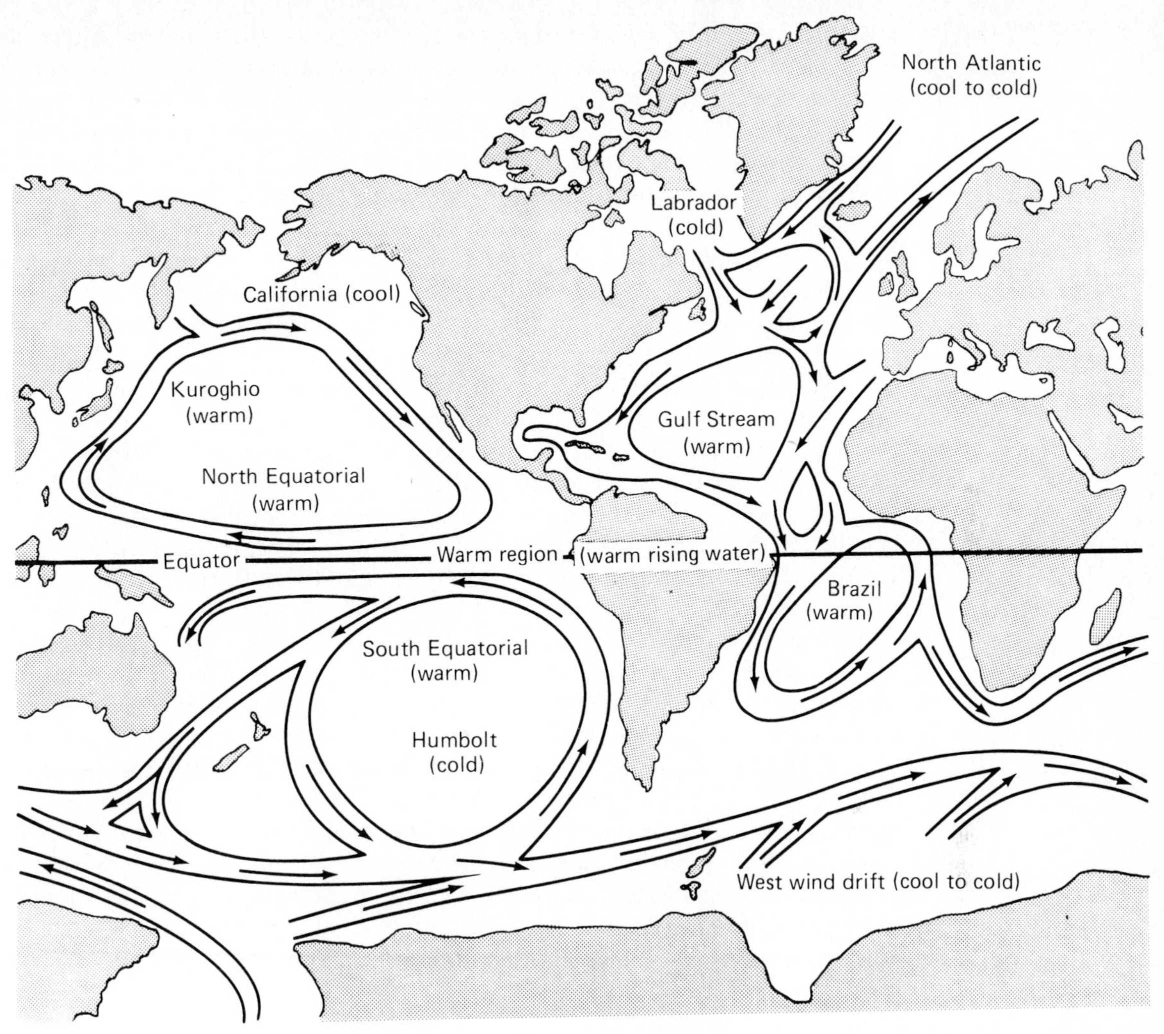

Figure 16-6. Main ocean currents. (Source: V. F. Lien, *Investigating the Marine Environment and Its Resources,* Texas A & M University, Sea Grant Program, 1979.)

west coast of the continental United States. Ask the students to explain how the differences between the two coasts are related to ocean currents near each coast. (*A cool current along the west coast tends to make summer temperatures lower; the warm Gulf Stream flows along the east coast.*) Have the students predict the effects of ocean currents upon adjacent land masses and obtain information to validate their predictions.

In addition to influencing temperature, currents can have other effects. Where a subsurface current runs into an obstacle, it brings rich nutrients with it. The nutrients support abundant water life, and such an area is usually a food source for people. Have students identify areas where subsurface currents come to the surface and determine whether fishing is an important industry of nearby countries.

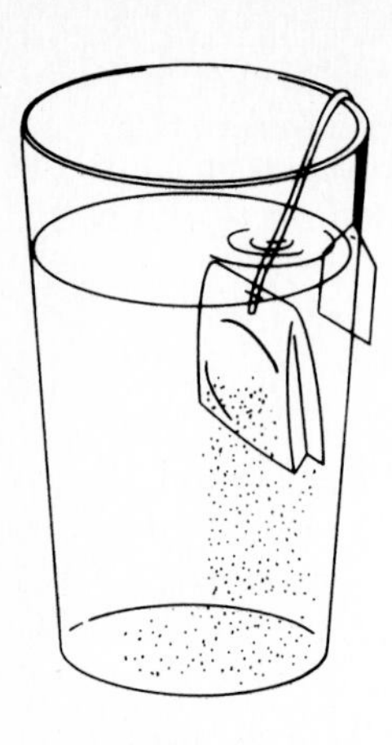

Figure 16-7. Effect of salt concentration on the density of saltwater.

Benjamin Franklin was among the first to note that the fastest way to travel between Europe and America was not the most direct route; he had discovered the effect of ocean currents on sailing vessels. Have the students suggest the fastest routes from the Americas to Europe and vice versa, making use of ocean currents. If they were planning to make a trip to Europe by raft, which route would be the fastest?

➤ DENSITY CURRENTS

Two factors which cause density differences in salt water are differing concentrations of salt and differences in temperature. The effects of both can be easily demonstrated by simple activities.

Differences in concentration of salt can be demonstrated by holding salt in a tea bag (make a slit in the top of an unused tea bag, empty out the tea leaves, and pour in salt) against the side of a glass of water just so that the salt in the bag is immersed in the water. If you watch carefully below the bag, you can see the concentrated salt water "flowing" to the bottom of the glass. (See Figure 16-7.) For more spectacular results, place sodium thiosulphate from a photography supply store instead of salt in the tea bag. After the sodium thiosulphate dissolves, carefully put a roasted peanut into the glass. The peanut may appear to "float" in the middle of the water; it will actually be floating on top of the sodium thiosulphate solution. Occasionally the dividing point between the water and sodium thiosulphate solution in this experiment can even be seen.

Density currents due to temperature differences in water can be demonstrated by using a rectangular cake pan, water, ice cubes, and food coloring. Fill the cake pan half full of water, and stack some ice cubes at one end. Then place a drop of food coloring on the surface of the water near the ice. (See Figure 16-8.) Have the students note the direction of movement of the coloring. The sinking effect of the cold water should be dramatically visible. Place another drop of food coloring on the surface of

Figure 16-8. Effect of temperature on the density of saltwater.

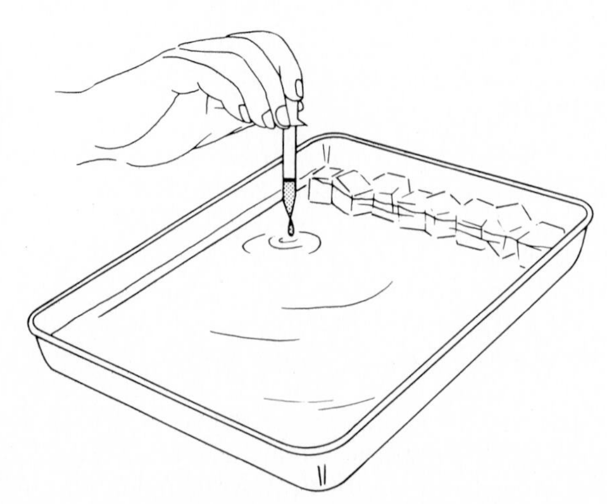

the water at some distance from the ice, and again have the students note the direction of movement. Ask the students to think of the oceans in relation to this activity. Where in the ocean are there "stacks of ice"? (*At the north and south poles.*) What do they think is happening to the water at the poles? Have them check out their guesses by consulting reference books.

➤ WHO LIVES NEAR LARGE BODIES OF WATER?

The chances are that you yourself live near a large body of water. It has been claimed that three-fourths of the people in the United States live within 160 kilometers (100 miles) of one of the saltwater coasts or the Great Lakes; and in fact, more than half live within 100 kilometers (62 miles) of a seacoast or one of the Great Lakes.

On a map of the United States, have your students draw different-colored lines 100 kilometers and 100 miles in from the coastlines and the shores of the Great Lakes. Then have them note the number and size of cities included in the areas they have marked off—or simply note the large metropolitan areas included. They can then subjectively compare the populations near and farther from the coasts and shores. What do they conclude? Do they think the above statements about people living near coasts and shores may be accurate?

➤ WATER ENVIRONMENT MURAL

Provide information on the marine physical environment and its inhabitants for your students, or have them do research. Then tell them to make a mural on marine life in a given environment. The mural can be temporary or permanent.

While the students are planning the mural, ask them to consider whether the food sources for the various animals in the mural are represented. Would the plants and animals shown normally be found in the same area? A good source of ideas is a set of charts available from the National Oceanic and Atmospheric Administration (see "Free and Inexpensive Materials").

➤ HOW MUCH DO WE KNOW ABOUT AQUATIC AND MARINE ENVIRONMENTS?

Only relatively small parts of the total aquatic and marine environment have been explored. To help students appreciate how limited and incomplete the available information is, have them plan an imaginary exploration of their own state from the viewpoint of "water people."

Ask the students to pretend that they have always lived under water. However, the population of their body of water has increased, and it is becoming necessary to find more living space. One possibility is the land which surrounds their home. After much discussion, it has been decided to send five "probes" to the land. What preparations will have to be made? What kind of exploration vessel will be needed? Remind the students that they normally live, eat, breathe, and perform all bodily functions under water.

Since the water people have no knowledge of land other than the first few meters (yards) around their body of water, they will be sending the five probes to random locations. The goal is, of course, to learn as much as possible about the land so as to determine its suitability for water colonies. Have the students work in groups of five, and give each student a small piece of paper.

Place a map of the state on the floor, and have each student in turn stand on a chair above the map and drop the paper onto the state map. The students are then to determine and describe the nature of the environment upon which their papers dropped. In case they are not already familiar with the locations, have almanacs (and perhaps other references) available for research.

After the individual students have written their descriptions, each group is to produce a composite description of the state, using *only* the information obtained at the five landing sites. Then the groups can share their descriptions with the class.

Have students draw an analogy between their sampling of their home state and scientists' exploration of oceans and other large bodies of water. What are some of the problems? (*Five locations will seldom give enough information to create a good overall description of the state. Scientists have studied even more widely spaced samples of the ocean. Of course, scientists do not sample only random underwater locations, but they probably are still unaware of much valuable information about ocean environments.*)

➤ HURRICANES

Hurricanes—and "typhoons," as they are called in some parts of the Pacific Ocean—receive a great deal of publicity. The energy in hurricanes places them among the most powerful phenomena on earth. Hurricanes can inflict a great deal of damage, especially on coastal areas.

Obtain hurricane-tracking charts from the National Oceanic and Atmospheric Administration (see "Free and Inexpensive Materials"; classroom quantities are available). During the hurricane season, have the class trace on the charts the paths of the hurricanes which occur. Table 16-2 shows a form for listing time and location reports for the devastating Galveston hurricane of 1900. After they plot the path of the Galveston hurricane, ask the students to comment upon the speeds at which hurricanes travel. (*Variable from hour to hour and day to day. Hurricanes can travel very fast at some times and very slowly at others.*) Can a hurricane affect inland areas?

If you live near a coast, it is important that your students and their families know what to do if a hurricane approaches. You can find out the particular requirements for your area by contacting your local Civil Defense Office.

TABLE 16-2 LOCATIONS OF HURRICANE IN GALVESTON, TEXAS, IN 1900

DATE	LONGITUDE	LATITUDE

➤ MUSIC RELATED TO AQUATIC AND MARINE ENVIRONMENTS

Because human history has been so closely involved with water navigation, a large amount of music deals with water environments. All forms of music are included—from classical to rock music. Ask your students to bring to class any records they have that refer to water environments; they will probably inundate you with records. Play as many as you can—at lunch time and recess, before and after school, etc.

Also bring in some of your own records. *Music of the Sea* is an album of beautiful mood music related to the marine environment.* This would be a good album to play either as background music during quiet study periods or for a few minutes of quiet relaxation. Pete Seeger's album *God Bless the Grass*† has songs such as "My Dirty Stream" and "San Francisco Bay." In a classical vein, Bedrich Smetana's *Moldau*, part of a long symphonic cycle, is enjoyable and will introduce your class to a musical form not frequently listened to by elementary school children.

Another exciting field for students is listening to river songs and sea chanteys.‡ They can learn a wealth of history as well as water lore from these songs. Check the card catalog of the school or community library for "sea chanteys" or "folk songs."

Ask your students to try to explain why so much music has been written and recorded which relates to water environments.

*RCA Camden CAL 639.

†Columbia CL 2432.

‡*Colonial and Revolutionary War Sea Songs and Chanteys*, by Cliff Haslam and John Miller, Folkways FH 5275; *Foc'sle Songs and Shanties*, by Paul Clayton and the Foc'sle Singers, Folkways FA 2429.

Have the class look at a map of the United States and locate the major cities in the country. Are they surprised to find that most all the major cities are located on a seacoast, on the shores of the Great Lakes, or along a major river? Ask them why they think this is. (*Transportation! Water transportation still carries a majority of the cargo shipped in this country and is by far the most energy- and cost-efficient means of shipping cargo.*)

FOLKLORE OF THE SEA

Humankind's first efforts to record and explain people's interactions with their environment are best known to us through folklore. Much of that interaction has taken place in aquatic and marine environments, and thus there is a great deal of folklore about water environments.

Recordings of river songs and sea chanteys were mentioned in the previous activity. There are also books full of sea chanteys (two are listed in "Building Your Own Library") which students can enjoy learning to sing. Often a sea chantey was used to set the cadence for some task such as hoisting a sail. Have your students act out a task to the cadence set by a chantey. (Remember, hoisting a sail is hard work.) After they listen to or sing several pieces of related folk music, ask the students to identify the important lessons taught by the music.

Folk art and literature about aquatic and marine environments are abundantly available. A visit to the local library should provide many examples. Have your class listen as you read some examples of folk literature which relate to aquatic and marine environments. Ask the students to describe what the environment was like and how the people living at that time seemed to feel about their environment.

Describe and show examples of folk art. Ask the students why they think the specific kinds of folk art seen were created. Have them make their own versions of some of the types of folk art. One form, scrimshaw, is described in the next activity.

SCRIMSHAW

Scrimshaw is the art of engraving or carving bone or ivory (especially from the whale, elephant, and walrus). It was started in New England in the 1800s by sailors on whaling boats. You may be able to find pictures to show your students in an encyclopedia; also look for actual pieces of scrimshaw in museums and shops. (Figure 16-9 shows an example of scrimshaw.)

Figure 16-9. Scrimshaw.

Your students can make scrimshaw by using beef bones—5-centimeter (2-inch) sections of bones which are 5 to 8 centimeters (2 to 3 inches) in diameter work very well. (Ask a friendly butcher to save some cross sections for you, as they become available.) Boil the bones in water for several minutes; drain and add clean water and soap, and then boil and drain again. After the bones have dried thoroughly, have the students sand them with fine sandpaper until they are smooth.

Tell the students to use a pencil to draw a simple picture on the bone—or better yet, make a scrimshaw yourself to show them how. Appropriate subjects you can suggest are starfish, simple boats, fish, and seashells; also encourage students to use their imaginations in thinking of other subjects. Next use a large nail to scratch into the bone over the pencil drawing. Sand lightly again to remove the pencil marks, and use artists' carbon black or a charcoal pencil to darken the carved lines. Wipe away any excess carbon black or charcoal.

Scrimshaw makes nice gifts from children to parents. And after your students get a little practice and make a few nice pieces, you may want to ask them to prepare a scrimshaw demonstration and exhibit for other classes in your school.

➤ A TRIP

Have your students pretend that they will be the first persons to explore ocean environments. What kinds of information will they try to gather? How will they attempt to gather that information? If they want to personally go into the water, what preparations will they have to make?

In other imaginative activities, you could tell the students to imagine they are fish, clams, shrimps, sea plants, pieces of sand, or other components of the aquatic environment and have them describe what happens to and around them.

Children's literature

Many children's books on aquatic and marine environments have been published, as evidenced by the list in "Building Your Own Library." Both fiction and reference books are included.

Read *The Cay** to your students or have them read it. Ask the students to identify some characteristics of the marine environment, as described in the book. What kind of knowledge was necessary to survive on the island?

Find other books on aquatic and marine environments in the school library and identify them for the students or place them in an interest center.

Activity cards

An example of an activity card is shown in Figure 16-10. Additional activity cards could be made for topics such as exploring a local aquatic environment (looking for insects, plants, water animals, large animal visitors), studying water quality, establishing an aquarium (freshwater or saltwater), making models of portions of the sea floor, determining temperature (or other physical) differences at various depths in a body of water, finding out about resources from the sea, cultivating fish and other water organisms (aquicultures), and finding means of exploring aquatic environments.

*Theodore Taylor, *The Cay*, Doubleday, Garden City, N.Y., 1969.

VISIT A LOCAL FOOD STORE AND FIND AS MANY KINDS OF FOOD AS YOU CAN WHICH COME FROM THE SEA.

WHICH WOULD YOU LIKE TO TASTE?

DID YOU CHECK CANNED FOODS? CRACKERS?

Figure 16-10. Activity card: Search for seafood.

Bulletin boards

RESOURCES FROM THE SEA
Collect pictures related to offshore oil production and the mining of various mineral resources such as manganese, and use the pictures to make a bulletin board. Include the question "What other kinds of resources do we get from the sea?"

THE NATURE OF THE AQUATIC ENVIRONMENT
Have students cut out magazine pictures of aquatic environments. Post the pictures on a bulletin board with the title "Aquatic Environments." After several pictures have been posted, ask students to begin arranging them in groups such as recreation, beauty, animals, plants, ecosystems, food for humans, mineral resources, and agriculture.

Field trips

If possible, take your class to observe an aquatic environment. In preparation, they will have to learn how to do sampling in water environments. Invite a local game warden or wildlife biologist to help you plan and conduct the field trip.

A visit to a local food store can show your students how many kinds of food do contain aquatic organisms or their derivatives. After sharing information from Table 16-3 with the class, make plans for searching in

TABLE 16-3 FOODS CONTAINING DERIVATIVES FROM ALGAE

Dairy	**Bakery**	**Dressings, sauces**
Ice cream	Bread doughs	Salad dressings
Milk shakes	Doughnuts	Syrups, toppings
Sherbets	Doughnut glazes	White sauces
Ice pops	Pie fillings	Mustards
Chocolate milk	Fruit pie fillings	Catsups
Puddings	Frozen pie fillings	**Meat, fish**
Cottage cheese	Meringues	Canned fish, meats
Cream cheese	Cookies	Sausage ingredients
Yogurt	Cake batters	**Miscellaneous**
Milk in cartons	Cake fillings and toppings	Jams, preserves
Beverages	**Dietetic foods**	Prepared cereals
Soft drinks	Starchfree desserts	Processed baby foods
Fruit juices	Salad dressings	Soups
Beer (foam stabilizers)	Jellies, jams	Frozen foods
Beer (clarification agents)	Puddings	Synthetic potato chips
Wines	Sauces	Fountain toppings
Spirits (aging agents)	Candies	Artificial cherries
Candy	Vegetarian and health foods	
Candy gels		
Caramels		
Marshmallows		

Source: V. F. Lien, *Investigating the Marine Environment and Its Resources*, Sea Grant College Program, Texas A. & M. University, College Station, Tex., 1979, p. 386.

specific areas of the store for aquatic foods. Divide the class into research teams and assign each team to a different section of the store. One group should talk with the store manager to find out about types of food which the class may not have considered. After visiting the store, have each group report on their findings. (How about planning a "tasting fair"?)

RESOURCES

Sources of assistance

FREE AND INEXPENSIVE MATERIALS

Here are some of the best sources of free and inexpensive materials for use in teaching about aquatic environments:

1. *State fish and wildlife (or parks and wildlife) departments.* These departments are usually located in the state capital and also have branch offices in various parts of the state. They can provide information on fish and wildlife in state waters and on recreation facilities.

2. *Nektonics.** A booklet on how to set up and maintain a marine aquarium.
3. *Tuna Research Foundation.* A kit including information on the history of the tuna industry and the nutritional value of tuna.
4. *Shell Oil Company.* A booklet, "Let's Collect Rocks and Shells," plus brochures on oil production.
5. *National Oceanic and Atmospheric Administration.* Several aids, including one entitled "Hurricanes." Ask for a list.
6. *University of Rhode Island.* A publication entitled "How to Build and Save Beaches and Dunes."
7. *U.S. Department of the Interior, Geological Survey.* Several informational booklets on water, including "Water of the World" and "Why Is the Ocean Salty?" Ask for a publications list.
8. *Cruise Lines International Association.* Several aids, including teaching suggestions for a unit on steamship travel.
9. *Metaframe Corporation.* A brochure entitled "Your First Aquarium."
10. *National Wildlife Federation.* A pamphlet entitled "Wildlife of Lakes, Streams, and Marshes." Ask for a publications list.
11. *Sea Grant College Program, Texas A. & M. University.* A number of marine resource teaching materials. Write for a list. Sea Grant College Programs in several other states also publish teaching aids.
12. *U.S. Department of Commerce.* Hurricane-tracking charts.

LOCAL RESOURCES

Local resources which can help in teaching about the oceans and their resources include:

1. A scuba diver who can explain how one type of exploration gear is used and may also be able to show the class underwater photographs.
2. A state fish and wildlife department employee may make a presentation to the class or may be able to give you teaching materials.
3. Fishing clubs may offer resource persons or materials on local bodies of water.
4. A sporting-goods store may have available information on fishing.
5. The manager of your grocery store may be willing to save promotional materials on aquatic and marine food products for you.

ADDITIONAL RESOURCES

1. Magazines such as *World, Ranger Rick, Currents, National Wildlife,* and *International Wildlife.*
2. Inexpensive activity and game books are often sold in discount and grocery stores.
3. Costeau Society, 930 West 21 Street, Norfolk, VA 23517. Current information on the marine environment.

*The addresses of specific agencies, organizations, and businesses are listed in the Appendix.

4. You may be able to find inexpensive aquarium equipment through garage and rummage sales and newspaper want ads.
5. Local streams and ponds contain aquatic organisms.

Strengthening your background

1. Go on a fishing trip with someone who is experienced in fishing.
2. Begin sampling many freshwater and marine foods. Save wrappers or containers with printed information on them.
3. Visit a large aquarium if possible, or study the fish at a pet shop.
4. Keep an aquarium in your home.
5. Make a collection of seashells.
6. Learn about sea urchins. Where do they live? What would you have to feed one to keep it alive? Would you want to keep one? Eat one?
7. Enroll in an oceanography or aquatic biology course at a nearby college or university. Special summer programs may be available for teachers, with assistantships.

Getting ready ahead of time

1. Make a collection of aquatic and marine articles from *National Geographic* and other magazines.
2. Construct a shoebox science kit on the marine environment. Include crossword puzzles, shell-identification activities, and activities that involve identifying "exotic" seafoods sold in local stores.
3. Make a list of aquatic and marine books available in the school library. Give the librarian the titles and authors of two more books (see "Building Your Own Library" and "Building a Classroom Library") and ask that they be added to the school library.
4. Gather the materials for the sand-painting activities described in this chapter. (Sand painting can be done in connection with many science studies.)
5. Find a health food store or other source of algae crackers.
6. Buy or have your librarian buy records of music about the sea.
7. Prepare a set of "Who's for Dinner?" cards for use with your students. Make photocopies of the cards and glue them on index cards. Laminate them if possible.
8. Sketch a plan for a mural of an underwater scene to be painted by your class.

Building your own library

Bagnall, Norma: *Children's Literature—Passage to the Sea,* Sea Grant College Program, Texas A. &. M. University, College Station, Tex., 1980.

Boyer, Robert: *The Story of Oceanography,* Harvey House, New York, 1975.

Brown, Joseph: *The Sea's Harvest: The Story of Aquaculture*, Dodd, Mead, New York, 1975.
Cooper, Elizabeth: *Science on the Shores and Banks*, Harcourt Brace Jovanovich, New York, 1960.
Klots, Elsie B.: *The New Field Book of Freshwater Life*, Putnam, New York, 1956.
Lien, Violetta: *Investigating the Marine Environment and Its Resources*, Sea Grant College Program, Texas A. & M. University, College Station, Tex., 1979.
McClung, Robert: *Treasures in the Sea*, Marine Biology Series, National Geographic Society, Washington, 1972.
Michelson, David: *The Oceans in Tomorrow's World: How Can We Use and Protect Them?* Messner, New York, 1972.
Morgan, Ann: *Field Book of Ponds and Streams*, Putnam, New York, 1930.
Read, Richard: *The Living Sea*, Puffin, New York, 1974.
Stratton, Leonard: *Your Book of the Seashore*, Faber, London, 1970.

Building a classroom library

Asimov, Isaac: *ABC's of the Ocean*, Walker, New York, 1970.
Barlowe, S.: *Oceans*, Follett, Chicago, Ill., 1969.
Blassingame, Wyatt: *The First Book of the Seashore*, Watts, New York, 1964.
Boston, Lucy: *The Sea Egg*, Harcourt Brace Jovanovich, New York, 1967.
Brindze, Ruth: *The Sea: The Story of the Rich Underwater World*, Harcourt Brace Jovanovich, New York, 1971.
Brown, Joseph: *Wonders of the Kelp Forest*, Dodd, Mead, New York, 1974.
Brown, Margaret Wise: *Seashore Noise Book*, Harper and Row, New York, 1941.
Buehr, W.: *Sea Monsters*, Norton, New York, 1966.
Cadbury, B. Bartram: *Fresh and Salt Water*, Creative Education Society, Minneapolis, Minn., 1967.
Carpelan, Bo: *Bow Island: The Story of a Summer That Was Different*, translated from the Swedish by Sheila LaFarge, Delacorte Press, Dell, New York, 1974.
Carson, Rachel: *The Sea around Us*, adapted by Anne Terry White, Simon and Schuster, New York, 1958.
Clemons, Elizabeth: *Shells Are where You Find Them*, Knopf, New York, 1960.
———: *Tide Pools and Beaches*, Knopf, New York, 1964.
———: *Waves, Tides, and Currents*, Knopf, New York, 1967.
Cortesi, Wendy: *Explore a Spooky Swamp*, National Geographic Society, Washington, 1978.
Creative Education Society: *Life in the Sea*, Minneapolis, Minn., 1971.
Dean, Anabel: *Exploring and Understanding Oceanography*, Benefic Press, Chicago, Ill., 1970.
———: *Men under the Sea*, Harvey House, New York, 1972.

———: *Submerge! The Story of Divers and Their Crafts*, Westminster, Philadelphia, 1976.
Dempsey, Michael: *By the Sea*, World Publishing, Cleveland, 1970.
Fenton, D. X.: *Harvesting the Sea*, Lippincott, Philadelphia, 1970.
George, Jean Craighead: *Spring Comes to the Ocean*, Thomas Y. Crowell, New York, 1966.
Goldin, A. R. : *The Bottom of the Sea*, Thomas Y. Crowell, New York, 1967.
Greenhood, David: *Watch the Tides*, Holiday, New York, 1961.
Hunt, Bernice Kohn: *The Beachcomber's Book*, Viking, New York, 1970.
Hurd, Edith Thacher: *The Mother Whale*, Little, Brown, Boston, 1973.
Jacobs, Francine: *Bermuda Petrel: The Bird that Would Not Die*, Morrow, New York, 1981.
Jacobson, Morris K. and David R. Franz: *Wonders of the Jellyfish*, Dodd, Mead, New York, 1978.
Kohn, Bernice: *The Beachcomber's Book*, Viking, New York, 1971.
Kumin, Maxine: *The Beach before Breakfast*, Putnam, New York, 1964.
Lubell, Winifred, and Cecil Lubell: *By the Seashore*, Parents Magazine, New York, 1973.
May, Julian: *Captain Cousteau, Underwater Explorer*, Creative Education Society, Minneapolis, Minn., 1972.
———: *The Land beneath the Sea*, Holiday, New York, 1975.
Milne, Lorus, and Margery Milne: *When Tide Goes Far Out*, Atheneum, New York, 1970.
Morris, Robert: *Dolphins*, Harper and Row, New York, 1975.
Olney, Ross: *Inquiring Mind: Oceanography*, Nelson, Camden, N.J.
———: *Men against the Sea*, Grosset and Dunlap, New York.
Pizer, Vernon: *The World Ocean: Man's Last Frontier*, World, Tarrytown-on-Hudson, N.Y., 1967.
Podendorf, Illa: *The True Book of Animals of the Sea and Shore*, Children's Press, Chicago, Ill., 1956.
Pringle, Laurence: *From Pond to Prairie: The Changing World of the Pond and Its Life*, Macmillan, New York, 1972.
Rabinowich, Ellen: *Seals, Sea Lions, and Walruses*, F. Watts, New York, 1980.
Ryan, Peter: *The Ocean World*, Penguin, Baltimore, 1973.
Selsam, Millicent: *Animals of the Sea*, Scholastic Book Services, New York, 1976.
Shannon, Terry, and Charles Payzant: *Windows in the Sea: New Vehicles that Scan the Ocean Depths*, Children's Press, Chicago, Ill., 1973.
Shepherd, Elizabeth: *Tracks between the Tides*, Lothrop, New York, 1972.
Simon, Seymour: *From Shore to Ocean Floor: How Life Survives in the Sea*, F. Watts, New York, 1973.
Sobol, Donald: *True Sea Adventure*, Seafaring Life series, Nelson, Camden, N.H., 1975.
Soule, Gardner: *New Discoveries in Oceanography*, Putnam, New York, 1975.

Spilhaus, Athelstan: *The Ocean Laboratory*, Creative Education Society, Minneapolis, Minn., 1967.

Stephens, William: *Come with Me to the Edge of the Sea*, Messner, New York, 1972.

———: *A Day in the Life of a Sandy Beach*, McGraw-Hill, New York, 1973.

———: *Life in the Open Sea*, McGraw-Hill, New York, 1972.

———: *Science beneath the Sea: The Story of Oceanography*, Putnam, New York, 1966.

Straker, Joan Ann: *Animals that Live in the Sea*, National Geographic Society, Washington, 1978.

Taylor, Theodore: *The Cay*, Doubleday, Garden City, N.Y., 1969.

Watson, Jane Werner: *Whales: Friendly Dolphins and Mighty Giants of the Sea*, Golden Press, New York, 1975.

Wildsmith, Brian: *Fishes*, F. Watts, New York, 1978.

Wong, Herbert, and Matthew Vessel: *Pond Life: Watching Animals Find Food*, Addison-Wesley, Reading, Mass., 1970.

Woodbury, David: *Fresh Water from Salty Seas*, Dodd, Mead, New York, 1967.

Yashima, Taro: *Seashore Story*, Viking, New York, 1967.

PART 4

TEACHING THE PHYSICAL SCIENCES

Fiberoptics—an artificial material so clear that a hundred miles of it is clearer than a quarter-inch window pane. Lasers used as scalpels to operate on delicate organs such as the human eye. Robots which come out in teams at night and perform a complex array of janitorial tasks in major industrial plants. Microcomputers which are capable of adding two numbers in a few millionths of a second by using an integrated circuit smaller than a dime which contains 20,000 discrete components. Bionics which make possible replacements for human limbs capable of greater strength and speed than the original limbs. A decade or two ago these were science fiction. Today they are reality.

The popularity of science fiction in the late 1970s and the 1980s has made it difficult to distinguish fact from fiction; and history has proved repeatedly that today's fiction can become tomorrow's fact. It seems that almost nothing is impossible and that humankind is capable of attaining almost any dream. But we must also realize that, as always, humans are capable of evil as well as good. Computers are just one example: in the wrong hands the computer can become an instrument of disaster, and it can enable even those with the best of intentions to make mistakes at an alarming rate. We believe that those who teach the physical sciences must present the dangers that can result from discoveries and inventions. Values are playing an increasing role in science; and simplistic ideas such as "bigger is better" are giving way to new concepts. Moreover, national pride must be balanced by a sense of world citizenship; national ecology must be accompanied by protection of the biosphere; and international competition must be tempered by willingness to cooperate.

In teaching the physical sciences—as in teaching the other sciences—we must always be concerned about improving the quality of life. This is not to imply that discoveries should be discouraged. On the contrary, students of the physical sciences should be encouraged to explore, tinker, and invent. Abraham Maslow reminds us of the need to focus on learners, not teachers:

"To understand the breadth of the role of the teacher, a differentiation has to be made between extrinsic learning and intrinsic learning. Extrinsic learning is based on the goals of the teacher, not on the values of the learner. Intrinsic learning, on the other hand, is learning to be and to become a human being, and a particular human being. It is the learning that accompanies the profound personal experiences in our lives. . . . As I go back in my own life, I find my greatest education experiences, the ones I value most in retrospect, were highly personal, highly subjective, very poignant combinations of the emotional and the cognitive. Some insight was accompanied by all sorts of autonomic nervous system fireworks that felt very good at the time and which left as a residue the insight that has remained with me forever."*

As you prepare to teach the physical sciences, it is our hope that you will consider your responsibilities in, and your own attitudes toward, this explosive and exciting field.

*Abraham Maslow, "What Is a Taoistic Teacher?" in *Facts and Feelings in the Classroom,* edited by Louis J. Rubin, Walker, New York, 1973, p. 159.

CHAPTER 17

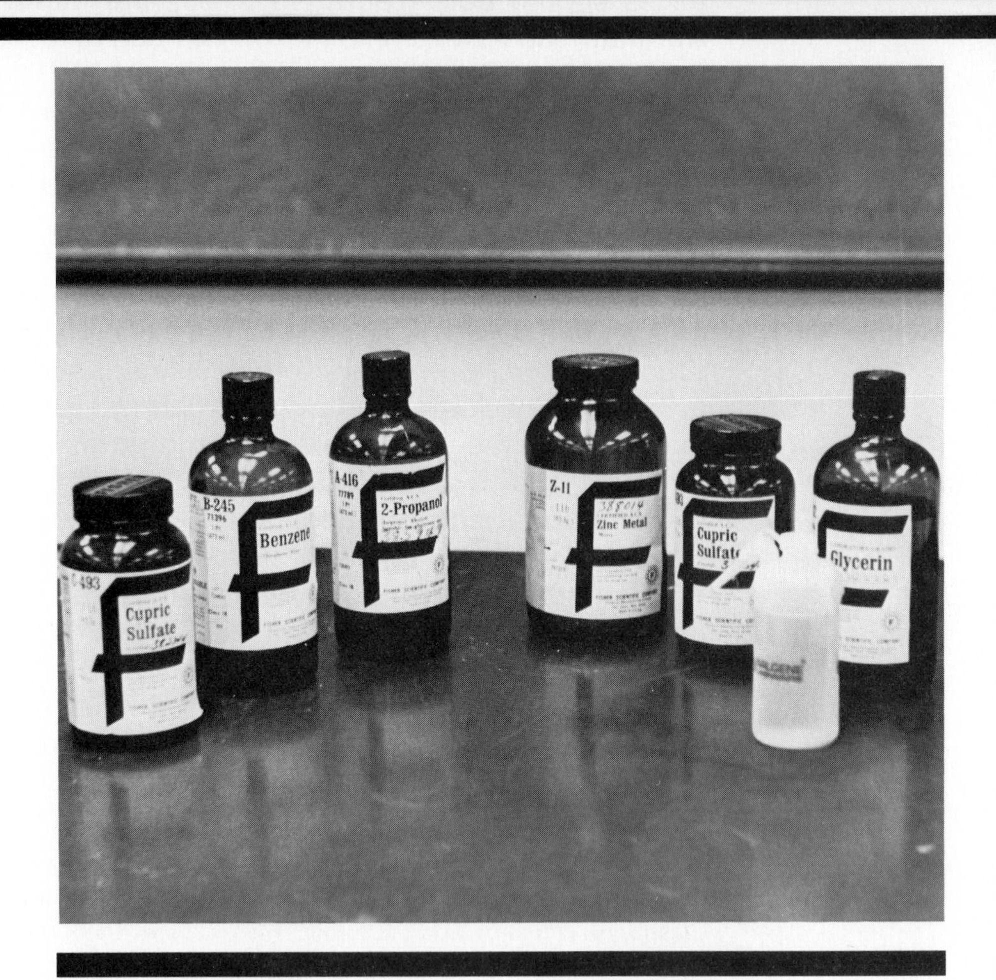

MATTER AND ENERGY

IMPORTANT CONCEPTS

Matter exists as solids, liquids, and gases.

Matter changes from one state to another when heat is added or subtracted.

Atoms are the building blocks of matter.

All matter is composed of atoms combined in differing ways.

An element is a substance composed entirely of one kind of atom.

Compounds are combinations of elements.

A molecule is the smallest particle of a substance that has the properties of that substance.

Chemical changes involve changes in the chemical composition of matter.

Physical changes involve changes in form.

When matter changes from one form to another, the total amount of matter is unchanged.

Energy is the capacity to do work.

Work is done when matter is moved.

Matter can be set in motion by forces.

Energy exists in many forms.

When energy changes from one form to another, the total amount of energy is unchanged.

The energy of matter in motion is kinetic energy.

Potential energy—the ability to do work because of position or condition—is "stored energy."

Friction slows down moving objects and makes it more difficult to set a nonmoving object in motion.

Although friction can be wasteful, it is necessary for many activities.

The sun is the ultimate source of most energy used by living things on earth.

In nuclear reactions matter is changed to energy; the combined total of energy and matter is unchanged.

Many of the things we do today are possible because people have invented ways to make use of great amounts of energy.

INTRODUCTION: STUDYING MATTER AND ENERGY

Matter and energy are two physical science topics which are readily observable in every student's environment. "Matter" is the real physical material of the world. Children can explore the states in which matter is found and how it can be changed. Students of all ages have long been fascinated by chemical reactions. In studying matter, your students will come to better understand the stuff of which their world is made.

Energy is important because we use such large amounts of it. Energy and its wise use are important to us. Basic information about energy—its changes from one form to another, the requirement of its use whenever motion changes, and other basic concepts—will help students better understand some of the complex energy issues in today's world.

In teaching matter and energy to young children it is very important to give them "hands-on" experiences, so that the information becomes real to them. Frequently ask students to look for additional examples of what they have just learned, and have them often relate concepts to everyday observations.

MATTER AND ENERGY FOR PRIMARY GRADES

Activities

➢ STATES OF MATTER

Ask students to name examples of solids, liquids, and gases which are found in the classroom. Tell them that these are three common states in which matter is found. Have them name additional examples of materials for each of the states.

Have students describe how it would feel to be a solid. A liquid. A gas.

➢ HEATING A SOLID

Give each student a plastic glass and an ice cube. Ask the students what state of matter the ice cube is in. *(Solid.)* Ask them what will happen to the ice cube if it sits in the glass. *(It will melt—change to a liquid.)* Have them try to find out. Ask what they could do to speed up the process.*

**Note:* This activity and the next three activities—"Heating a Liquid," "Cooling a Gas," and "Cooling a Liquid"—present the concept that changes in the state of matter are produced by adding or subtracting heat energy.

Water is one of the few compounds which we rather commonly experience in three states: solid, liquid, and gaseous.

➢ HEATING A LIQUID

Place a Pyrex glass container of water on a hotplate. Ask the students to tell what is happening to the liquid water. *(It is boiling and "disappearing" from the container.)* Ask what has happened to the water; several students will probably say that they saw the water vapor (gas) escaping over the container. What was needed to change the water from a liquid to a gas? *(Heat.)*

➢ COOLING A GAS

Place some ice in a tin can. Have students touch the can and note that it is cool. Hold the can over a pan of boiling water. Ask the students what is happening. *(Water vapor condenses on the cool can.)* What does this illustrate? *(When a gas is cooled, it changes to a liquid.)* Ask students to name other examples of this phenomenon which they have seen.

➢ COOLING A LIQUID

This activity can be done at school if a refrigerator-freezer is available, or the students can be asked to do it at home. Have each student fill a small plastic container half full of water and place it in a freezer. Ask what happens to objects that are placed in a freezer. *(They become cooler.)* What happens to the liquid water? *(It changes to ice—solid water.)* What does this illustrate? *(When a liquid is cooled, it changes to a solid.)*

➢ SMALL PARTICLES OF MATTER

Show the class a bottle of a very fragrant perfume, and place a few drops of the perfume in a large container. As it evaporates, have the students come up to the container and gently wave their hands over the container to move air toward their noses (see Figure 17-1). Ask them what they notice. *(They should smell the perfume.)*

Give each student a glass of water and a sugar cube. Have the students

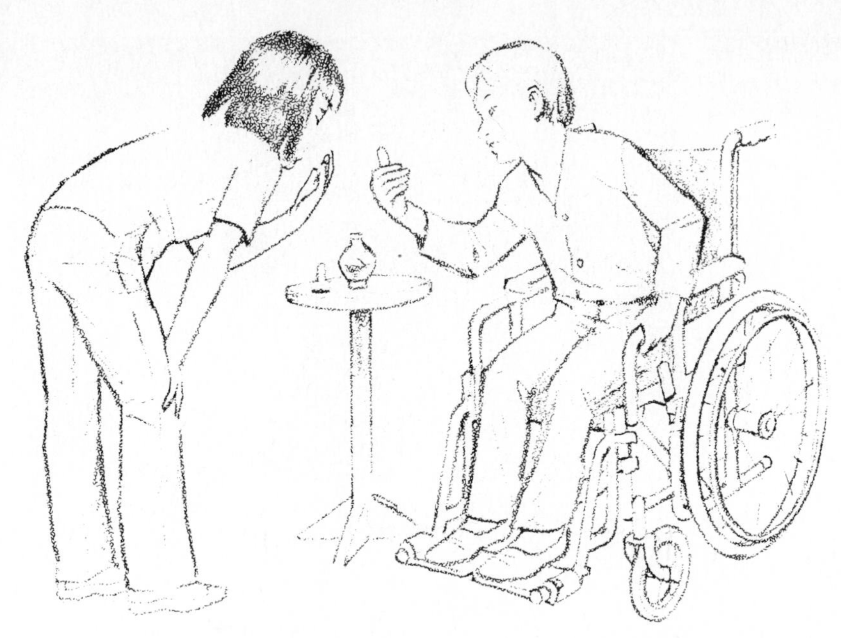

Figure 17-1. Demonstration of the behavior of small particles of matter.

take a small taste of the water and then place the sugar cube in the water. After the sugar dissolves, have the students taste the water again. Ask them what they can taste. *(Sugar.)* Can they see the sugar in the water? *(No.)*

Explain to the students that these two activities illustrate that the perfume and the sugar are made up of particles so small that, under the conditions of the experiment, they cannot be seen, but that these small particles are still the same original substances, as evidenced by smell and taste.

➢ PROPERTIES OF MATTER

Have students observe selected pieces of matter (simple objects such as pencils, oranges, shirts or dresses, chalk, books), and use their senses to describe what they observe. They should note properties such as color, shape, texture, taste, and smell. As part of this activity, you can have students order objects by selected properties—for example, texture, lightest to darkest yellow objects, least intense to most intense smell.

➢ THE BUTTON BOX

Another sorting activity which causes students to focus on describing properties of objects is to have them group buttons from a box which contains several types of buttons. Ask the students to separate the buttons into piles which contain buttons that are alike in some way. Then have them look at their neighbors' piles and try to tell on what basis they are grouped. Have them repeat the activity, this time regrouping the buttons on the basis of a different property.

Sorting buttons can help students focus on several properties of objects.

➢ MATTER TAKES UP SPACE

Show the class a glass full of water and some marbles. Ask them what would happen if a few marbles were placed in the glass. *(The water would overflow.)* Explain that this activity illustrates that matter takes up space and that if a space is "full" of one kind of matter, other matter cannot be placed in that space unless some of the original matter is removed.

Give each student an empty glass and some marbles, and tell the students to "fill" their empty glasses with marbles. Ask them if their glasses are "full." *(The glasses are "full" of marbles. The spaces between the marbles are "full" of air.)* Ask if they think it would be possible to put anything other than a gas into the glass full of marbles. Give them some sand and have them carefully pour it in around the marbles. Is the glass "full" now? Would it now be possible to place anything other than a gas in the "full" glass? Give the students some water to pour into the glass of marbles and sand. Again ask them if the glass is "full." *(Now it is "full" of matter other than gases, for all practical purposes.)*

➢ AIR IS MATTER

To demonstrate that air is matter and takes up space, have your students inflate a large, heavy-duty plastic trash bag and tie the opening shut. Then they can gently sit on the bag (see Figure 17-2), and they will be supported by the bag full of air.* Explain that air is occupying the space between them and the floor.

*If your students are too heavy to be supported by a plastic trash bag, use a plastic air mattress.

Figure 17-2.
Air is matter.

➢ WORK

To help students learn that work is done only when something is moved, have them apply a push or a pull (a force) to several small objects on their desks or slightly larger objects around the room. Explain that they are doing work because they are moving something.

Ask the students if they would be doing work if they pushed against the outside walls of the school building. *(Work would not be done unless they moved it.)*

➢ REDUCING FRICTION

Place a large box full of books on the classroom floor. Ask two students to move the box by sliding it across the floor. Then place the box in a wagon and have the same students move it again. Which is easier? *(Moving it in the wagon.)* Why? *(In sliding there is much more friction than in rolling.)* Explain that friction is a force which keeps stationary objects from moving and which slows down and stops moving objects.

➢ WHAT STOPS IT?

Have students roll a toy vehicle or a ball across a surface. What eventually happens? *(The object stops.)* Why? *(Because friction acts on the object to stop it.)* Have students try out the toy or ball on several different types of surfaces—hard floor, rug, sidewalk. Upon which surface do the objects stop fastest? What does that tell us about the relative amounts of friction on the different surfaces? *(The surfaces upon which the objects stop fastest have the greatest friction.)* Have the students rub their hands on each of the surfaces. Ask them what kind of surfaces have the least friction. *(The surfaces with the least friction feel hard and smooth.)*

➢ STORING ENERGY
Explain to the class that people and animals store energy to do work by eating and that we may also store energy in other ways. Have them wind a clock or a wind-up toy. Ask them what will now happen. *(The clock's hands should move and the toy should run; when energy is released, objects move.)* Ask the students to name other ways in which we store energy.

➢ BECOME A PARTICLE
Have the students pretend that they are particles of liquid water. Tell them that they are in a pan with other water particles and all at once the pan is placed over a fire. They become hotter and hotter until they make a big change. Ask them what that change is. *(They become a gas.)*

You can then have them return to liquid and finally to solid as they are cooled and then frozen.

Children's literature

Read *Push! Pull! Stop! Go!** to your students or have them read it, and then discuss with them the concepts presented.

Find other books on matter or energy in the school library, and identify them for the students or place them in an interest center.

Activity cards

An example of an activity card is shown in Figure 17-3 (page 402). Additional activity cards could be made for topics such as heating and cooling a balloon full of air, using lubricants to reduce friction, and making "smelling jars" (jars containing materials which can be identified by smell).

Bulletin boards

STATES OF MATTER
Divide a bulletin board into three sections labeled "Solid," "Liquid," and "Gas", and place in a pocket nearby several examples or pictures of matter in each of the states. Tell the students to place the matter and pictures in the appropriate places on the bulletin board. (See Figure 17-4.)

*R. J. Lefkowitz, *Push! Pull! Stop! Go!* Parents Magazine, New York, 1975.

- MAKE A BUTTON COLLECTION FROM WORN-OUT CLOTHING. TRY TO GET AT LEAST 10 DIFFERENT KINDS OF BUTTONS.
- DIVIDE YOUR BUTTON COLLECTION INTO TWO GROUPS WHICH HAVE BUTTONS THAT ARE ALIKE IN SOME WAY. HOW ARE THEY ALIKE? DIVIDE EACH GROUP INTO TWO MORE GROUPS. HOW ARE THE BUTTONS IN EACH GROUP ALIKE?

WHAT ARE YOUR FAVORITE KINDS OF BUTTONS LIKE?

WHAT ARE BUTTONS MADE OF?

Figure 17-3. Activity card: Buttons.

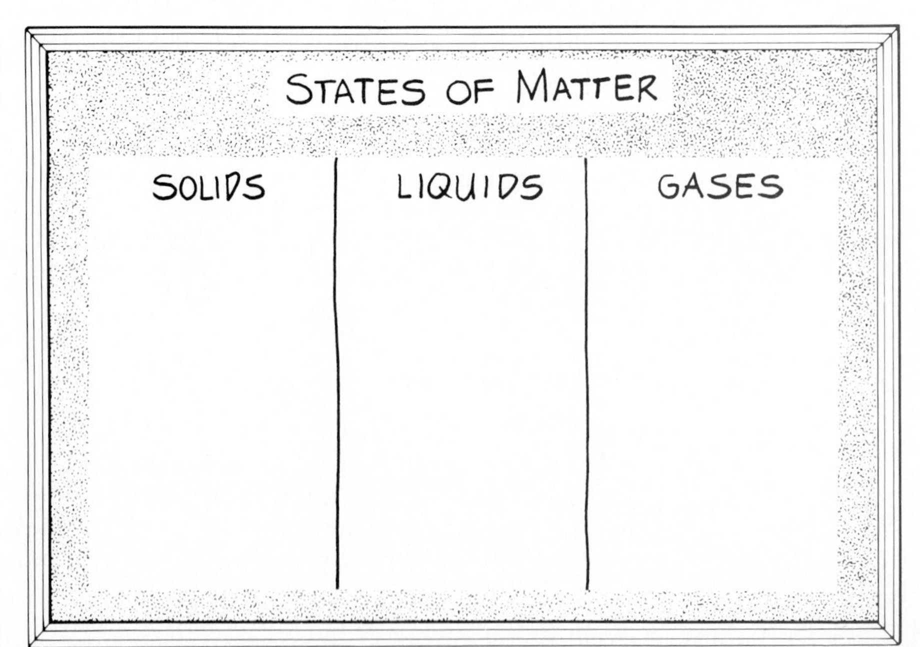

Figure 17-4. Bulletin board: States of matter.

WORK BEING DONE
Ask students to cut out magazine pictures which show work being done and attach them to a bulletin board.

Field trips

Take the class just outside the school building and tell them to look around and find three examples of work being done. *(Possibilities include a person carrying a package, an ant moving a small piece of food, an automobile traveling, a bird flying, wind moving the trees.)*

Have students also look for three examples of how friction is useful. *(Examples include an automobile stopping, a person climbing a rope, a child riding a tricycle.)*

MATTER AND ENERGY FOR INTERMEDIATE GRADES

Activities

➢ ATOMS: BUILDING BLOCKS
In this activity the students can make simple models of atoms out of gumdrops and toothpicks. They should use three colors of gumdrops—one color to represent protons, one to represent neutrons, and one to represent electrons. On the basis of information from their textbook or a reference book, they can construct models of some atoms. Protons and neutrons are closely packed into the nucleus of the atom, whereas electrons are held at a distance from the nucleus by the toothpicks. (See Figure 17-5.)

Tell the class that each chemical element has a specific number of protons and the atomic number of the element indicates how many protons that is. The number of electrons is equal to the number of protons, and the number of neutrons is determined by subtracting the atomic number from the mass number.

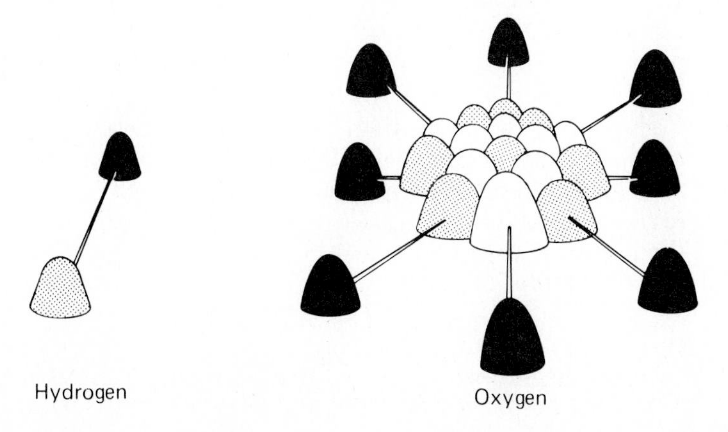

Figure 17-5. Gumdrop atom models.

After completing this activity, the students will be happy to clean up the protons, neutrons, and electrons.

➢ MOLECULAR MOVEMENT

Have students fill three glasses with water—one glass with very cold water, one with room-temperature water, and one with very hot water. Ask the students which glass contains the fastest-moving molecules. *(They should expect the molecular movement to be fastest in the hot water.)* Ask what would happen if a drop of food coloring were placed in each glass. Let the glasses sit still long enough for the water to stop moving, and then add a drop of food coloring to each glass of water (all three drops should be the same color). What can be seen? *(The coloring should spread evenly throughout the hot water the fastest.)* Have the students explain what has happened. *(The molecular movement in the hot water is fastest.)*

➢ PHYSICAL CHANGES AND CHEMICAL CHANGES

Tell students to make physical changes in several kinds of matter—they can, for example, freeze water, hammer a soft metal, tear paper, break up a sugar cube, change the shape of a piece of clay. Explain that the new material is still like the old except for its size, shape, or state. Ask the students to name additional examples of physical changes.

Explain that in a chemical change new materials are produced. One example of a chemical change that can be observed in the classroom is the burning of a small piece of paper.* After burning, what is left is no longer paper. What does remain after this chemical change? *(Carbon dioxide, water, and the residue left as ash.)* Another chemical change which students enjoy seeing is the combining of vinegar and baking soda. The chemical reaction is highly visible (a vigorous bubbling action). What are the bubbles? *(An escaping gas.)* How does this indicate that the change may be a chemical change? *(Apparently, new materials are being produced.)*

➢ PROPERTIES OF MATTER

Give your students samples of three to five similar-looking white "mystery powders"—perhaps salt, sugar, baking soda, cornstarch, and plaster of paris—and ask them to name as many differences or similarities as they can.

Have them do more than just simply taste and smell. They might try adding a little water, a lot of water, and a little vinegar; dropping the substances from a height of 10 centimeters (about 4 inches); feeling textures; and looking at the substances through a hand lens. Encourage them to investigate powders in a variety of ways.†

**Note:* This demonstration is to be performed by the teacher only. Never let students use matches or fire.

†For many suggestions related to this activity, see Del Alberti, *Mystery Powders,* Elementary Science Studies (ESS), Webster, McGraw-Hill, Manchester, Mo.,1974.

Students enjoy discovering the properties of "mystery powders."

➢ FINDING ELEMENTS

Elements are substances which are composed entirely of one kind of atom. The names of some elements will be familiar to your students—gold, silver, copper, aluminum. Have the students look at a list or a periodic table of the elements and then collect samples or pictures of several elements.

➢ ACIDS AND BASES

Chemical compounds are grouped in several ways. Two well-known groups are acids and bases.* Mild acids and bases are interesting to explore because they can be detected with an indicator referred to as "litmus paper" (available from a science supply house—see the Appendix). Blue litmus paper placed in an acid turns pink, and pink litmus paper placed in a base turns blue. Some toothpastes and shampoos are advertised as being basic or neutral or acid. The accuracy of these claims can be checked with litmus paper. Challenge your students to determine whether many of the products they use (food, drinks, toothpaste, etc.) are acid, base, or neutral.

➢ DISSOLVING SUBSTANCES FASTER

Challenge students to find ways that they can cause a common substance—e.g., salt—to dissolve in water more quickly. For example, does heating or cooling affect the rate? Stirring? Adding some other substance to water? What else?

**Note:* Warn your students that strong acids and strong bases are dangerous to handle.

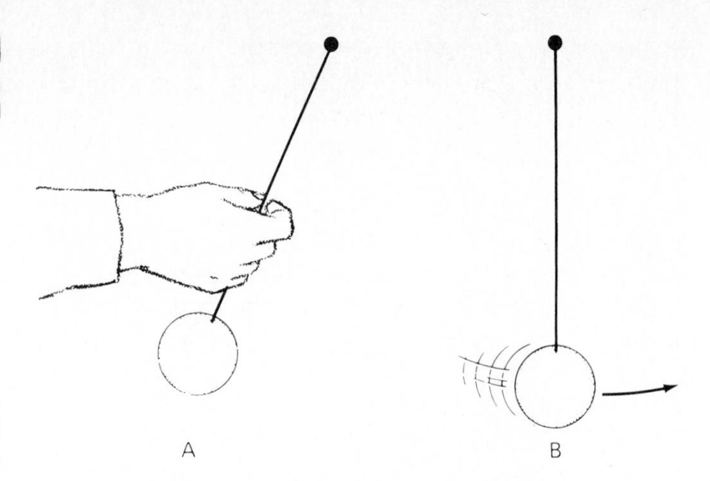

Figure 17-6. Making a pendulum.

⪢ POTENTIAL AND KINETIC ENERGY

Potential and kinetic energy can be illustrated with a simple pendulum, constructed of a piece of string and a weight, such as a fishing weight. (See Figure 17-6.) Tie one end of the string to the weight (bob), and attach the other end to a bulletin board or some other convenient surface in the classroom.

Pulling the bob to the side and holding it there demonstrates potential (stored) energy. The bob has the "potential" to move if released. What will happen if the bob is released? *(Once the bob is released, it moves.)* The moving bob demonstrates kinetic energy, since kinetic energy is the energy of motion.

Have the students identify the points in the arc of swing where the potential energy is greatest. *(At the outside points, where the bob stops briefly, and then swings back in the other direction.)* Where is kinetic energy (movement) the greatest?

⪢ SMELLING

One sense which is not well developed in most people is smell. Make a few "smelling jars" (or "smelling boxes") by placing small amounts of various common substances in small containers. Ask students to smell the containers and identify the contents. Have the students make additional smelling jars.

Help the students learn to identify the smells of several spices. Then cook samples of food using the spices and see if the students can detect the spices by smelling the samples.

In a similar activity, students could develop their sense of taste more fully.

⪢ DRAWING MOLECULES

If some of your students become interested in finding out which atoms make up some common substances, have them do a little library research and then make drawings of the atomic makeup of the substances. They may have heard that water is H_2O—drawing the structure could be very meaningful. Encourage them to find out about some slightly more complex but still common chemicals such as sugar and gasoline. Display the drawings in the classroom or in a hallway of the school.

NOTE ON THE ENERGY WE USE

Beginning in the late 1970s, people have become increasingly aware of the use of energy—an awareness which came about because some kinds of energy resources became scarce. The next several activities relate to use of energy in homes, while traveling, and at work.

ELECTRICAL ENERGY

Where does the electricity we use come from? Have a team of your students visit or telephone the local electrical utility office and find out the answer to this question. Then the team can prepare an illustrated report on how electrical energy is generated in your area.

Nationwide, what are the different ways of generating electricity? What percentage of the nation's electricity is generated in each manner? The utility office may be able to answer these questions also. If not, consult a recent reference book on energy or contact the federal agency charged with planning our country's energy policy. (In 1983 the U.S. Department of Energy was responsible for energy policy.)

In order to use energy wisely, it is important not to waste energy. Have

The distribution or generating plant of a local electric utility is a site for a possible field trip.

students find out how much electrical energy their families have used each month for the past 3 months. Help each student to make a plan of how his or her family could use 10 percent less energy next month. Tell them to ask for their families' cooperation in trying out the plans, and to report the results to the class at the end of a month. *(It will probably be necessary to determine electrical consumption for the same month 1 year earlier, to make a fair comparison; and the average temperatures of the two months should also be compared, in order to make the amounts of electrical consumption really comparable.)*

FOSSIL FUELS

Much of the energy which people use comes from fossil fuels such as petroleum products, natural gas, and coal. Why are these substances called "fossil fuels"? How large a supply of each of the major fossil fuels, in terms of years of available supply, remains in the earth? Ask the students to find out the answers to these questions, and then have them make a chart showing the amounts that remain.

How do we obtain these fuels? Where are these resources located in the United States? What environmental problems occur when we remove fuels from the earth? Have the students work in groups to compile reports that attempt to answer these questions, as well as other pertinent questions which you and they think of.

WIND ENERGY

Some people have stated that we should obtain and use more of the energy generated by the wind. Windmills have been used for many years. For what have they been used? What kind of research is presently being done on wind-generated energy? Have students write to the U.S. Department of Commerce, National Oceanic and Atmospheric Administration (see the Appendix), to request information on current research.

Hold a windmill construction contest. The winning windmill will be

Windmills such as this one have harnessed wind power for many years.

the one whose rotating shaft makes the greatest number of revolutions per minute when all are placed side by side in the wind. Tell students to do some library research to find out how to construct windmills and how to determine revolutions per minute.

➢ SOLAR ENERGY

Often when we think of solar energy, we think of converting sunlight to electricity. That kind of conversion is being done and may become an important energy source. However, there are many other ways of using solar energy. Houses and other buildings can incorporate "passive" solar designs, which use the sun for heat, instead of using other forms of energy which must be paid for. In warmer climates homes can be constructed in ways that allow avoidance of extra solar heating that is unwanted in summer.

Invite an architect to discuss passive solar designs with your class. Also obtain pictures and information on passive solar designs from the architect or from the U.S. Department of Energy (or other agency responsible for energy policy).

Have small groups of students list ways that a new home can be designed to use passive solar heating and ways in which an existing home can be modified to use passive solar heating. Students can make drawings or models to illustrate their design suggestions. Have a "Solar Day" and exhibit the illustrations and models for other students in your school. Invite news reporters from local newspapers and radio and television stations.

Many homes now are planned to use available solar energy efficiently.

▸ VISITING MOLECULES

Ask the students to relax and close their eyes. Tell them that they have become very small—so small, in fact, that they are able to travel between molecules in a piece of solid matter. Describe some of the motions of the molecules. Tell them that they are inside an ice cube. The temperature is rising, and the solid water changes to liquid water. Have them note the distance between molecules and the speed at which they are traveling. The heat continues to rise, and the liquid water changes to gaseous water. Have the students note molecular spacing and speed of travel.

After the fantasy visit with molecules is over, discuss with the class the molecular spacings, speed of travel, and effect of adding heat energy which they encountered during the visit.

Children's literature

Read *How Did We Find Out about Energy?** to your students or have them read it, and then discuss with them the concepts presented.

Find other books on matter or energy in the school library, and identify them for the students or place them in an interest center.

Activity cards

An example of an activity card is shown in Figure 17-7. Additional activity cards could be made for topics such as investigating many kinds of energy (wind, water, electrical, chemical, solar, nuclear), doing safe chemical activities, finding out whether some shapes of ice cubes melt faster than others, causing and observing physical and chemical changes, and conserving energy.

Bulletin boards

BUILD AN ATOM

Post on a bulletin board a periodic table of the elements or a list which includes at least the names, symbols, atomic numbers, and mass numbers. In pockets nearby place a large number of circles in three different colors—say, red to represent protons, yellow for electrons, and blue for neutrons. Allow students to go the bulletin board and construct an atom of any element they choose. For elements with more than 15 or 20 protons, tell the students to place perhaps 20 to 30 proton and neutron circles in the nucleus and write on one proton and one neutron the number of protons and neutrons present in that atom. (See Figure 17-8.)

NUCLEAR ENERGY

Obtain pictures and diagrams which explain the operation of a nuclear generating plant (see "Free and Inexpensive Materials"), and make an informational bulletin board. Ask some questions which will cause students to begin thinking about the pros and cons of nuclear power plants.

*Isaac Asimov, *How Did We Find Out about Energy?* Walker, New York, 1975.

BUILD A MODEL OF THE NUCLEUS OF A URANIUM-238 ATOM (REMEMBER—THE NUCLEUS HAS PROTONS AND NEUTRONS. LOOK IN YOUR TEXTBOOK OR A REFERENCE BOOK IF YOU NEED TO.). YOU CAN USE DRIED PEAS AND DRIED CORN TO REPRESENT THE PROTONS AND NEUTRONS.

SOME PEOPLE SAY NUCLEAR REACTORS ARE VERY SAFE. SOME SAY THEY MAY BE DANGEROUS. DO YOU THINK THEY ARE SAFE? SOMETIMES WE ARE AFRAID OF THINGS WE DO NOT KNOW MUCH ABOUT. HOW COULD YOU FIND OUT ENOUGH ABOUT NUCLEAR GENERATORS SO THAT YOU COULD MAKE UP YOUR MIND ABOUT WHETHER OR NOT THEY ARE SAFE?

FIND OUT HOW URANIUM-238 PRODUCES ENERGY. MAKE MORE MODELS OR A DRAWING TO SHOW HOW IT HAPPENS.

Figure 17-7. Activity card: Build an atom.

Figure 17-8. Bulletin board: Build an atom.

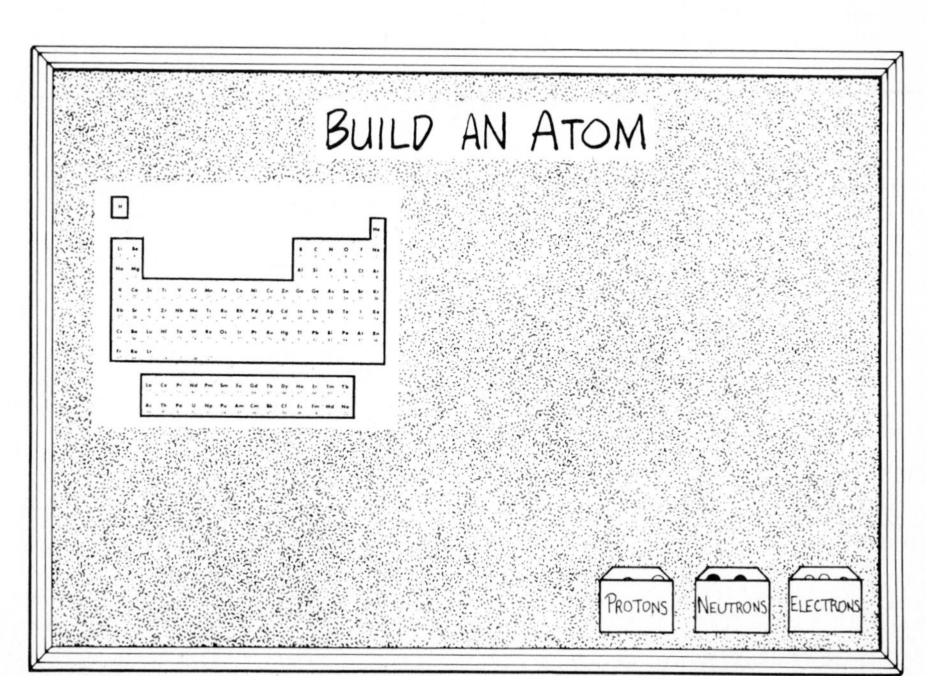

Field trips

The study of energy production and conservation can provide several opportunities for field trips. Visits can be made to a variety of generating plants—solar, nuclear, wind, hydroelectric, and fossil fuel.

Other trips can be made to businesses or homes which use energy-saving features, devices, or practices. A visit to a chemical plant and a presentation by a staff scientist could also be very rewarding for students.

RESOURCES

Sources of assistance

FREE AND INEXPENSIVE MATERIALS

Here are some good sources of free and inexpensive materials for use in teaching matter and energy:

1. *American Museum of Atomic Energy.** Several folder sets of energy activities.
2. *American Gas Association.* Information on gas as an energy source.
3. *American Petroleum Institute.* Information on the exploration, production, refining, and use of petroleum products.
4. *Edison Electric Institute.* Comic presentations on the atom, electricity, and fuel use, along with many other teaching aids.
5. *Energy Research and Development Administration (ERDA).* Many teaching aids on all aspects of human use of energy.

LOCAL RESOURCES

Local resources which can help in teaching matter and energy include the following:

1. Publications and other information from energy companies or utility companies.
2. Resource persons and education specialists from energy and utility companies.
3. Companies that generate scrap material which happen to be elements—such as copper and aluminum.

ADDITIONAL RESOURCES

1. Magazines such as *SciQuest, Science News, World,* and *Ranger Rick.*
2. Almost all major oil companies produce teaching materials.
3. The departments of education in many states have energy education specialists who can tell you about teaching materials which have been produced within the state and nationally.

*The addresses of specific agencies, organizations, and businesses are listed in the Appendix.

4. The state geologic survey, department of mineral resources, or equivalent may be able to identify sources of minerals (including elements) within your state.

Strengthening your background

1. Read the Time-Life books *Matter* by Ralph Lapp and *Energy* by Mitchell Wilson (see "Building Your Own Library").
2. Do an energy conservation analysis of your home; a checklist should be available from your local electrical or gas utility company.
3. Find a suitable recipe and prepare a food treat which illustrates the concept of heat changing the state of matter. (Your students would enjoy helping you make the treat and sharing it with you.)
4. Collect examples of as many elements as you can.
5. Visit a large museum of natural history and spend half a day in the section on matter and energy.
6. Read a short article on how nuclear generating plants operate.

Getting ready ahead of time

1. Prepare the materials for the bulletin board "Build an Atom."
2. Request to be placed on the mailing list for the Energy Research and Development Administration (see the Appendix), so that you can receive new teaching materials as they are produced.
3. Make a list of matter and energy books in your library. If the supply is small, give the librarian a copy of "Building a Classroom Library" and ask him or her to order some of the books.
4. Ask for a set of the energy activities available from the American Museum of Atomic Energy (see the Appendix). The activities cover many kinds of energy, not just nuclear.
5. Put together a set of the "mystery powders" described in the activity "Properties of Matter" for intermediate grades. Design activities using the powders which your class can perform.

Building your own library

Asimov, Isaac: *Inside the Atom,* 2d ed., Abelard-Schuman, New York, 1974.

Barker, Eric, and W. Millard: *Machines and Energy,* Arco, New York, 1972.

——— and ———: *Nature and Energy,* Arco, New York, 1972.

Blackwood, Paul: *Push and Pull—The Story of Energy,* rev. ed., McGraw-Hill, New York, n.d.

Booth, Verne, and Mortimer Bloom: *Physical Science: A Study of Matter and Energy,* 3d ed., Macmillan, New York, 1972.

Elementary Science Study (ESS): *Balloons and Gases,* Webster, McGraw-Hill, Manchester, Mo., 1970.

———: *Clay Boats,* Webster, McGraw-Hill, Manchester, Mo., 1976.

———: *Colored Solutions,* Webster, McGraw-Hill, Manchester, Mo., 1968.
———: *Gases and "Airs,"* Webster, McGraw-Hill, Manchester, Mo., 1967.
———: *Kitchen Physics,* Webster, McGraw-Hill, Manchester, Mo., 1967.
———: *Mystery Powders,* Webster, McGraw-Hill, Manchester, Mo., 1974.
———: *Pendulums,* Webster, McGraw-Hill, Manchester, Mo., 1976.
Lapp, Ralph: *Matter,* Time-Life, New York, 1969.
Wilson, Mitchell: *Energy,* Time-Life, New York, 1969.

Building a classroom library

Adler, Irving: *Energy,* John Day, New York, 1971.
Ardley, Neil: *Atoms and Energy,* F. Watts, New York, 1976.
Asimov, Isaac: *Building Blocks of the Universe,* Abelard-Schuman, New York, 1974.
———: *How Did We Find Out about Energy?* Walker, New York, 1975.
Bendick, Jeanne: *Solids, Liquids, and Gases,* F. Watts, New York, 1974.
Bronowski, J., and M. E. Selsam: *Biography of an Atom,* Harper and Row, New York, 1965.
Coombs, Charles: *Coal in the Energy Crisis,* Morrow, New York, 1980.
Fermi, Laura: *The Story of Atomic Energy,* Random House, New York, 1961.
Fuchs, Erich: *What Makes a Nuclear Power Plant Work?* Delacorte Press, Dell, New York, 1972.
Gallant, Roy: *Explorers of the Atom,* Doubleday, Garden City, N.Y., 1974.
Lefkowitz, R. J.: *Matter All around You,* Parents Magazine, New York, 1972.
McDonald, Lucile: *Windmills: An Old-New Energy Source,* Elsevier-Nelson, New York, 1980.
Metos, Thomas H., and Gary G. Bitter: *Exploring with Solar Energy,* Messner, New York, 1978.
Pine, Tillie: *Friction All Around,* McGraw-Hill, New York, 1960.
——— and Joseph Levine: *Energy All Around,* new ed., McGraw-Hill, New York, 1975.
Pringle, Laurence: *Nuclear Power: From Physics to Politics,* Macmillan, New York, 1979.
Veglahn, Nancy: *The Mysterious Rays: Marie Curie's World,* Coward-McCann, New York, 1977.
Watson, Jane W.: *Alternate Energy Sources,* F. Watts, New York, 1979.
Weiss, Ann E.: *The Nuclear Question,* Harcourt, Brace and World, New York, 1981.

CHAPTER 18

HEAT

IMPORTANT CONCEPTS

Heat is molecular motion in matter.

Temperature is a measure of the rate of molecular motion within matter.

Thermometers are used to measure the average amount of heat in a substance.

Heat is one form of energy.

Heat can be changed into other forms of energy and vice versa.

Heat energy travels by conduction, convection, and radiation.

Increasing the heat of a substance increases its rate of molecular motion and can change a solid to a liquid and a liquid to a gas.

Reducing the heat of a substance reduces its rate of molecular motion and can change a gas to a liquid and a liquid to a solid.

Increasing the heat of most matter causes it to expand, whereas reducing heat (cooling) causes most matter to contract.

Heat energy can be produced by chemical reactions—including combustion—and from electrical energy.

Radiant energy coming from the sun may be absorbed by solids, liquids, or gases and changed to heat energy.

Insulators can slow down the movement of heat energy.

INTRODUCTION: STUDYING HEAT

Heat is one of the many forms of energy which students commonly encounter in their environments. Several very interesting activities can familiarize children with the basic nature of this form of energy and also can teach them to use a thermometer to determine the average heat of a substance.

HEAT FOR PRIMARY GRADES

Activities

➤ MEASURING TEMPERATURE

To introduce students to thermometers, have them fill three glasses with water—hot, room-temperature, and cold—and insert a thermometer into each glass. What happens to the red line in the thermometers in the various glasses? *(It goes up or down, or becomes longer or shorter.)* Which water causes it to go up the most? *(Hot.)* Down the lowest? *(Cold.)* Does the thermometer in the room-temperature water change? *(Very little, or not at all.)*

Ask them to predict where the red line would stop if they mixed all the room-temperature water with the hot water. Have them check their predictions. Let them try other combinations.

What happpens if the thermometers are placed against their warm skin? Against an ice cube? In sunlight coming through a window? What is indicated if the red-colored liquid in the thermometer goes up? *(Warming.)* What is indicated if it goes down? *(Cooling.)*

➤ TEMPERATURE OF FREEZING WATER

Have students place a thermometer in a clear glass, half full of room-temperature water. Have them read the temperature. Then place the glass and thermometer in a freezer, and read and record the temperature every 15 minutes. How long does it take the water to freeze? What is the temperature of the water when it begins to freeze? *(Should be 0° C, or 32° F.)* Make a graph showing data all the way from room temperature to frozen solid (or colder). Would it make any difference if we added 2 teaspoons of salt to the water before putting it into the freezer? Sugar? Antifreeze?* Make predictions and try each mixture. Graph the results of each of these experiments.

**Warning:* Antifreeze is poisonous. Only the teacher should handle it or a mixture to which it has been added.

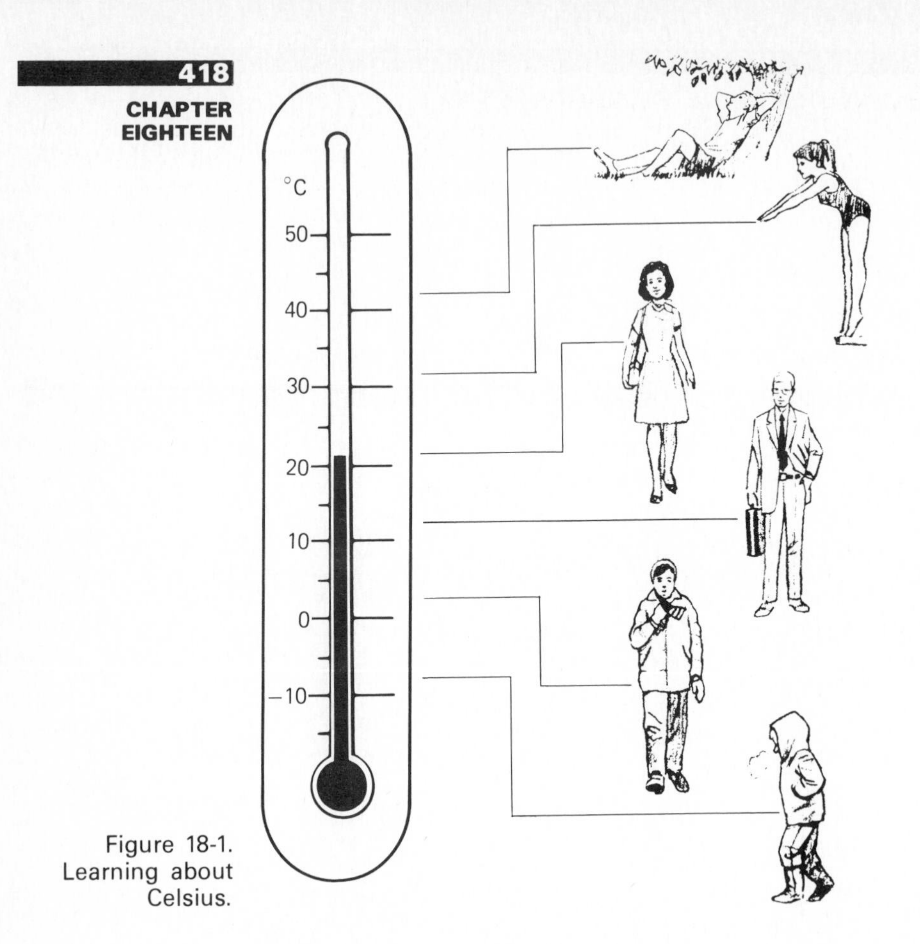

Figure 18-1. Learning about Celsius.

▶ CELSIUS SCALE

To help students become familiar with the Celsius scale, have them take readings of Celsius temperatures, throughout the school year, which can be used as references points—the freezing point of water, the boiling point of water, their body temperatures, an average classroom temperature, the temperature outside on a cool or cold winter day, the temperature on a sunny spring day, and others. At various times during the year, have them make drawings of objects, scenes, and people and animals at various temperatures, and post the pictures on a bulletin board beside a large diagram of a thermometer, similar to the one shown in Figure 18-1.

Ask these and other questions: "What would the temperature be on a hot summer day? A cold winter day? In your home? In a freezer? In a tropical fish aquarium?" Have the class check out their ideas about these temperatures.

▶ CHANGING TEMPERATURE CHANGES VOLUME

Have one student inflate a balloon slightly. Ask the class what would happen if the balloon were placed on a bed of ice cubes. Try it. *(The air in the balloon will take up less volume, and the balloon will shrink.)*

Ask them what would happen if the balloon were placed in a dish of very hot water. Try it. *(The air will expand, and the balloon will become larger.)* Most materials shrink when cooled and then expand when heated. Ask students to tell examples from their experience which illustrate this principle. How about a basketball on a cold day? What happens when you open a can of soda which has been sitting in the sun on a hot day?

HEAT CAN BE CHANGED TO LIGHT

Ask students what would happen if a nail (or you can use a bobby pin, a straight pin, or some other small metal object) were placed in a flame. The teacher should use a gas burner to perform this demonstration.* Hold the nail with a pair of pliers and place 1 to 1½ centimeters (about ½ inch) of the tip of it into the flame. After a while, the nail will glow a bright red. What kind of energy is being emitted by the nail? *(Light.)* Tell the students that this is one way of changing heat energy to another type of energy (light energy). Ask the students to name additional examples of changing heat energy to some other form. *(Burning gasoline to produce heat drives an automobile–mechanical energy.)*

HEATING SPEEDS EVAPORATION

Have the students place an equal number of drops of water on two similar dishes. Set one of the dishes in direct sunlight and the other in a shaded place nearby. Check periodically to determine whether one dish of water evaporates before the other. *(The water in the sunlight should evaporate considerably faster.)* Ask the students to explain the difference. *(The dish in direct sunlight was heated more.)* How much faster does the heated water evaporate? If we added more heat by placing the dish over a candle, would it evaporate still faster? What practical applications are made of evaporation by heat in the home? *(Clothes dryers, among others.)*

HEAT CAN BE CONDUCTED BY A SOLID

To demonstrate that heat can be transmitted by a solid (in a process called "conduction"), melt some wax and place small dots of it at four or five points along the bottom of a wire clothes hanger, as illustrated in Figure 18-2 (page 420). An easy way to do this is to light a candle and allow drops of wax to fall onto the hanger. Ask the class to predict what will happen if you place a heat source (a candle or can of Sterno) at one end of the hanger (see Figure 18-2). Try it and ask the students to describe what happens. *(The wax dot nearest the candle melts first, then the second dot melts, and so on.)* Have students offer explanations for their observations. *(The heat is conducted down the wire.)* How is this like touching the metal handle of a frying pan or a pot while it is over a burner? Several students will probably be able to tell the class about personal experiences with hot pots.

**Warning:* Never let children use fire or flames in the classroom.

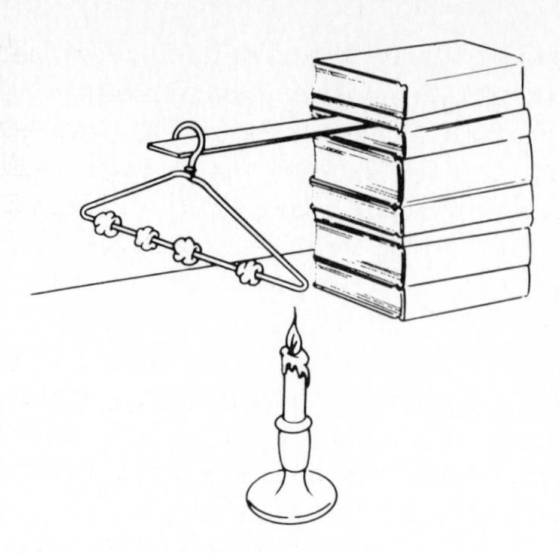

Figure 18-2. Transportation of heat by conduction. (Source: Adapted from M. K. Blecha et al., *Exploring Science,* Laidlaw, New York, 1979, grade 3, p. 77. By permission of Laidlaw Brothers, a Division of Doubleday and Company, Inc.)

➢ BECOMING COLD
Have the students write or tell about what it would be like to lose all sources of heat. What would happen as all things moved toward a temperature of absolute zero? *(Even molecular motion would stop at that temperature.)*

Children's literature

Read *A Book about Heat** to your students or have them read it, and then discuss with them the concepts presented.

Find other books on heat in the school library, and identify them for the students or place them in an interest center.

Activity cards

An example of an activity card is shown in Figure 18-3. Additional activity cards should be made for topics such as finding out how homes are heated, measuring temperatures in various places, determining the effect of heating on the rate at which ice cubes melt, and finding out what people can do to keep excess heat out of their homes in summer.

Bulletin boards

MEASURING TEMPERATURE
Make a large "ribbon" thermometer (see Figure 18-4) and place it on a bulletin board, along with a regular themometer. Have students take the real thermometer outside to measure the temperature several times a day, and then move the ribbon to indicate the outside temperature on the ribbon thermometer.

*Harlan Wade, *A Book about Heat,* Raintree, 1979.

CAN GLASS BE USED TO TRAP HEAT?

PLACE A THERMOMETER IN A SMALL OPEN BOX IN AN AQUARIUM–THE BOX IS USED ONLY TO KEEP DIRECT LIGHT FROM SHINING ON THE THERMOMETER.

PUT A GLASS TOP ON THE AQUARIUM.

READ THE TEMPERATURE WHEN THE AQUARIUM IS IN A SHADED PLACE.

PLACE THE AQUARIUM IN DIRECT SUNLIGHT AND READ THE TEMPERATURE AGAIN.

IS THIS WHY A CAR GETS HOT IN THE SUMMER?

Figure 18-3. Activity card: Trapping heat.

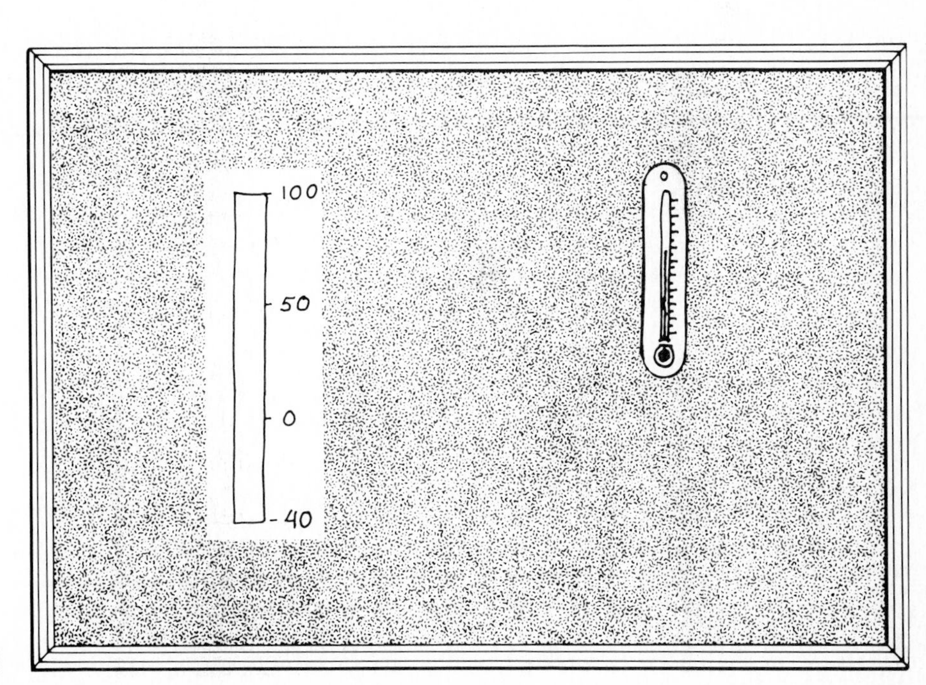

Figure 18-4. Bulletin board: Measuring temperature.

HOME HEATING
Obtain illustrated information on home heating from local electrical companies and natural gas and coal companies. Post the information on a bulletin board, and ask students to cut out and post magazine pictures that show means of heating homes—for example, advertisements for fireplaces, space heaters, and solar heating devices.

Field trips

Take the students outside on the school grounds and have them find where it is warmest and where it is coolest. Then analyze the differences between the warmest and coolest places. Which place do they enjoy most? Would they have the same preferences in 6 months? Have students estimate the temperatures in the warmest and coolest places and then check them with a Celsius thermometer.

HEAT FOR INTERMEDIATE GRADES

Activities

▶ HOMEMADE THERMOMETER
Construct a home-made thermometer similar to the one shown in Figure 18-5, using a small throwaway soft-drink bottle, a rubber stopper or a snap-on plastic cap from a soft-drink bottle, and a glass tube or a clear plastic drinking straw. Fill the bottle a little over half full with room-temperature colored water. Cut a hole in the stopper or cap and insert the tube or straw through the stopper or cap into the bottle so that the bottom of the tube or straw is 1 to 1½ centimeters (½ inch) from the bottom of the bottle. Make a tight seal between the tube or straw and the stopper or cap by using clay or melted wax.

Place the thermometer in hot water, in cold water, in direct sunlight, and in a refrigerator. Have the students note what happens. *(The liquid moves up the tube in warm places and down in cool places.)* Ask them to explain what is happening. *(Most of the effect is produced by the increase and decrease in volume of the air above the water in the thermometer as it is heated and cooled.)*

If the students would like to, they can calibrate this homemade thermometer by comparing it with a standard thermometer and marking the straw or tube. What range does the thermometer have? Will it work below 0° C or above 100°C? What kind of thermometer could be used outside the 0° to 100°C range?

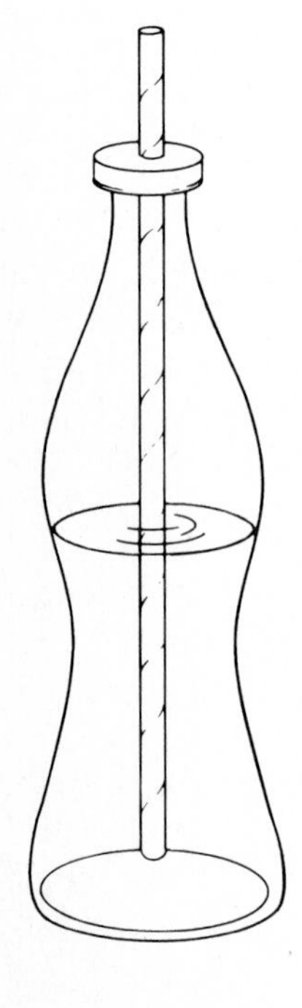

Figure 18-5. Making a thermometer.

Placing 16 students on a 4-by-4 square of floor tiles is a good way to demonstrate what happens when heat is added to a substance.

➢ EFFECT OF HEATING ON MOLECULES

The purpose of this activity is to demonstrate the effect of heating on molecules, as well as its role in changing solids to liquids and liquids to gases. Have 16 students each stand still on a floor tile in your room (or mark off a grid of 16 squares, each 30 centimeters or about 12 inches square). Explain that they represent a solid, and each "molecule" (each student) is doing relatively little moving. Then tell them that heat is being added: each molecule (each student) is to begin moving about a little, while still remaining in her or his square. Ask them how their slight movement is similar to that of molecules in a solid—a metal, for example. *(Adding heat does cause increased movement of the molecules.)*

Tell them to keep increasing their movement gradually, and then suddenly say, "Stop! Do not move at all." What do the students observe about their positions? *(Undoubtedly several will have their feet outside the original grid.)* How is this like a heated metal? *(The material expands when it is heated.)* What would happen if "heating" continued and the students' movement became even more vigorous? *(The "molecules" would become even more widely spaced; by analogy, the metal would eventually become liquid and then gaseous.)*

⋗ HEAT FROM CHEMICAL REACTIONS

Into each of four (or more) clear plastic drinking glasses pour 125 milliliters (about ½ cup) of room-temperature water. Have on hand at least the following chemicals: baking soda, sodium thiosulphate, washing soda, and copper chloride.* Ask the class what, if anything, would happen if each chemical were added to water? Have students measure the temperature of the water, add 1 teaspoon of a chemical to a glass, stir the mixture for approximately 1 minute, and take another temperature reading. Ask the students to describe what happened. Repeat the procedure with the other chemicals.

The changes in temperature are due to the reaction of the chemicals with the water or to the dissolving of the chemicals in the water. Most often when a chemical is added to water, the temperature increases. However, there are a few chemicals—sodium thiosulphate is an example†—which cause water to become cooler.

⋗ HEAT FROM ELECTRICITY

Have students collect pictures and advertisements of electrical devices which are used to generate heat. What different types of heaters are available? *(Glowing filament, quartz, radiant-panel—others?)* Bring an electrical space heater to class to demonstrate. Have students check the temperatures at successive ½-meter (about 1½-foot) distances from the heater. How close would they like to sit to a heater?

⋗ BECOMING FAMILIAR WITH TEMPERATURE READINGS

Have the students guess the temperatures (on the Celsius scale, of course) of various locations—inside a refrigerator, inside a freezer, in the classroom, outside in shade, and outside in sun, for example. Then have them measure the temperatures and evaluate their estimates. How many trials are necessary before the students become accurate (to some predetermined extent) at estimating the temperature?

After the class acquires skill in estimating, hold a "Celsius Olympics." Students who choose to participate will be taken to several locations and asked to estimate the temperatures. The student with the least average difference from the measured temperature is named "Best Guesser" and is awarded a thermometer as a trophy.

**Warning:* Stress the importance of treating these and all unfamiliar chemicals with respect. Tasting is not permitted. If you tell the students about this activity in advance, they may wish to bring other substances from home to try in the experiment. Tell them that they must have your approval before trying each chemical. Carefully read the information on each package, and approve the substance only if you are sure there are no hazards.

†Sodium thiosulphate is inexpensive and readily obtainable at photographic supply shops.

▶ INSULATING MATERIALS

Students are familiar with special insulators such as styrofoam picnic boxes and insulated containers for cold or hot drinks. However, they may not realize that rather ordinary materials can be used as insulators. Actually one of the best insulators is air—but there is a catch; it must be trapped air that is not free to circulate. This activity gives the class a chance to find out how well various other insulating materials work.

Give each student a 500-milliliter (1-pint) heat-proof jar with a lid and a box (with or without a cover). The boxes must be of uniform size—for convenience, use boxes of less than 30 centimeters, or about 12 inches, per side in length. Tell the students that they are to try using several different kinds of materials as insulators. You may wish to make a few suggestions—sand, crumpled newspaper, crushed leaves, styrofoam, grass—just to get the students thinking. Help them gather the insulating materials ahead of time.

On the day of the experiment, tell the students to put the jars in the center of the boxes, and fill the space around the jars with the insulating materials they have selected. After the boxes are ready, the teacher should heat some water to approximately 50°C and fill the jars.* The students can add the lids to the jars and the covers, if used, to the boxes. Have them take temperature readings every 30 minutes and plot the readings on a graph for each type of insulator. After deciding which insulator tested works best, ask the students what other insulators might be even better. Let them try the new suggestions if they want to.

Ask the students if the insulator which kept the water warmest for the longest period of time would also be the best for keeping a container of crushed ice coolest for the longest time. Let them repeat the experiment using ice, if they want to.

▶ A MOLECULE OF WATER

Have students write descriptions of what it would be like to be a molecule of water in an ice cube while heat is being applied. They are to eventually change to water vapor and escape into the atmosphere. Have a committee of students select the best parts or ideas from all the descriptions, and make a tape recording of one combined description.

Children's literature

Read *Heat and Its Uses*† to your students or have them read it, and then discuss with them the concepts presented.

Find other books on heat in the school library, and identify them for the students or place them in an interest center.

**Caution:* Do not let students heat the water or pour hot water; they could be badly burned.

†Irving Adler and Ruth Adler, *Heat and Its Uses,* John Day, New York, 1973.

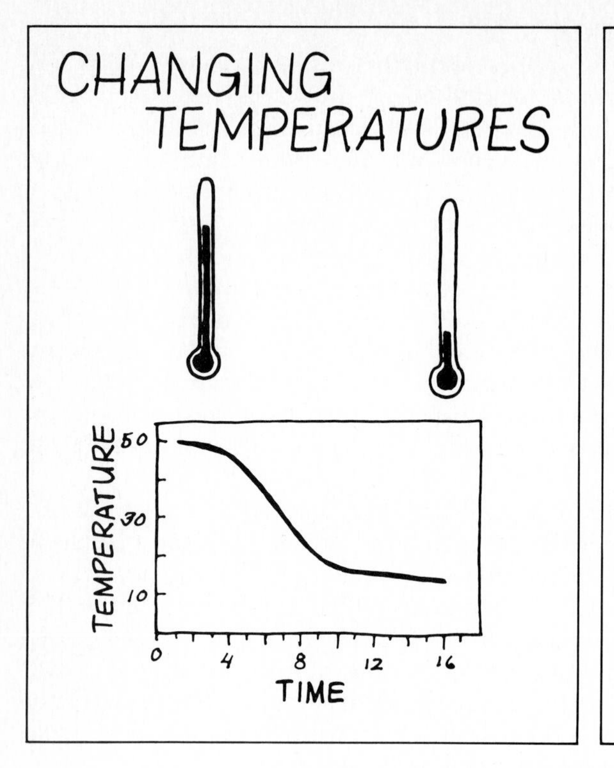

·PLACE A THERMOMETER IN EACH OF THE FOLLOWING PLACES: (1) A GLASS OF HOT WATER (2) A GLASS OF COLD WATER (3) A GLASS OF WARM WATER (4) THE AIR.

·READ AND RECORD EACH OF THE TEMPERATURES EVERY 2 MINUTES.

·DRAW A GRAPH FOR EACH LOCATION WHICH SHOWS HOW THE TEMPERATURE CHANGES.

CAN YOU DO SOMETHING TO CAUSE THE TEMPERATURES TO CHANGE MORE SLOWLY?

WOULD SALT WATER DO THE SAME?

Figure 18-6. Activity card: Changing temperatures.

Activity cards

An example of an activity card is shown in Figure 18-6. Additional activity cards could be made for topics such as the effectiveness of different kinds of home insulation materials, the amount of heat given off by different kinds of light bulbs, and the production of heat from other forms of energy.

Bulletin boards

TYPES OF THERMOMETERS
Have students make drawings for a bulletin board to illustrate the development of the modern thermometer and the diverse types of thermometers in use. They can do some research to find the information and pictures they need.

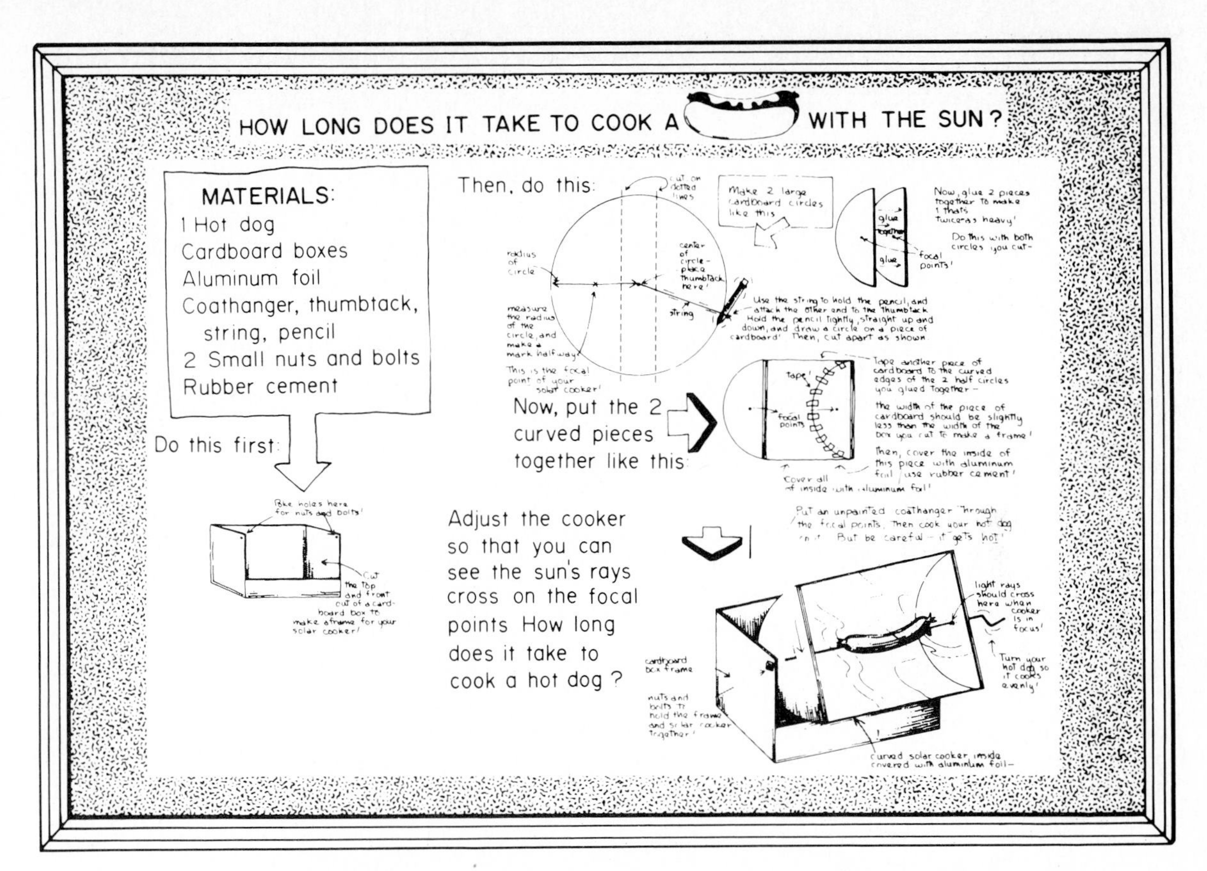

Figure 18-7. Bulletin board: Making a solar cooker. (Source: *Science Activities in Energy: Solar Energy,* U.S. Department of Energy, activity 4. Used by permission.)

SOLAR COOKERS

Post on a bulletin board instructions for constructing a small solar heater (a hot dog cooker), as illustrated in Figure 18-7. After students who choose to build cookers have completed theirs, hold a hot dog cookout for lunch for the class.

Field trips

Take the class to visit either a local house which is solar-heated or a demonstration project on solar heating.

On another field trip, have students place equal-sized ice cubes in various locations outside the school building. They are to determine which cubes melt fastest and then try to explain the reasons for the different melting rates.

RESOURCES

Sources of assistance

FREE AND INEXPENSIVE MATERIALS
Here are some good sources of free and inexpensive materials for use in teaching heat:

1. *American Museum of Atomic Energy.** A folder of activities on solar energy.
2. *American Gas Association.* Information on production of gas and on its uses in the home.

LOCAL RESOURCES
Local resources which can help in teaching about heat include the following:

1. Resource persons from electrical or gas utility companies.
2. A civil engineer who is willing to visit your class or school to discuss how the expansion and contraction of building materials in response to heating and cooling must be taken into consideration in construction work.

ADDITIONAL RESOURCES
1. Magazines such as *SciQuest* and *National Geographic World.*
2. Pamphlets and articles on heating and insulation from local utilities, newspapers, and building supply stores.

Strengthening your background

1. Do a heating and insulating energy check on your home, using a checklist obtained from your state energy advisory commission or from an electrical or gas utility company.
2. Talk with an older person about home heating in her or his youth. Ask especially about recollections of old pot-bellied stoves.
3. Read the chapter on heat in one of the junior high school textbooks listed in "Building Your Own Library" or in a textbook used in your school system.
4. Make a list of five chemicals which could be added to water to produce heat and which would be safe for use in an elementary school classroom.

*The addresses of specific agencies, organizations, and businesses are listed in the Appendix.

Getting ready ahead of time

1. Collect newspaper and magazine articles and pictures on home heating.
2. Write a guided-imagery activity for use in teaching heat.
3. Construct a "ribbon thermometer" for use with your class.
4. Make a collection of home insulating materials.*
5. Locate a source of supply of inexpensive thermometers for use in your classroom.

Building your own library

Adler, Irving: *Hot and Cold,* rev. ed., John Day, New York, 1975.
Blecha, Milo, et al.: *Exploring Matter and Energy,* Laidlaw, River Forest, Ill., 1980.
Brandwein, Paul: *Energy: A Physical Science,* Harcourt Brace Jovanovich, New York, 1975.
Elementary Science Study (ESS): *Heating and Cooling,* Webster, McGraw-Hill, Manchester, Mo., 1971.
———: *Ice Cubes,* Webster, McGraw-Hill, Manchester, Mo., 1974.
Feravolo, Rocco: *Easy Physics Project: Air, Water, and Heat,* Prentice-Hall, Englewood Cliffs, N.J., 1966.
Podendorf, Illa: *The True Book of Energy,* Children's Press, Chicago, Ill., 1963.
Ramsey, William, et al.: *Physical Science,* Holt, New York, 1978.

Building a classroom library

Adler, Irving: *Heat,* John Day, New York, 1964.
——— and Ruth Adler: *Heat and Its Uses,* John Day, New York, 1973.
Barratt, D. H.: *Heat,* World Publishing, Cleveland, 1972.
Bendick, Jeanne: *Heat and Temperature,* F. Watts, New York, 1974.
Cobb, Vicki: *Heat,* F. Watts, New York, 1974.
Feravolo, Rocco: *Junior Science Book of Heat,* Gerrard, Champaign, Ill., 1961.
Liss, Howard: *Heat,* Coward-McCann, New York, 1965.
Pine, Tillie, and Joseph Levine: *Energy All Around,* new ed., McGraw-Hill, New York, 1975.
Ruchlis, Hy: *The Wonder of Heat Energy,* Harper and Row, New York, 1961.

**Note:* Be careful when handling fiberglass. Wear gloves and long sleeves; you might even tie a scarf or handkerchief over nose and mouth.

Shimek, William: *The Celsius Thermometer,* Lerner, Minneapolis, Minn., 1975.
Simon, Seymour: *Hot and Cold,* McGraw-Hill, New York, 1972.
———: *Let's Try it Out: Hot and Cold,* McGraw-Hill, New York, 1972.
Stone, A. Harris, and Bertram Siegel: *The Heat's On!* Prentice-Hall, Englewood Cliffs, N.J., 1969.
Wade, Harlan: *A Book about Heat,* Raintree, Milwaukee, Wis., 1979.
Wilson, Mitchell: *Energy,* Time-Life, New York, 1969.
Wohlrabe, R.: *Exploring Solar Energy,* World Publishing, Cleveland, Ohio, 1966.

CHAPTER 19

LIGHT

IMPORTANT CONCEPTS

The sun is directly or indirectly the source of most of the light on earth.

Other forms of energy can be converted to light.

Light travels in straight lines.

Light may be reflected or absorbed.

Materials may be transparent, translucent, or opaque to light.

White light may be separated into the colors of the spectrum.

Light energy has behavior which is explained by both wave and particle properties.

Lenses are used to bend light.

Objects are seen by light which they produce or which they reflect.

A colored object appears to be the color or mixture of colors of the light which it reflects.

Light is scattered by rough surfaces.

Shadows are produced by objects which block light.

INTRODUCTION: STUDYING LIGHT

Light can be readily investigated by students in many ways. Several children's toys are based upon properties of light—periscopes, kaleidoscopes, and other toys which use mirrors. Much of what is known about the properties of light is too complex and difficult for young children, but there is no shortage of light phenomena with which they can experiment and from which they can learn.

In teaching light listen closely to children's questions or statements about light. These expressions of interest may give you ideas of ways to channel their interests into investigations which they can make individually or in groups.

One especially important aspect of the study of light relates to eyesight and eye care. Your students should become aware of the importance of proper eye care and of specific practices which they should follow in order to care for and protect their eyes.

LIGHT FOR PRIMARY GRADES

Activities

➢ SOURCES OF LIGHT

Show students a day and a night photograph or drawing of the same place. Ask them to name the sources of light in the scene. (Figure 19-1 shows an example.) They should note the dominance of sunlight in the

Figure 19-1. Sources of light. (Source: Adapted from Harold E. Tannenbaum et al., *Experiences in Science: Light and Shadow,* Webster Division, McGraw-Hill, New York, 1966, pp. 2 and 3.)

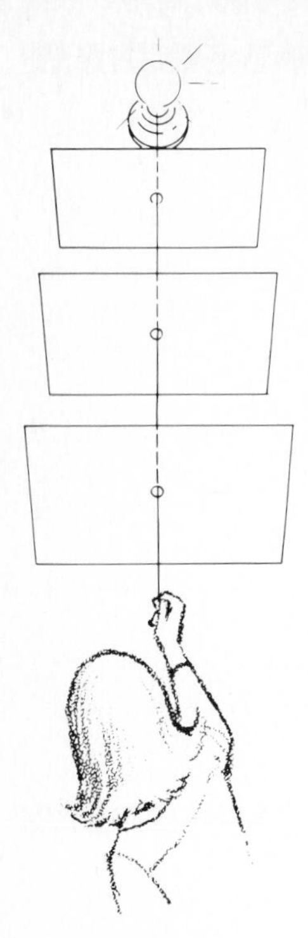

Figure 19-2. Demonstration that light travels in a straight line.

day scene. List on the blackboard all the light sources which the students can see in the scene. Ask them also to think of other light sources, and ask how the light is produced in each instance. *(For example, electricity produces the light of lamps and automobile lights, whereas burning produces the light of candles, lanterns, and fireplaces.)*

➢ LIGHT TRAVELS IN STRAIGHT LINES

Give each student three cards with holes punched in the centers. (All the holes must be the same distance from the bottoms of the cards.) Have the students arrange the cards in a row, one behind the other, so that they can see through all the holes. They can bend the bottom part of each card to make it stand up, or card stands or pieces of clay can be used to hold the cards erect. After the students have aligned the cards to enable them to see through all the holes, tell them to thread a piece of string through the three holes and use it to pull the cards together. (See Figure 19-2.) What does this demonstrate? *(The holes must be in a straight line in order for people to be able to see through them.)*

Ask two students to stand around a corner from each other. Can they hear each other talk? Can they see each other? Why not? *(Because light travels in straight lines and there are no reflecting surfaces such as mirrors in the proper locations.)* Continue the class discussion until you are sure the students understand that light travels in straight lines.

➢ WHAT WILL LIGHT PASS THROUGH?

Give students samples of transparent, translucent, and opaque objects, such as clear plastic, cellophane, waxed paper, thin tissue paper, dark-colored paper, and a piece of wood. Ask them to hold the objects up to sunlight to determine which will let light pass through and which will not. *(The transparent and translucent samples will allow light to pass through.)* Ask the students which materials they can see through clearly. *(Transparent ones.)* The translucent objects (waxed paper, thin tissue paper) will allow light to pass through, but sharp images cannot be seen through them. Which objects will not allow light through?

Ask the students to identify other materials in the classroom or their homes which they see through clearly, materials which allow light to pass through but do not transmit sharp images, and materials which block light. How would each of the kinds of materials be used? If appropriate for your grade level, introduce the terms "transparent," "translucent," and "opaque."

➢ REFLECTING LIGHT WITH A MIRROR

In this activity, the students will begin to develop some understanding of the concept of angles of incidence and reflection of light. However, they will not be formally addressing this concept until the upper elementary grades.

Give the students small mirrors and have them stand in the sunlight and reflect sunlight onto a wall. Tell them to move the reflections around. Can they make them shine on a spot which you specify?

Place a mirror where the class can see it, and shine a flashlight into it. Have the students locate the spot where the light is reflected. Move the flashlight and ask them what happens to the reflected light. Ask the students where they would have to hold the flashlight to get the light to shine on a designated spot. Have them try this several times until they become proficient at it.

➢ LIGHT IN A BAG

Have each student divide a 3- by 5-inch index card into quarters, then color or paint three of the quarters red, yellow, and blue, while leaving the remaining quarter white. Tell them to put their cards inside very heavy (preferably black) paper bags which exclude as much light as possible. Then they should put their faces into the bags and close the bags about their heads so that no light enters. Ask what they can see. *(Nothing.)* Tell them to let just a little light in, and again ask what they can see. *(They should be able to see the cards, but colors will not be distinguishable in low light.)* Tell them to keep letting more light in, until finally all the colors become distinguishable.

Ask them what is necessary for seeing. *(Eyes and light.)* Would the colors be visible if a lighted flashlight could be placed in the sack in such a way that all other light was excluded?

➢ SHADOWS

Have students identify some objects which cast dark shadows and some which do not. What is one characteristic of the objects which cast dark shadows? *(They block light which strikes them.)*

Have students make shadows with the following (or similar) objects: a coin, a cylinder such as a flashlight battery, a rectangular box such as a cigar box, a cube such as a building block, and a spherical ball. Ask them if they can use the coin to make shadows which look like a line, a circle, and an egg. Ask them to see how many different-shaped shadows they can make with the other objects.

➢ DIRECTION OF SHADOWS FROM THE SUN

Have students place a golf tee or a similar small object on a piece of paper and trace around its base. Early in the morning on a clear day they should place the paper at the point of intersection of two cracks on a section of the sidewalk which will remain in sunlight all day. Tell them to place the golf tee on its mark and trace around the shadow. Repeat at noon and again late in the afternoon.

Have students note where the sun is when the shadow is in a certain position. *(The sun is on the opposite side.)* When is the shadow shortest? Longest? If the grade level is appropriate, tell students that the sun rises in the east and sets in the west.

If a window of the classroom is exposed to direct sunlight, paint or tape an X on the window. Have the students mark on the floor or wall where the shadow falls at selected hours—say, nine o'clock, noon, and three o'clock. Will the shadow be in the same place the next day? A week

The shadow of this golf tee may be observed at several different times during the day to learn about the "sun's movement."

later? A month later? Ask them to explain any differences which occur. *(Change in the sun's elevation during the seasons will cause differences, and so will changing to or from daylight savings time.)*

➤ SHADOW PLAY

Have students cut out shapes of animals, people, clouds, and other objects, and attach them to sticks. Place a bright light behind a white bedsheet on a stage, as shown in Figure 19-3. The students can then make up a story and present a shadow play to the class.

Ask the students: "How can you make the 'characters' appear larger? Smaller? What is the relationship between light source, object, and shadow?"

Figure 19-3. Making shadows.

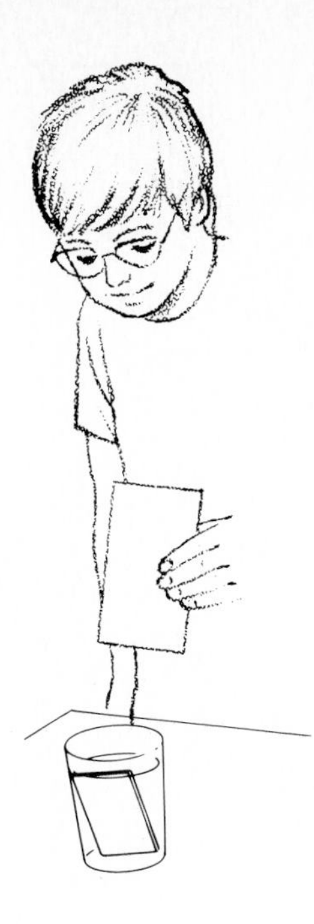

Figure 19-4. Sunlight has all colors. (Source: Adapted from Milo K. Blecha, P. G. Gega, and M. Green, *Exploring Science: Green Book,* 2d ed., Laidlaw, New York, 1979, grade 5, p. 161. By permission of Laidlaw Brothers, a Division of Doubleday and Company, Inc.)

COLORS OF THE RAINBOW

Use a prism to separate a beam of sunlight into its component colors. Allow the students to use the prism, and ask them to name the major colors. Tell the students that they can make a "prism" by using a glass of water and a small flat mirror as illustrated in Figure 19-4. Let them try out this idea, and have them show you when they have been successful. Ask what prisms tell us about light from the sun. *(It contains all the colors.)*

FAVORITE COLORS

Ask each student to name his or her favorite color and then to make a painting using chiefly that favorite color. Have students describe how it would feel to be their favorite colors. If they want to, ask them to explain how they chose their favorite colors.

SIGHT

Invite the school nurse or a doctor to visit the class and make a presentation on proper eye care, possible eye problems, and how eyes are checked. If possible, the visitor should bring along a model of the eye and should point out the major parts to the class and also compare the model with a diagram in the students' textbook. If the nurse or doctor can also bring the appropriate equipment to the classroom, have students use the equipment to look into each other's eyes. How do we properly care for our eyes? How often should they be checked?

BLINDNESS

Invite a representative of a society for the blind to make a presentation to the class, and ask the person to present information on the causes of blindness and on how blind persons live. How should the students treat a classmate who is blind? A blind person whom they meet on the street or in a shop?

I'M A SHADOW?

Ask the students to pretend that they have become their own shadows for a short while. Later have them describe what it was like being a shadow during periods of activity, when they came into bright light, when the sun began to get low in the sky, when the sun went behind a cloud, and so forth.

·CUT OUT A CARDBOARD PERSON.
·USING A LAMP OR A FLASHLIGHT, MAKE A SHADOW OF THE "PERSON."
·MAKE THE SHADOW LARGER.
·MAKE IT SMALLER.

CUT OUT OTHER FIGURES AND PRODUCE A SHADOW PLAY.

Figure 19-5. Activity card: Shadows.

Children's literature

Read *The Day We Saw the Sun Come Up** or *What Could You See?*† to your students or have them read one or both books, and then discuss with them the concepts presented.

Find other books on light in the school library, and identify them for the students or place them in an interest center.

Activity cards

An example of an activity card is shown in Figure 19-5, above. Additional activity cards could be made for topics such as playing shadow games, following shadow movement throughout the day, identifying the kinds of days on which shadows are visible, listing various sources of light, and using mirrors to signal.

Bulletin boards

LIGHT
Have students cut out and bring to school magazine pictures which illustrate different sources of light. Post the pictures on the bulletin board with a suitable heading, such as "Light Sources."

*Alice Goudey, *The Day We Saw the Sun Come Up*, Scribner, New York, 1961.

†Jeanne Bendick, *What Could You See?* Whitt House, 1957.

Figure 19-6.
Bulletin board: Pin on the shadows.

SHADOWS
Set up a bulletin board similar to the one illustrated in Figure 19-6. Attach colored cutouts of a few objects to the bulletin board, and a "sun" in the "east." In a nearby pocket place "shadows" (black cutouts) of the objects. Have one student pin the shadows where they belong. *(On the opposite side of each object from the sun.)* Have another student verify placement of the shadows. Put the shadows back in the pocket, move the sun halfway toward "noon," and assign two more students to place and verify the shadows. Keep repeating the activity until the sun "sets" in the "west."

Field trips and outdoor activities

On a sunny day take your students outside and look at the shadow of a building—the taller the better. Ask the students how far they think the shadow will move in 10 minutes. Have them mark a point or edge of the shadow on the ground. Then, for 10 minutes, let the students play shadow tag—whoever is "it" must step on someone else's shadow.

After 10 minutes, have the students mark the same point or edge of the shadow on the ground. Ask the students if they are surprised by the amount of movement. Ask them where they think the shadow will be in another 10 minutes. Again have them check the accuracy of their predictions.

Challenge the students to do some of these activities:

1. Make the shadows of two students shake hands without the students actually touching each other.
2. Find some way to separate themselves temporarily from their shadows. Point out that when they are standing on the sidewalk, their feet and their shadows' feet touch—how can they be separated?
3. Do something which causes them to cast no shadows. *(Step into a larger shadow.)*

Divide the students into pairs and give them blindfolds. One member of each pair is to wear a blindfold for several minutes, while the other leads him or her around a designated area to learn about the objects located in that area. Then have the pairs exchange roles; the other student wears a blindfold and is led around a different designated area. Back in the classroom, ask the students to discuss their experiences in exploring without use of the sense of sight.

LIGHT FOR INTERMEDIATE GRADES

Activities

➢ SOURCES OF LIGHT

Students in intermediate grades already know that the sun produces most of the light people use in the daytime. But what other light sources do we use? In this activity the class can learn two things: first, that the sun is the ultimate source of most of the light which we use; second, that one form of energy can be changed to another.

Have the students list as many sources of light other than the sun as they can. *(Lamps, flashlights, lanterns, lightning bugs, etc.)* Then have them name the immediate energy which produces each light. *(Electricity, burning fuel, chemical reactions, etc.)* Finally have them trace that energy source back as far as they can. *(In almost all cases they should be able to trace energy back to the sun as the ultimate source.)*

➢ ANGLES OF REFLECTING LIGHT IN AN AQUARIUM

In this activity, the students should note the similarity of the angles at which a beam of light strikes and leaves two surfaces—the top of the water in an aquarium and a mirror placed on the bottom of the aquarium. (This is a good introduction to the activity below.)

To prepare for the activity, place a small mirror flat on the bottom of an aquarium, as shown in Figure 19-7. If the water is not perfectly clear, the light beam will be seen more clearly; add a *little* soil or a few drops of milk to create a slight cloudiness. Blowing some chalk dust into the air over the aquarium will also improve the visibility of the light beam in the air. Then darken the room and shine a bright flashlight into the aquarium. Ask the students to discuss and explain what they observe.

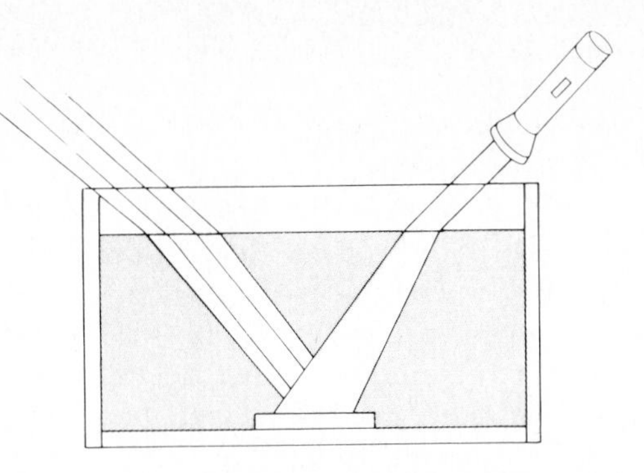

Figure 19-7. Angles of reflection in an aquarium. (Source: Adapted from P. F. Brandwein et al., *Concepts in Science: Orange Level,* Curie ed., Harcourt Brace Jovanovich, New York, 1980, p. 49.)

ANGLES OF REFLECTING LIGHT IN A MIRROR

Give the students a piece of cardboard or styrofoam approximately 40 centimeters (16 inches) square, and tell them to construct a setup like the one shown in Figure 19-8. The line on the cardboard is drawn between the midpoints of the two sides (see Figure 19-8*a*). Use a highly polished piece of metal as a mirror if possible; or use a thin glass mirror.

First have the students stick a pin into the right-hand side of the cardboard. Then they should place another pin so that the bases of the two pins will be in a straight line with the place where the midline meets the mirror. To do this, they should place one eye at the level of the cardboard and sight until all three points are lined up. Next have students look into the mirror from the other side of the midline and determine along what line of sight the bases of the two pins and their images in the mirror appear to be a single pin (see Figure 19-8*b*).

Then they should insert two more pins into the left side of the cardboard so that they are in a straight line with the "single-pin" image. Tell them to draw lines from the original two pins to the midpoint at the mirror and from the other two pins which were lined up with the "single-pin" image to the midpoint. Measure the angle between each of the lines and the edge of the cardboard. What is true about the sizes of the two angles? *(Allowing for slight errors in measuring, the angle of incidence of light—the angle between the line drawn from the first two pins and the edge of the cardboard—should be equal to the angle of reflection—the angle between the line drawn from the last two pins and the edge of the cardboard).*

Have the students try the activity again, placing the pins in different locations. Are the angles still equal?

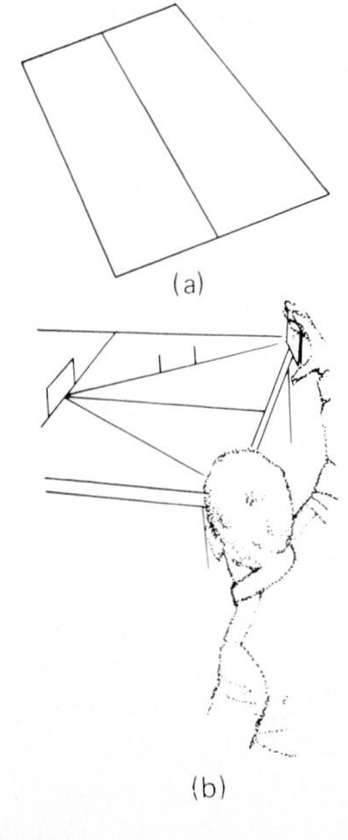

Figure 19-8. Angle of incidence equals angle of reflection. (Source: G. C. Mallinson et al., *Science: Understanding Your Environment,* Silver Burdett, Morristown, N.J., 1975, level 4, p. 171.)

HOW TALL IS A "FULL-LENGTH" MIRROR?

Have students stand approximately 4 meters (12 feet) from a full-length mirror (defined as one which runs from the floor to the top of the head of the tallest person in the household). Using pieces of cardboard, cover as much of both the top and the bottom of the mirror as possible until each student can just see both head and feet in the mirror. Compare the height of the "full-length" mirror with the student's actual height.

Is there a difference in the heights of "full-length" mirrors for tall and short persons? Ask the students what height would be required for "full-length" mirrors for their families. Have them find out from a store the cost per square foot (meter) to have a mirror made and then calculate the savings on getting a real "full-length" mirror as compared with what people usually think of as a full-length mirror.

Why would some people still choose the longer mirror? Which would the students prefer to have? Ask them to explain why a "full-length" mirror does not have to extend to the floor. *(They should relate their answer to angle of incidence and angle of reflection.)*

IMAGE IN A MIRROR

Hang a 20- by 25-centimeter (8- by 10-inch) mirror on the classroom wall at a height such that a student can see the top of his or her head in it. Have a student stand approximately 3 feet (1 meter) from the mirror. Ask the student to "point out how low on your body you can see in the mirror." Then ask the student, "How far down on your body do you think you will be able to see if you back up 4 meters (12 feet) farther

Students can discover something quite surprising by using a wall mirror. To find out what that "surprising" discovery is, do the activity "Image in a Mirror."

away from the mirror?" Have the student back up. Ask, "Are you surprised by what you see?" (Do not let the other students know what is seen.) Allow all the students, in turn, to do the same activity. Finally, ask the students to explain what they saw in terms of the angles between their eyes, the mirror, and the lowest point on their bodies which they could see.

➤ SCATTERED LIGHT

Darken the room, blow some chalk dust into the air, and shine a bright flashlight on several different surfaces. First try a mirror. What can be seen? Next try a light-colored, rough surface, such as a ceiling tile. Is there any difference between this observation and the previous one? *(The light beam may be seen shining onto the tile, but then the light scatters.)* What does this demonstrate? *(Rough surfaces scatter light.)* Have the students try surfaces with varying degrees of roughness. Does the reflected light appear the same with all degrees of roughness?

➤ ABSORPTION OF LIGHT

In this activity the students can examine the relative degrees to which dark and light surfaces absorb light. Again, darken the room and blow some chalk dust into the air. Place first white and then black paper of the same texture in front of a flashlight. Does the chalk dust show up any differences in the amount of light being reflected from the two papers? *(The white will reflect more.)*

Try to obtain one or more shades of gray paper of the same texture and a light meter (perhaps from an amateur or professional photographer). Ask the students whether they think there will be any differences in the meter readings for the different papers. If so, why? Hold the meter the same distance from the black, white, and gray papers, and also keep the flashlight at a constant distance from the papers. Take readings for the different papers and discuss the results with the class, comparing them with the students' expectations. If the light meter is not too delicate or expensive, have the students take readings in several places around the classroom.

➤ COLORS

Students can perform this activity to demonstrate that an object appears to be the color of light which it reflects—i.e., that a red ball looks red because it reflects red light. In a dark room shine a flashlight on a red ball (or some other red object). What color is it? *(Red.)* Place green cellophane over the light, and shine the green light on the ball. What color is it? *(It appears dark, but the color cannot easily be identified.)* Why is this so? *(The green cellophane blocks the red component of light, and thus there is no red light to be reflected.)*

Have the students try the same activity with different objects and cellophane of other colors. What hapens with primary colors? Is there a difference when they use secondary colors?

A piece of wax paper and a drop of water can make a "magnifying glass."

➢ MAKING YOUR OWN MAGNIFYING GLASS

Most intermediate-grade students have had some experience with a magnifying glass. If there are any who have not, have them use one. Ask them, "Where is the lens thickest?" *(At the middle.)* "What happens when you look at distant objects?" *(The lens inverts them.)*

After you are sure all the students are familiar with magnifying glasses, tell the class that they can build their own "magnifying glasses" by simply using newsprint, waxed paper, and water. They should put a piece of waxed paper on a piece of newsprint and put a drop of water on top of the waxed paper. What happens? *(The drop of water balls up, and the print under the drop is magnified.)* How is the drop of water like a magnifying glass? *(Thicker at the middle—thinner at the edges.)* Could they make larger lenses of different shapes by using ice? Let them try it. One problem they will encounter is how to freeze water so that it remains clear rather than becoming frosty. Let them solve that problem on their own.

Figure 19-9. Pinhole camera. (Source: Adapted from Milo K. Blecha, P. G. Gega, and M. Green, *Exploring Science: Green Book,* 2d ed., Laidlaw, New York, 1979, grade 5, p. 171. By permission of Laidlaw Brothers, a Division of Doubleday and Company, Inc.)

PINHOLE CAMERA

Tell your students to bring to class some oatmeal boxes, or other cylindrical boxes, so they can make pinhole "cameras" (see Figure 19-9). They will also need some tissue or waxed paper, a small amount of aluminum foil, and masking tape.

Cut a 2- by 2-centimeter-square (1- by 1-inch-square) hole in the bottom of the box, place aluminum foil over the opening, and make a pinhole in the foil. Tape the tissue or waxed paper over the opening on the other end, making sure the surface is smooth. Point the aluminum-foil end of the camera at a bright scene—an outdoor view on a bright day, or a light bulb or other light source indoors. What do students notice about the image which is on the tissue or waxed-paper screen? *(It is inverted.)*

If any of your students are amateur photographers and have darkrooms, challenge them to make the pinhole cameras into actual working cameras. They can experiment to determine exposure times and the effect of the size of the hole upon the picture.

Bring in or have a student bring in a camera with a B (bulb) setting, place a piece of tissue or waxed paper over the place where the film usually rests, and compare the image with that of a pinhole camera.

LIGHT DIMS WITH DISTANCE

Cut a 2-centimeter-square (or 1-inch-square) hole in a piece of tagboard or dark paper. Draw a grid of 2-centimeter (1-inch) squares on a piece of thin paper (onionskin or tissue). (See Figure 19-10, page 446.) Set up the tagboard in front of a very bright light—a floor lamp, if possible.

Have the students move the grid paper until the light fills a 2×2 grid and then measure the distance from the grid to the hole in the tagboard. Ask the students what happens as they move the grid farther from the hole. *(The light covers a larger area but is less bright.)*

Now have the students move the grid twice as far away from the hole as when the light covered a 2×2 grid. Ask them to compare the amounts of light at the different locations. Have the students note the

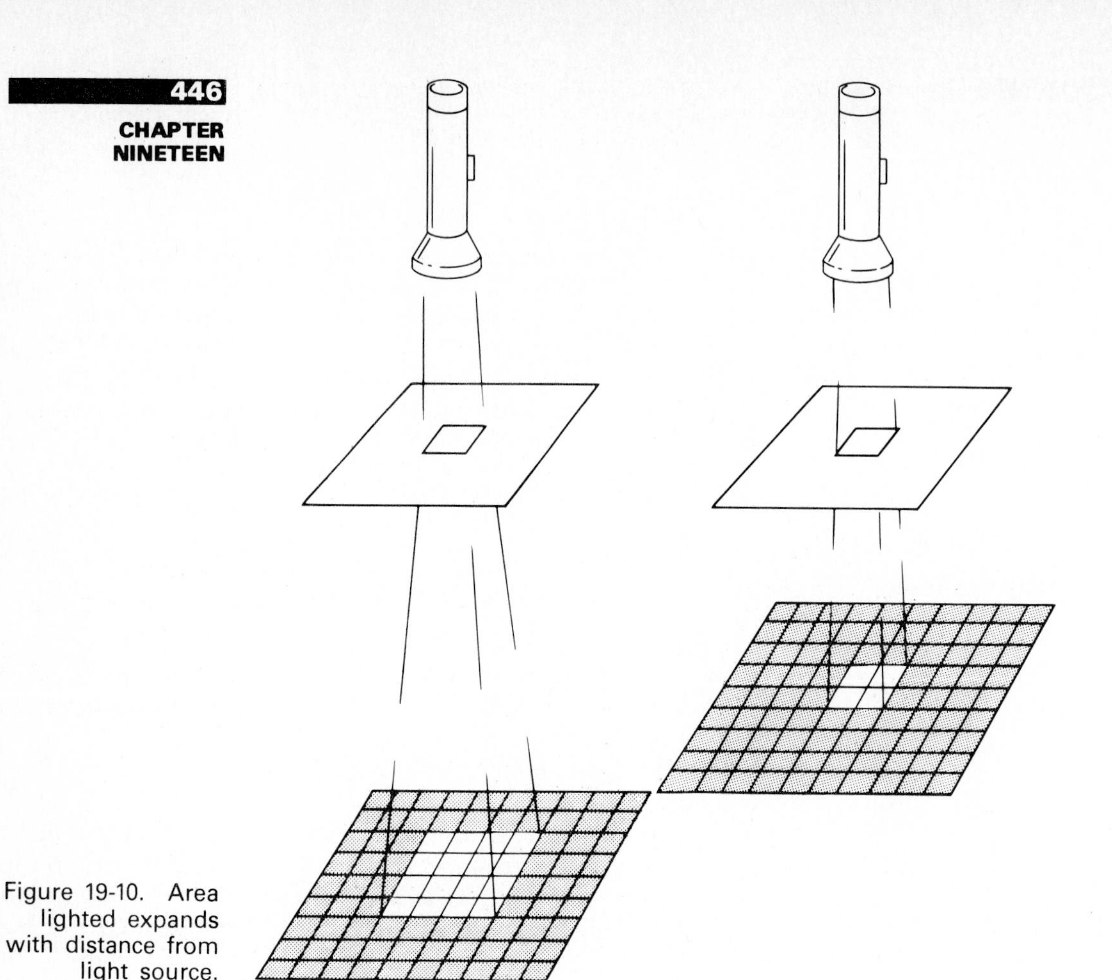

Figure 19-10. Area lighted expands with distance from light source.

size of the grid which is lit at the new location. *(It should be a 4 × 4 grid.)* How many square centimeters (inches) are now being lit? *(16.)* How many were lit on the 2 × 2 grid? *(4.)* This activity illustrates the concept that when the light source is twice as far away, the intensity of the light is only one-fourth as great for a given area. *(When the distance was doubled, the same amount of light covered 16 squares rather than 4.)*

Try to bring to class or borrow a 110 camera, and have the students note at what distance a flash picture may be taken (many cameras indicate about 2 meters or 7 feet). If they attempted to take a flash picture at 4 meters (or about 14 feet), would it turn out? *(Not very well—there would only be one-fourth as much light as recommended. And if they were 20 meters or about 70 feet away, there would be only one-hundredth as much light as recommended.)* Ask them what they think would happen if they tried to take a picture in a large auditorium with a 110 camera. Tell them to remember this exercise when they want to take photographs at concerts, graduation exercises, or other large gatherings, and to make sure they use an appropriate camera and film.

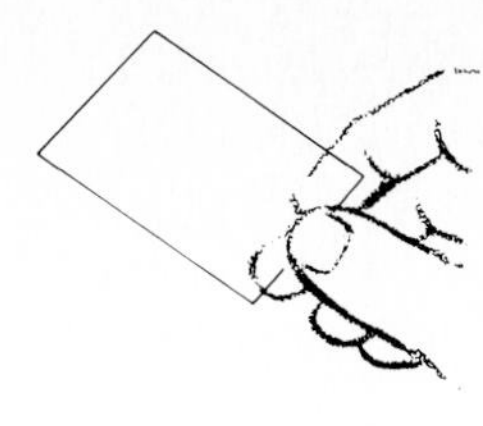

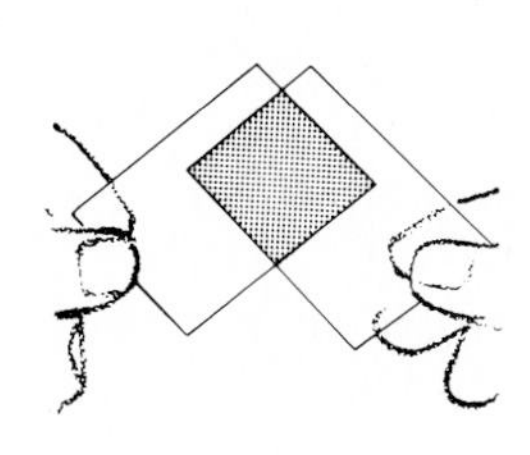

Figure 19-11. Polarizing plastic sheets. (Source: Adapted from P. F. Brandwein et al., *Concepts in Science: Orange Level,* Curie. ed., Harcourt Brace Jovanovich, New York, 1980, p. 53.)

➢ POLARIZING LIGHT

Allow students to play with two pieces of "polarizing plastic"—the lenses from a pair of polarizing sunglasses will do nicely. Make sure the students note that when the lenses are oriented in certain directions in front of one another, no or little light passes through. (See Figure 19-11.) This is a demonstration of the wave nature of light.

An analogy can be made by stretching a rope between the back rungs of two chairs. If they are properly aligned, the students will be able to send "waves" back and forth along the rope. Then have someone hold one chair sideways, as in Figure 19-12. Now the waves cannot pass.

Explain to the class that polarizing plastic lenses do something similar. Waves coming from a light source are vibrating in all directions. The first piece of polarizing plastic stops all waves except those vibrating in one direction. The second piece, when properly oriented, stops the waves coming in the direction permitted by the first piece, and thus no light passes through. Have the students further use the polarizing lenses to look at shining surfaces, water surfaces that reflect light, etc.

➢ OPTICAL ILLUSIONS

On or just before the day of the full moon (look at the weather page of a local newspaper to find out the right day), ask the students if they have ever noticed the "giant" size of the full moon as it is just rising. Tell them to think of reasons why the moon is (or appears to be) larger at that time. After hearing their suggestions, tell them that you have heard that the moon is not really larger at that time but rather that it "just looks that way." One way to determine whether or not it is really larger is to hold a ruler with your arm stretched out so that the top of the ruler is aligned with the top of the moon and mark with your thumb the position

Figure 19-12. Using chairs to show the wave properties of light. (Source: Adapted from P. F. Brandwein et al., *Concepts in Science: Orange Level,* Curie ed., Harcourt Brace Jovanovich, New York, 1980, p. 52.)

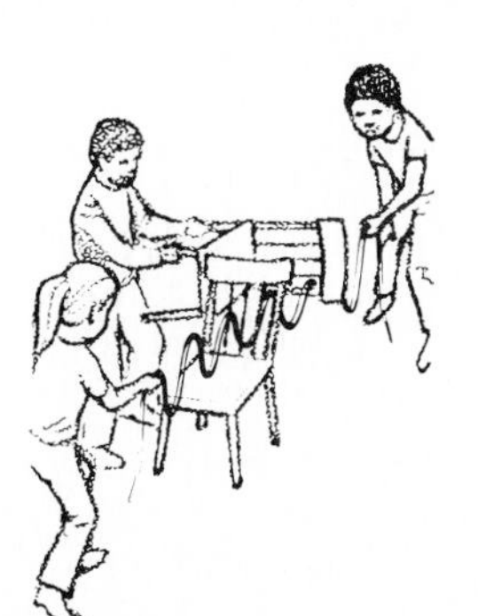

of the bottom of the moon on the ruler. Then 2 or 3 hours later, when the moon appears smaller, stand in the same spot and repeat the process. Is there really a difference in size? *(There is no difference. When the moon is close to the horizon, we have reference points, such as trees or buildings, which "mislead" us. When the moon is higher, we lose the reference points.)*

Have students do some library research to find some examples of optical illusions which they can illustrate. Place the illustrations on a bulletin board or in an interest center.

➢ NO LIGHT THIS YEAR

Tell the students to imagine that all light is to be eliminated from earth for a year. Have them think about how this would affect people such as assembly line workers, farmers, doctors, teachers, and others who live in your area. (Give them a few seconds to think after you mention each group of people.) How would it affect entertainers such as baseball players, singers, and musicians? What about flowers, birds, frogs, plants? (Again, give them a few seconds to think.)

Then have the class discuss the possible effects of such an event. What would happen to the temperature? Have students name other persons or things which they think would be affected. Ask them to discuss some way in which they personally would be affected. Could life even survive if earth were deprived of light for such a long time?

Children's literature

Have your students read all or parts of *Color in Your Life** or *Light and Radiation,*† and then discuss with them the concepts presented.

Find other books on light and radiation in the library, and identify them for the students or place them in an interest center.

Activity cards

An example of an activity card is shown in Figure 19-13. Additional activity cards could be made for topics such as investigating rainbows, lenses, or optical illusions; changing other forms of energy to light energy; making a microscope or telescope; and investigating the influence of various colors on people's behavior.

Bulletin boards

OPTICAL ILLUSIONS

Have students produce examples of optical illusions and place them on a bulletin board.

*Irving Adler, *Color in Your Life,* John Day, New York, 1962.

†Ira Freeman, *Light and Radiation,* Random House, New York, 1968.

"RAINBOWS" CAN BE PRODUCED BY BLOWING BUBBLES, PLACING A DROP OR TWO OF OIL ON A PUDDLE OF WATER, OR SPRAYING THE LAWN ON A SUNNY DAY.

- MAKE A RAINBOW IN ONE OF THE ABOVE WAYS.
- FIND SOME OTHER WAY TO PRODUCE A RAINBOW.

WHAT IS THE ORDER OF THE COLORS IN THE RAINBOW?

Figure 19-13. Activity card: Rainbows.

RADIANT ENERGY
Make an informative bulletin board which illustrates the whole range of radiation and points out the rather narrow band of visible light.

Field trips and outdoor activities

Arrange a visit to a home-decorating firm, and ask to have a staff member make a presentation on color coordination, how different colors or shades make rooms appear smaller or larger, how mirrors are used to create special effects, and other ways in which interior decorators use light and color in their work.

If there is a wooded area near the school, visit it and have the students notice what types of plants grow in shady places. What types in direct sun? Ask what would happen if a plant growing in shade were moved to a sunny spot. *(It might die.)* Why? *(Plants are usually found growing in the conditions which are best for them. If moved, they often will not grow as well, or they may not survive.)*

RESOURCES

Sources of assistance

FREE AND INEXPENSIVE MATERIALS

Here are some good sources of free and inexpensive materials for use in teaching light:

1. *Edison Electric Institute.** Several books on electricity which include information on light. Informational material plus descriptions of activities for students.
2. *Power and light company.* Information on proper lighting.
3. *County medical society.* Information on eyes and sight.

LOCAL RESOURCES

Local resources which can help in teaching light include:

1. The school nurse or a doctor may be able to give you information and materials on eyes and sight.
2. A special education consultant may be able to supply information on blindness.
3. An architect may give you information on ways in which lighting is considered in constructing homes and other buildings.

ADDITIONAL RESOURCES

1. Magazines such as *SciQuest,* and *National Geographic World.*
2. Books on light, sight, eyes, and energy from the school or municipal library.

Strengthening your background

1. Determine proper lighting levels for several things which you normally do at home and at school—reading, craft work, watching television, etc. Check your home and school to determine whether the lighting you use is at the recommended levels.
2. Find out which colors are helpful in creating certain moods (e.g., which colors are relaxing, exciting, etc.).
3. Figure out how a slide projector works. (Remove the cover and look at the parts of the machine.)

Getting ready ahead of time

1. Obtain some of the materials available from the Edison Electric Institute (see the Appendix).

*The addresses of specific agencies, organizations, and businesses are listed in the Appendix.

2. Collect several lenses which might otherwise be thrown away—from old eyeglasses, discarded projectors, etc.
3. Prepare a bulletin board for use in teaching a unit on light. Examples include sources of light and the history of lighting used in homes.
4. Collect a few sayings, cartoons, and jokes about light to show your students as examples which may inspire them to develop others.
5. Write the script for a guided fantasy on light in which students imagine they are particles of light.

Building your own library

Beeler, Nelson, and Franklyn Branley: *Experiments with Light,* Harper and Row, New York, 1957.
Blecha, Milo, et al.: *Exploring Matter and Energy,* Laidlaw, River Forest, Ill., 1980.
Brandwein, Paul: *Energy: A Physical Science,* Harcourt Brace Jovanovich, New York, 1975.
Elementary Science Study (ESS): *Light and Shadow,* Webster, McGraw-Hill, Manchester, Mo., 1976.
———: *Optics,* Webster, McGraw-Hill, Manchester, Mo., 1970.
Feravolo, Rocco: *Junior Science Book of Light,* Gerrard, Champaign, Ill., 1961.
Gregg, James: *Experiments in Visual Science,* Ronald, New York, 1978.
Heuer, Kenneth: *Rainbows, Halos, and Other Wonders,* Dodd, Mead, New York, 1978.
Levin, Edith: *The Penetrating Beam, Reflections on Light,* Rosen Press, New York, 1979.
Oxenhorn, Joseph, et al.: *Sound and Light,* Globe Book, New York, 1968.
Ramsey, William, et al.: *Physical Science,* Holt, New York, 1978.
Ruchlis, Hyman: *Wonder of Light: A Picture Story of How and Why We See,* Harper and Row, New York, 1960.

Building a classroom library

Adler, Irving: *Color in Your Life,* John Day, New York, 1962.
———: *Story of Light,* Harvey House, New York, 1979.
Alexenberg, Melvin: *Light and Sight,* Prentice-Hall, Englewood Cliffs, N.J., 1969.
Asimov, Isaac: *Light,* Follett, Chicago, 1970.
Barratt, D. H.: *Light,* Collins, New York, and World Publishing, Cleveland, 1972.
Beeler, Nelson, and Franklyn Branley: *Experiments in Optical Illusion,* Thomas Y. Crowell, New York, 1951.
Branley, Franklyn, M.: *Color: From Rainbows to Lasers,* Thomas Y. Crowell, New York, 1978.
———: *Light and Darkness,* Thomas Y. Crowell, New York, 1975.
Bulla, Clyde: *What Makes a Shadow?* Thomas Y. Crowell, New York, 1962.

Bursill, Henry: *Hand Shadows to Be Thrown upon a Wall,* Dover, New York, 1967.

———: *More Hand Shadows,* Dover, New York, 1971.

Colby, C. B.: *Modern Light: New Uses in Protecting and Improving Life,* Coward-McCann, New York, 1957.

Cooper, Miriam: *Snap! Photography,* Messner, New York, 1981.

Davis, Edward E.: *Into the Dark: A Beginner's Guide to Developing and Printing Black and White Negatives,* Atheneum, New York, 1979.

DeRegniers, Beatrice, and Isabel Gordon: *Shadow Book,* Harcourt, Brace and World, New York, 1960.

Feravolo, Rocco: *Light,* Gerrard, Champaign, Ill., 1961.

Freeman, Ira: *Light and Radiation,* Random House, New York, 1968.

Gardner, Robert, and David Webster: *Shadow Science,* Doubleday, Garden City, N.Y., 1976.

Goor, Ron, and Nancy Goor: *Shadows: Here, There, and Everywhere,* Thomas Y. Crowell, New York, 1981.

Heuer, Kenneth: *Rainbows, Halos, and Other Wonders,* Dodd, Mead, New York, 1978.

Highland, Harold: *How and Why Wonder Books of Light and Color,* Grosset and Dunlap, New York, 1969.

Lewis, Bruce: *What is a Laser?* Dodd, Mead, New York, 1979.

Meyer, Jerome: *Prisms and Lenses,* World Publishing, Cleveland, Ohio, 1972.

Myers, Bernice: *Come Out, Shadow, Wherever You Are,* Scholastic Book Services, New York, 1971.

Podendorf, Illa: *Shadows and More Shadows,* Children's Press, Hill, New York, 1971.

Schneider, Herman: *Laser Light,* McGraw-Hill, New York, 1978.

———: *Science Fun with a Flashlight,* McGraw-Hill, New York, 1975.

Schwalberg, Carol: *Light and Shadow,* Parents Magazine, New York, 1972.

Simon, Seymour: *Light and Dark,* McGraw-Hill, New York, 1970.

———: *Mirror Magic,* Lothrop, New York, 1980.

Ubell, Earl: *World of Candles and Color,* Atheneum, New York, 1969.

CHAPTER 20

SOUND

IMPORTANT CONCEPTS

Sound is the result of vibrations.

Sound travels in waves.

Sounds travel in all directions from their point of origin.

The pitch of a sound is determined by the vibration rate.

Sound travels through solids, liquids, and gases.

Sound waves may be reflected or absorbed.

An echo is a reflected sound.

Hearing is the result of the reception of sound waves by the ear and the processing of those waves by the ear and the brain.

Sound may be used to identify objects and events.

Sounds may be pleasing or disturbing.

INTRODUCTION: STUDYING SOUND

An in-depth study of sound would involve some rather complex concepts which are not appropriate for most elementary school students. However, your students can easily acquire an understanding of the basic concepts of sound by participating in activities which they can not only relate to but also enjoy.

Hearing is one of our most important senses. Students should understand not only how the hearing process operates, but also how to care for and protect this valuable sense. Music, an important part of most people's lives, is partially based on an understanding of how to combine pleasant sounds. Musical instruments can become an integral part of the study of a unit on sound.

Encourage students to follow up on their questions about sound. Ask them questions which will help them design investigations to answer their questions.

SOUND FOR PRIMARY GRADES

Activities

➤ VIBRATING RULER OR HACKSAW BLADE

Have the children place a ruler or hacksaw blade* on a table as illustrated in Figure 20-1. At first, the blade should extend as far as possible over the edge of the table. They should pull the blade down just a short distance and release it.

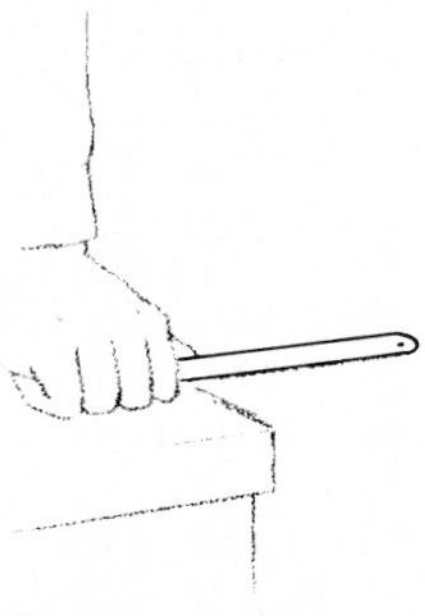

Figure 20-1. Effect of length on vibrations. (Source: Adapted from Harold E. Tannenbaum et al., *Experiences in Science: Sound,* Webster Division, McGraw-Hill, New York, 1967, p. 10.)

Have the children describe what happened. Tell them that the up-and-down motion of the blade can be called "vibration." Have them count the number of times the blade vibrates in 15 seconds. Ask them what they think would happen if a shorter length of blade were extended over the edge of the table. Have them try it and again count the number of vibrations in 15 seconds. What has happened? Try making the blade even shorter.

Sound is produced in a similar manner. However, the vibration rate must be greater than that of the blade in order for us to hear it. Ask the students to name some examples of sound linked with vibration. How about when a radio is turned up very loud? How about the strings of a guitar? How about a drum?

*It is our experience that a strip of "spring" steel or a hacksaw blade works much better than a ruler. Caution the students against bending the blade back too far.

➢ VIBRATING TUNING FORK

All the experiments in this activity demonstrate that a vibrating object—a tuning fork—produces sound. The vibrations are barely visible but can be detected in several ways. An ordinary table fork will work; however, a tuning fork, which can be obtained relatively inexpensively,* works much better. Strike the tuning fork with your hand or a plastic or rubber instrument and have the students look closely at it. Does it look "fuzzy"? Have the students touch it. How does it feel?

Strike the tuning fork again and have the students place it close to their ears. If more than one length of tuning fork is available, have them compare the sounds. *(Shorter tuning forks vibrate faster and produce a higher-pitched sound.)*

Strike the tuning fork and place it in a glass or bowl of water. What happens?

Have one student hold a piece of paper with a few grains of sand on it while another student touches a vibrating tuning fork to the bottom of the paper. What happens? *(The grains of sand bounce around.)* Why does this happen?

➢ VOCAL CORDS

Have the children hold their hands over their throats while singing some favorite songs. They will feel their vocal cords vibrate. They should recognize this as further evidence that all sounds are created by vibrations. What happens when they change pitch? Volume?

➢ PHONOGRAPH RECORDS VIBRATE

Run a sewing needle through a 3- by 5-inch or 4- by 6-inch index card. Place an old record which you are willing to sacrifice (it will probably be damaged by the activity†) on a phonograph turntable. Start the turntable, but do not place the phonograph stylus on the record. Instead, have each child hold the index card so that the needle is in a groove of the record and listen to the music. Ask the children if they can feel the vibrations. Does a large index card make the sound louder? Why might that be so? Have the students look at the grooves in the record under a hand lens or through a microscope. What can be seen?

*Check the catalogs of several publishing companies and science supply companies. One source of relatively inexpensive tuning forks is the Webster Division of McGraw-Hill Book Company (see the Appendix), as mentioned in the *Sound* unit of the Experiences in Science Program.

As an alternative, you could use the type of tuning forks which musicians use. These may be available at local music stores but are usually quite expensive. A music teacher or high school teacher may be willing to lend you a few tuning forks.

†*Caution:* Tell the students not to try this activity at home because it may damage their records. If they are very eager to do it at home, however, tell them to have their parents help them select a record which can be sacrificed.

An index card can serve as an amplifier to play a record.

➢ RADIO SPEAKERS VIBRATE

Try to find an old radio or phonograph of little value, and have the students *gently* touch the speaker to feel it vibrate as it is playing.* What happens if the volume is turned up?

➢ STRINGED INSTRUMENTS

Bring a stringed instrument—perhaps a banjo, guitar, or dulcimer—to class, or ask someone who plays such an instrument to visit your class. Pluck the strings and have the students touch them to feel the vibrations; then have a sing-along session. Have students make simple musical instruments—or start a class band. Can vibrations be detected in all the instruments? For more suggestions see the ESS unit *Musical Instrument Recipe Book.*†

➢ SOUND TRAVELS THROUGH WATER

Ask the students if they think that sound could travel through water. Tell a few students at a time to place their ears against the side of an aquarium, and then click two pebbles together under the water. They should detect the sound clearly. Ask them to do similar experiments if they go swimming and then to report the results to the class. Can the sound be heard more or less clearly than if it were traveling through air?

**Caution:* Be sure that electronic parts cannot be touched—a severe shock could result.

†Elementary Science Study (ESS), *Musical Instrument Recipe Book,* Webster, McGraw-Hill, Manchester, Mo., 1971.

Figure 20-2. Sound travels through solids.

SOUND TRAVELS THROUGH SOLIDS

Have the students place their ears against one end of a long table. Scratch on the other end with a needle. (See Figure 20-2.) Can the sound of the scratching be heard? Better or less well than through air? What other solids would transmit sound? Have them test several solids. Can some be found which will not transmit sound?

SOUND TRAVELS LIKE WAVES

Drop small pebbles into a large pan of water, or touch the water with one finger. What happens to the waves? In what directions do they travel? *(In all directions.)* Ask the students whether everyone in the classroom can hear when you stand in the middle of the room and talk. How is that similar to the waves in the pan of water?

ECHOS

Some of the students may have noticed the waves of water "bouncing back" from the sides of the pan in the previous activity. If not, repeat the activity and make sure they notice the bounce-back effect. Ask them, "Could sound waves do something similar?" Then ask if they have ever heard echos. On a still day, take them outside and have them produce echos: Tell them to stand at various distances away from a flat surface such as the side of a large building. Then they should clap their hands, fire a cap pistol, or make some other noise. What is the closest distance at which they can still hear the echo? *(If they are closer than about 17 meters or 56 feet, the echo comes back so fast that it cannot be distinguished from the original sound.)*

➢ TAPE RECORDING OF VARIOUS SOUNDS
One very important use of sound is that it helps us to identify objects and events in our environment. Tape-record several sounds: a clock, a police siren, a bird chirping, a food mixer running, an automobile passing, children playing, etc. Play the tape for the students and have them identify the sounds. Ask them to describe warning sounds which indicate danger to them. Allow students to use a tape recorder to record various sounds around the school or in their homes. Then they can play the recordings and let the other students try to name the sounds.

➢ PLEASING AND DISTURBING SOUNDS
Ask the class to name some sounds which they enjoy and some sounds which they do not like. Have the students talk about places where they like to be because of the sounds they can hear in those places. What kind of music makes them feel relaxed? What kind makes them feel energetic? Play some of each kind. Also play some music you like and some you dislike. Allow them to do the same.

➢ VISIT A LAKE
Describe a scene from the ballet *Swan Lake.* Ask the students to try to visualize that scene as they close their eyes and listen to a recording of Tchaikovsky's music for the ballet. Discuss their reactions to the music. What did they "see"?

Children's literature

Read *Sounds of a Summer Night** to your students or have them read it, and then discuss with them the concepts presented.

Find other books on sound in the school library, and identify them for the students or place them in an interest center.

Activity cards

An example of an activity card is shown in Figure 20-3. Additional activity cards could be made for topics such as stopping sound, using a megaphone, and testing hearing range.

Bulletin boards

SOUNDS WE LIKE—SOUNDS WE DO NOT LIKE
Set up a bulletin board as illustrated in Figure 20-4 (page 460). Have the students make drawings or cut out magazine pictures which portray sources of sounds they like and sounds they do not like. Then they can write their names on the pictures and place them in the appropriate spaces on the bulletin board.

*May Garelik, *Sounds of a Summer Night,* Scott, New York.

- MAKE A "GUITAR" FROM A CIGAR BOX AND RUBBER BANDS.
- HOW CAN YOU CHANGE THE PITCH OF THE RUBBER BANDS?

FORM A BAND FROM ALL IN YOUR CLASS WHO PLAY ANY MUSICAL INSTRUMENT. HAVE A CONCERT.

HAVE SOME FRIENDS MAKE OTHER INSTRUMENTS AND PLAY SOME TUNES TOGETHER.

Figure 20-3. Activity card: Musical instruments.

Figure 20-4. Bulletin board: Sounds we like and dislike.

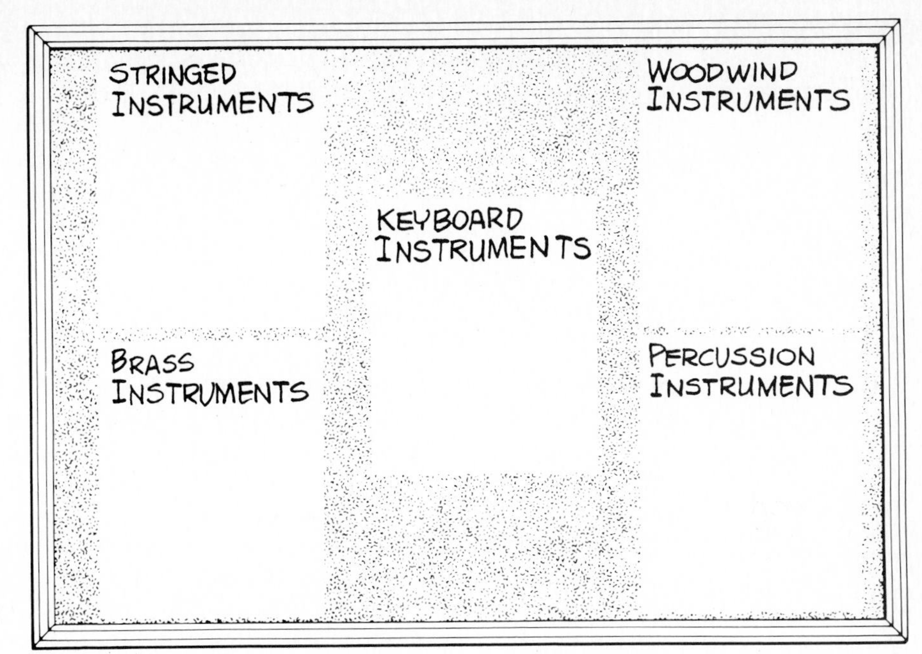

Figure 20-5. Bulletin board: Classes of musical instruments.

MUSICAL INSTRUMENTS
Set up a bulletin board as illustrated in Figure 20-5. Have the students locate and place photographs or drawings of instruments in the appropriate places.

Field trips

Take the class outside on the school grounds or to any other convenient outdoor location. Have the students sit down, close their eyes, and listen quietly to identify as many sounds as they can hear. Later have them name the sounds they heard. Which were made by people? Which sounds do they like? Dislike?

Arrange for your class to visit the school music room. Ask the music teacher to identify and play several instruments for your students.

SOUND FOR INTERMEDIATE GRADES

Activities

➤ A MODEL FOR SOUND WAVES
Tie one end of a rope to a desk leg near the floor; have a student hold the other end loosely, so that the rope is slack; and then move the rope rapidly back and forth across the floor. The resulting wave pattern is a model of sound moving through the air. Any individual particle of air

(analogous to a "particle" of the rope) does not travel from the source of the sound (the student's hand) to the desk; it is the wave (the disturbance in the rope) that does the moving.

Use a Slinky toy as another model. Stretch the Slinky and after pinching one segment of it together, release it. What happens? *(A "compression" wave travels the length of the Slinky. Again, it is the compression that travels, not the actual "pieces" of the Slinky.)*

➢ SOUND VIBRATION DETECTOR

Have students construct a "vibration detector" by cutting one end from a round salt box and cutting a 2-centimeter-square (1-inch-square) hole in the other end. Glue a small piece of mirror to the center of a piece of heavy tissue paper or waxed paper, and tie the paper over the open end of the box. (See Figure 20-6.)

While one student talks into the small hole, have another look at the mirror to see if any movement can be detected. Stand in the sunlight holding the vibration detector, or use a flashlight to shine a reflection on some surface. Have the students talk again. What happens to the reflected light on the surface? Place the detector in front of a radio or a musical instrument. What are the results? *(When sound waves strike the paper, it vibrates. This in turn causes the mirror to move, and the reflected light from the mirror also moves.)*

Are the results different in response to different pitches of sound? Does loudness make a difference? Ask the students what else could cause changes and check out their various suggestions.

➢ CHANGING PITCH: HOMEMADE GUITAR

Obtain several pieces of wood approximately 2 by 5 by 15 centimeters (¾ by 2 by 6 inches). (One of your students would probably like to get them for you—just ask!) Show the students how to wrap two rubber bands of nearly the same length but different thicknesses around each board and place a pencil between the rubber bands and the board. Have them make the rest of the "guitars" themselves. Let them strum their homemade guitars, and then ask if there is any difference between the sounds from the two rubber bands. *(The thicker one should be lower in pitch—but if it is shorter and therefore stretched tighter, it may not be lower.)*

Ask the students to find out how many different ways they can change the pitch of their guitar strings. *(Tightening the rubber bands, holding down the rubber bands as if they were guitar strings, and so forth.)* Make sure they understand that three things affect pitch—the thickness, tightness, and length of the string.

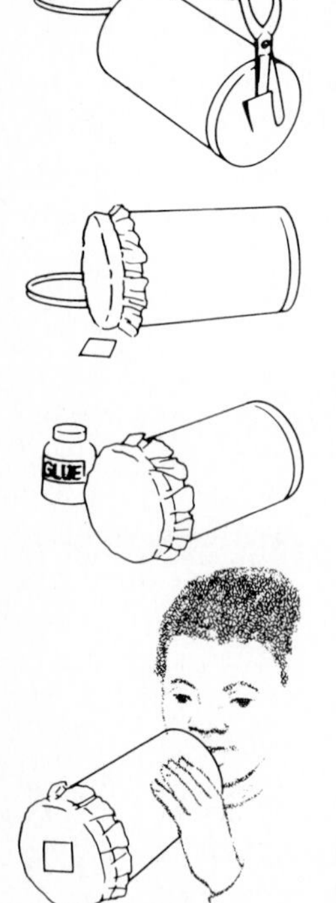

Figure 20-6. Sound vibration detector. (Source: Adapted from G. C. Mallinson et al., *Science: Understanding Your Environment,* Silver Burdett, Morristown, N.J., 1975, level 5, p. 226.)

➢ CHANGING PITCH: STRINGED INSTRUMENT

Obtain a stringed instrument such as a banjo or a guitar which the students can handle.* Demonstrate how the pitch is affected by the diameter, tightness, and length of the strings.

Allow the students to play the instrument. Ask which diameter of string produces the lowest-pitched sound. The highest? Tell them that they may change the tension on the strings *slightly;* but warn them that if they tighten the string too much, it may break.†

Which raises the pitch—increasing or decreasing the tension? Have them "stop" the string with a finger. What are they actually doing? *(Changing the length of the amount of string that can vibrate.)* As they shorten the length of the string that can vibrate, what happens to the pitch? *(It goes up.)* What happens when the length of vibrating string is increased? *(The pitch goes down.)*

Invite someone to play a stringed instrument for the class. Ask the students to explain how the pitch is changed.

If there is a piano in the school, have the children look inside it. Can they figure out how the pitch is changed in a piano?

➢ ABSORBING SOUND

Point out to the class that sometimes people need to make some sounds quieter—for example, traffic noise from a freeway. Tell them that it has been claimed that a thick growth of shrubs or bushes will cut down the noise level.

To test this claim, have students stand at a known distance from a radio and play it at a selected volume. Then, keeping the radio at the same distance, have them stand behind thick shrubs or bushes and play the radio at the same volume. Is the sound level different? *(It is much lower behind the vegetation.)*

If possible, obtain a sound-level meter and make measurements with it. What could be done to make our homes quieter with reference to outside noises?

➢ DOES SOUND TRAVEL BETTER THROUGH AIR OR WOOD?

Move a watch which makes a quiet ticking sound away from a student's ear until the student indicates that it can no longer be heard. Then touch one end of a piece of wood—a meter stick or yardstick—to the student's ear and rest the middle of the stick against the watch. Can the ticking be heard now? *(It should be.)* Move the watch to the far end of the stick, but make sure it still touches the stick. Can it still be heard? (See Figure 20-7, page 464.)

* *Note:* Tell the students to handle it carefully, reminding them that a musical instrument is not a toy.

†*Caution:* Supervise the students closely during this part of the activity to avoid broken strings and possible eye injuries.

Figure 20-7. Another demonstration that sound travels through solids.

Is sound transmitted audibly a greater distance through air or through wood? Ask the students if they have ever heard of the Indian practice of placing an ear to the ground to hear buffalo or horses in the distance even before they can be seen. Why would this be possible? How does it relate to the activity they have just completed?

➢ EAR MODEL

Ask the school nurse or a doctor to show a model of the human ear to your class and to explain how the ear functions and proper procedures for ear care. The nurse or doctor may also be able to show the class a short movie on the functioning of the ear.

If your students have not already had hearing tests, it may be possible for the nurse to conduct them. First the procedure for conducting the tests should be explained to the class.

➢ MOOD MUSIC

Ask students to tell you about types of music which have different effects upon them. What kinds of music would they play to relax and fall asleep by? What kind would make them want to move faster? How does music without lyrics affect them? Have students select music which they would enjoy hearing during a half-day of school and which would not disturb their learning—and play it for them.

These hearing-impaired students "listen" by watching their teacher signing.

➢ DEAFNESS

Invite a special education consultant or other expert on hearing problems to explain to your class how deaf persons can communicate. Ask the person to demonstrate as many procedures as possible. If a speaker is not available, show a film on the topic and have the students research the ways that deaf persons learn to communicate. Have students learn how to communicate by "signing."

The students should learn that, though deaf persons do have a handicap, they can lead quite normal lives and do almost all the things other people can do. Point out that we all have handicaps of various types. Ask the class how they think they should treat a deaf student if one were in the class.

➢ NOISE LEVELS

Obtain a sound-level meter and have students measure noise levels in several places where they enjoy the sounds and in several places where the sounds annoy them. Do they notice a trend in the differences in loudness at these locations?

Have the students check loudness levels in many places where they spend time: school hallways, music room, shop, homes, street corners, in front of their radios, at concerts, etc. Then compare the noise levels with those in Figure 20-8 (page 466), which indicates the effects of various common sound levels on people.

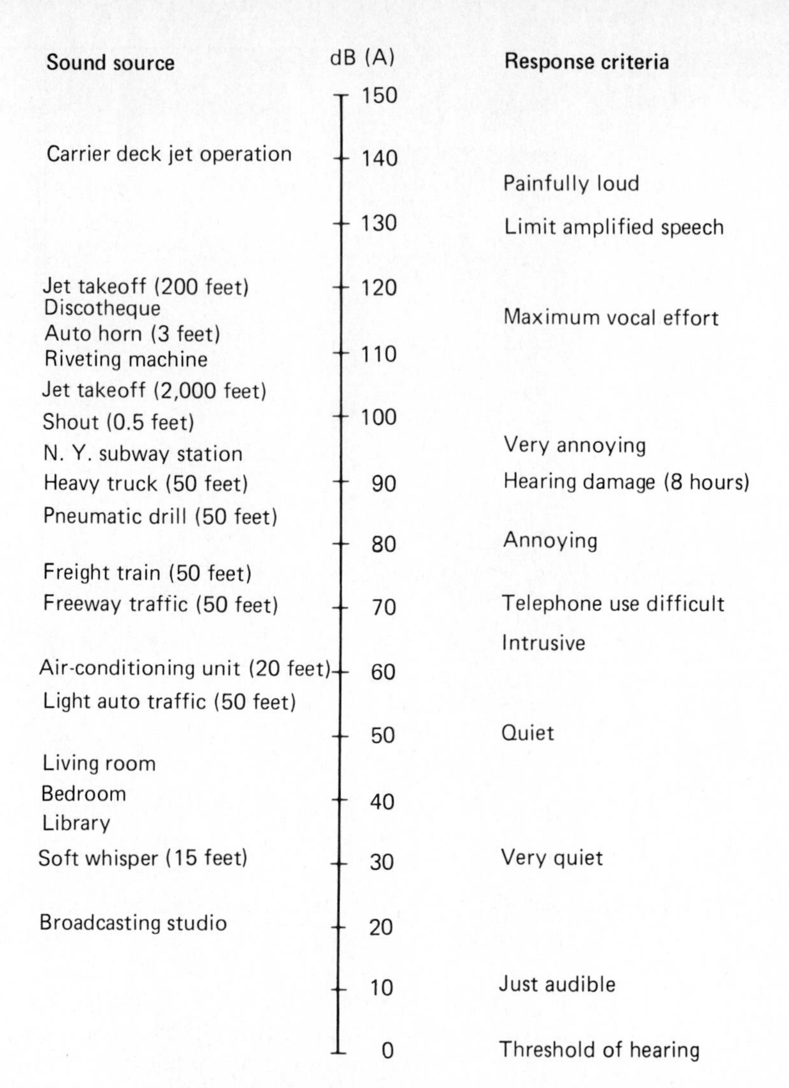

Typical A – weighted sound levels taken with a sound-level meter and expressed as decibels on the scale. The "A" scale approximates the frequency response of the human ear.

Figure 20-8. Weighted sound levels and human responses. (Source: *Environmental Quality: The First Annual Report of the Council on Environmental Quality,* U.S. Government Printing Office, Washington, D.C., 1970, p. 125.)

➤ DOWN THE RIVER WITH SOUND

Prepare the students for this activity by telling them that they will be listening to a musical composition, part of a symphonic cycle, which portrays a trip down a river. Then have them close their eyes and listen while you play a record or tape of Bedrich Smetana's *The Moldau.* Discuss with the students their interpretations of the various parts of the composition. Ask them to bring to class classical or other music which creates a picture by using instrumental music only.

- WHAT SOUNDS DO YOU ENJOY?
- WHAT SOUNDS DO YOU NOT LIKE?
- WHAT KIND OF MUSIC DO YOU LIKE?
- WHAT MUSICAL INSTRUMENTS MAKE THAT MUSIC? (COULD YOU PLAY ANY OF THEM?)

BRING YOUR FAVORITE RECORD TO SCHOOL FOR OTHERS TO HEAR.

DRAW A PICTURE OF A PLACE WHERE YOU WOULD ENJOY THE SOUNDS.

Figure 20-9. Activity card: Sounds.

Children's literature

Read *The Listening Walk** or *Who Will Drown the Sound?*† to your students or have them read one or both of these books, and then discuss with them the concepts presented.

Find other books on sound in the school library, and identify them for the students or place them in an interest center.

Activity cards

An example of an activity card is shown in Figure 20-9. Other activity cards could be made for topics such as sound pollution, usual and unusual musical instruments, how animals make sounds, and the speed of sound.

Bulletin board

Make a bulletin board which shows what people can or must do to avoid harm from loud sounds—mufflers on persons who work around airplanes, sound barriers surrounding freeways, etc.

*Paul Showers, *The Listening Walk,* Thomas Y. Crowell, New York, 1961.

†Carleen Mausy Hutchins, *Who Will Drown the Sound?* Coward-McCann and Geoghegan, New York, 1972.

Field trips

Arrange a field trip to a radio station, and have a station employee describe to your class the process by which the radio signals are transmitted. Ask the person also to point out to the students the soundproofing tile in special locations at the station, and to comment on the purposes of the tile.

As another field trip, you could arrange to visit an auditorium and have an acoustical engineer explain its acoustical design. A third informative visit could be made to a music shop, where you could ask the manager to explain how stereo speakers should be arranged to produce the best sound.

RESOURCES

Sources of assistance

FREE AND INEXPENSIVE MATERIALS

Here are some good sources of free and inexpensive materials for use in teaching sound:

1. *U.S. Environmental Protection Agency.** Information on sound pollution.
2. *County medical society.* Pamphlets and other materials on ears and hearing.
3. *A lumber company or home-decorating establishment.* Information on sound control, acoustical tile, and other materials designed to control sound.

LOCAL RESOURCES

Local resources which can help in teaching sound include these:

1. The school nurse or a doctor can give you information and materials on ears and hearing.
2. A special education consultant should have information on deafness.
3. A radio station can supply information on broadcasting.
4. A record shop can supply information on recording and perhaps special recordings which demonstrate stereophonic sound.

ADDITIONAL RESOURCES

1. Magazines such as *SciQuest,* and *Science World.*
2. Books from the school or municipal library on sound, music, sound pollution, and hearing.

*The addresses of specific agencies, organizations, and businesses are listed in the Appendix.

Strengthening your background

1. Find out how stereophonic recording systems operate.
2. Identify pieces of classical music which portray special places, animals, etc.
3. Find out how to improve the acoustics in your home and classroom.

Getting ready ahead of time

1. Obtain tuning forks (see the Appendix).
2. Collect pictures of several musical instruments.
3. Prepare a shoebox science kit on sound.
4. Obtain a phonograph to use permanently in your classroom.
5. Identify six children's books on sound.
6. Obtain blocks of wood for making homemade guitars (see the activity "Changing Pitch: Homemade Guitar").

Building your own library

Blecha, Milo, et al.: *Exploring Matter and Energy,* Laidlaw, River Forest, Ill., 1980.
Brandwein, Paul: *Energy: A Physical Science,* Harcourt Brace Jovanovich, New York, 1975.
Elementary Science Study (ESS): *Musical Instrument Recipe Book,* Webster, McGraw-Hill, Manchester, Mo., 1971.
———: *Whistles and Strings,* Webster, McGraw-Hill, Manchester, Mo., 1971.
Feravolo, Rocco: *Wonders of Sound,* Dodd, Mead, New York, 1962.
Oxenhorn, Joseph, et al.: *Sound and Light,* Globe Book, New York, 1968.
Ramsey, William, et al.: *Physical Science,* Holt, New York, 1978.
Stevens, S., and *Life* editors: *Sound and Hearing,* Time-Life, New York, 1965.

Building a classroom library

Anderson, Dorothy: *Junior Science Book of Sound,* Gerrard, Champaign, Ill., 1962.
Bendick, Jeanne: *The Human Senses,* F. Watts, New York, 1968.
Branley, Franklyn, M.: *High Sounds, Low Sounds,* Thomas Y. Crowell, New York, 1975.
Catherall, E. A., and P. N. Holt: *Working with Sounds,* A. Whitman, Chicago, 1969.
Darwin, Len: *How the Telephone Works,* Little, Brown, Boston, 1970.
Elgin, Kathleen: *The Ear,* F. Watts, New York, 1967.
Freeman, Ira: *Sound and Electronics,* Random House, New York, 1968.
Friedman, Jay: *Look around and Listen,* Grosset and Dunlap, New York, 1974.

Heuer, Kenneth: *Thunder, Singing Sands, and Other Wonders: Sounds in the Atmosphere,* Dodd, Mead, New York, 1981.
Kavaler, Lucy: *The Dangers of Noise,* Thomas Y. Crowell, New York, 1978.
Knight, David C.: *Silent Sound: The World of Ultra-Sonics,* Morrow, New York, 1980.
Koen, Martin: *The How and Why Wonder Book of Sound,* Grosset and Dunlap, New York, 1962.
Kohn, Bernice: *Echos,* Coward-McCann, New York, 1965.
Miller, Lisa: *Sound,* Coward-McCann, New York, 1965.
Pine, Tillie, and Joseph Levine: *Sound All Around,* McGraw-Hill, New York, 1961.
Scott, John: *What's Sound?* Parents Magazine, New York, 1973.
Showers, Paul: *How You Talk,* Thomas Y. Crowell, New York, 1975.
———: *The Listening Walk,* Thomas Y. Crowell, New York, 1961.
Sootin, Harry: *Science Experiments with Sound,* Grosset and Dunlap, New York, 1970.
Spier, Peter: *Gobble, Growl, and Grunt,* Doubleday, Garden City, N.Y., 1971.
Thomas, Anthony: *Things We Hear,* F. Watts, New York, 1976.
Victor, Edward: *Exploring and Understanding Sound,* Benefic Press, 1960.
White, Lawrence: *Investigating Science with Rubber Bands,* Addison-Wesley, Reading, Mass., 1969.

C H A P T

ELECTRICITY AND MAGNETISM

IMPORTANT CONCEPTS

Magnetism is a basic property of some matter and is related to the arrangement of atoms and molecules within that matter.

Some metals are magnetic.

Magnets have two poles.

Magnets are strongest at their poles.

Like poles of magnets push each other apart (repel), whereas unlike poles pull toward each other (attract).

Magnetic fields exist around magnets.

Iron and steel become magnetized if they are placed in a magnetic field.

Magnetic force passes through nonmagnetic materials.

Earth has magnetic poles and a magnetic field.

Magnets which are free to move line up with the magnetic field of earth.

Earth's magnetic poles are located near its geographical north and south poles, but do not coincide with them.

Electricity occurs because the electrons in matter can move.

Static electricity is produced by rubbing some kinds of materials together.

An electric current can occur when a circuit (path) is completed.

Electric energy is one kind of energy and can be changed to other kinds.

Electric energy can be produced by using other kinds of energy.

Electric energy can be used to do work.

Electrons flow easily through conductors.

It is difficult for electrons to flow through insulators.

Electricity can travel through an unbroken circuit of conductors.

Cells, bulbs, switches, and other electric equipment may be arranged in series circuits or parallel circuits.

In a series circuit the electricity flows through all the equipment in order.

In a parallel circuit the electricity divides into two or more paths, and not all the electricity has to pass through each piece of electrical equipment.

Fuses and circuit breakers are safety devices, used in electric circuits to prevent fires and damage to electric equipment.

Electricity can be generated by moving a wire through a magnetic field.

Magnets may be made by using electricity.

INTRODUCTION: STUDYING ELECTRICITY AND MAGNETISM

Electricity and magnetism are especially fascinating to young minds. As has already been mentioned, your students are naturally inquisitive and a little nudge can really get them going in the right direction. In a given class there are usually at least half a dozen students who will want to investigate electricity from the day it is introduced until the last day of school—and well beyond. All you have to do is ask a few open-ended questions, or suggest that the class look at some books which you just happen to know about, or ask students to help you construct some materials which you would like to use with next year's class.

Teaching magnetism is rather straightforward in the sense that there are many interesting activities which can demonstrate it, but it becomes more difficult when the children start asking questions such as "How does magnetism work?" and "What causes magnetism?"

Before you begin to teach magnetism, especially in third grade, ask the students about the magnetism activities you are considering using; they may have already done many of the activities in grades K, 1, and 2.

The teaching of electricity requires some special cautions. *Never work with the electricity that comes from wall plugs!* The activities in this book use only D-size *flashlight* cells. If four of those cells are hooked together, they add up to 6 volts of electrical pressure, whereas the electricity from wall plugs is usually 110 or 220 volts—a very large difference indeed! Using cells is more expensive than using electricity from wall plugs, but safety in the classroom is essential—and one of the goals of teaching electricity is to have children develop a healthy respect for household electricity. They should learn that electric current is dangerous, and that only people with special training can work with it safely. Question: Can you get a shock from a D-size flashlight cell? Answer: No. Try it. *Use D cells for all activities.*

We have, with few exceptions, placed magnetism studies in the activities for primary grades and electricity studies in the activities for intermediate grades. You may wish to deviate from this arrangement, depending upon the characteristics of your students.

Remember to listen to your students while they are doing the activities. Ask probing and divergent questions whenever opportunities arise. Some questions are suggested in this chapter; and as your students make observations and attempt to understand what is happening, you should be able to think of many more.

ELECTRICITY AND MAGNETISM FOR PRIMARY GRADES

Activities

➤ WHAT DOES A MAGNET ATTRACT?
Ask your students whether they know about magnets; even very young children may have had some experience with magnets at home. Give each student a small bar magnet, and ask all of them to find objects in the classroom to which the magnet will stick. After a few minutes, ask the class to identify the kinds of materials a magnet sticks to. (*They will probably respond that magnets stick to metals. Some may note that magnets do not stick to all metals.*) Ask the students to place their magnets against a piece of aluminum, such as a throwaway food container. The students should conclude that magnets stick to some metals, but not to all metals. You may wish to inform them that the common metals which are attracted to magnets are iron, steel, and nickel.

Reinforce what your students have learned by giving them a box which contains magnetic and nonmagnetic materials. Have them predict which will be attracted by their magnets and then test their predictions.

➤ WHAT HAPPENS WHEN MAGNETS TOUCH EACH OTHER?
To begin this activity, give each student a small bar magnet on which the north pole is identified. (You can mark the north poles with paint or nail polish if you like.) Tell the students to work together in pairs to find out what happens when the magnets touch each other. First, let them touch their magnets together at random, and ask them what happens. (*They will get different results depending upon how the magnets were touched together.*) Tell the students to bring the two marked ends together. What do they notice? (*The magnets push apart.*) Tell them to bring the two unmarked ends together. What happens this time? (*Again the magnets push apart.*) Then ask the students to bring one marked and one unmarked end together. What happens? (*The magnets pull together.*) Have the students exchange partners to see whether the same results occur. (See Figure 21-1.) Tell the class that all magnets have two "poles," named the "north" pole and the "south" pole, and that the poles are so named because they can be used to find earth's north and south poles. Point out that on their magnets, the north pole is the marked end.

➤ WHICH END IS THE NORTH POLE?
Give each student a bar magnet on which the north pole is not marked and also another magnet, such as a U-shaped magnet. Tell the students that these magnets are like the ones in the previous activity except that the north poles are not marked. Ask them to find out which end is the north pole. Allow the students to work with their unmarked magnets and the U-shaped or other magnets until they can demonstrate how they know which end is the north pole.

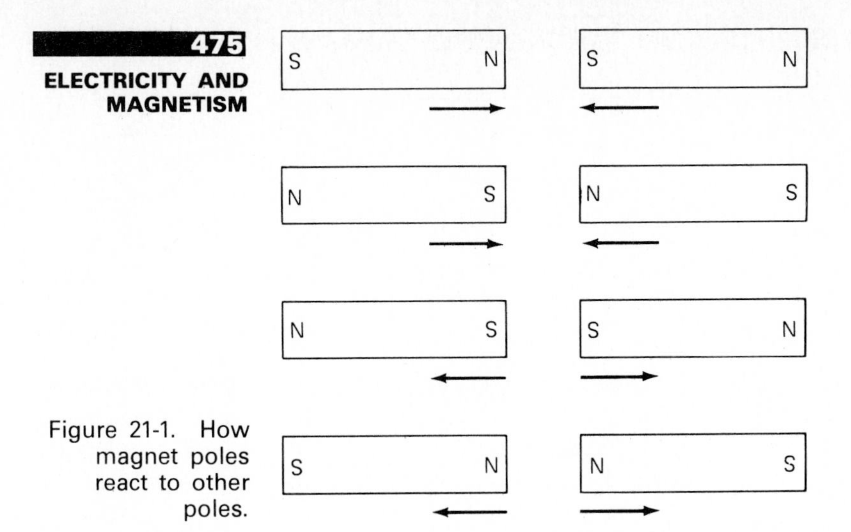

Figure 21-1. How magnet poles react to other poles.

➢ ARE BOTH ENDS OF THE MAGNET OF THE SAME STRENGTH?
Have each student make a mark on a smooth nonmetal surface (such as a piece of paper on a wood surface), and place one tip of a paper clip on the mark with the other tip pointing away from the mark. Then the students should slowly move the north pole of their magnets toward the paper clip until the clip "jumps" to the magnet, and mark the positions of the end points of the magnet. Have them do this three or four times using the north pole. Then tell them to repeat the process, but this time to use the south pole of the magnet. Is the magnet at the same point or nearly the same point when the paper clip moves to the south pole? (*It should be.*) Make sure the students understand that this is evidence that both poles are of the same strength. (See Figure 21-2.)

Ask the students if they can think of other ways to show that both poles of a magnet are of the same strength. (*One other way is to first place one pole into a box of paper clips or straight pins and count how many stick to it, then put in the other pole and count again.*)

➢ ARE ALL MAGNETS OF THE SAME STRENGTH?
Ask your students how they could determine which of a number of magnets was strongest. Have them work in pairs to test two or three magnets and explain to the class how they determined which was strongest. (*Two ways of testing are described above, in "Are Both Ends of the Magnet of the Same Strength?"*)

Figure 21-2. Relative strength of opposite poles. (Source: Adapted from H. Tannenbaum et al., *Experiences in Science: Magnets*, Webster Division, McGraw-Hill, New York, 1966, p. 19.)

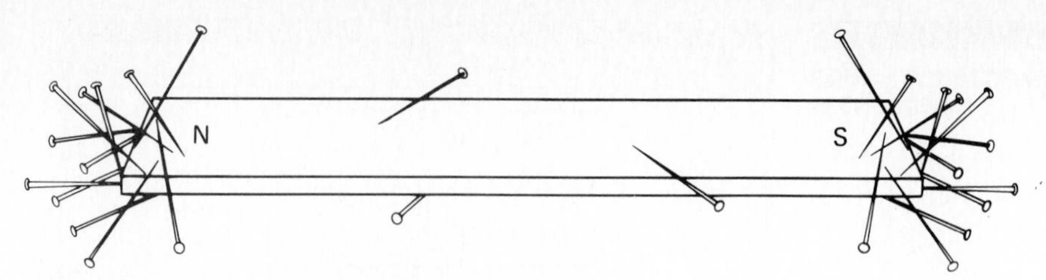

Figure 21-3. The strongest parts of a magnet.

➢ WHERE IS A MAGNET STRONGEST?

Ask your students where on the magnet the pull is strongest, or whether it is the same strength over the whole magnet. Give them an adequate supply of steel paper clips, straight pins, small nails, or other magnetic materials, and allow them to work in pairs or small groups to find the answer. (See Figure 21-3.)

After they have had time to solve the problem, ask each group to explain how they answered the question. *(Here are two possible ways: First, hold a steel paper clip and approach the magnet at different points, which makes it possible to sense the force of the pull at various points on the magnet. Second, place the magnet into a box of pins or other metal items and see what parts of the magnet attract the most pins. The magnet is strongest at its poles.)*

➢ DOES MAGNETIC FORCE WORK THROUGH NONMAGNETIC MATERIALS?

From this activity, the students will learn that a magnetic force does indeed pass through nonmagnetic materials. Show the class a plastic glass with a small bar magnet in it, and ask them if they think another magnet, outside the glass, will attract the magnet in the glass through the plastic. (See Figure 21-4.) What do they think would happen if the glass contained water also? Give your students plastic glasses, magnets, and a supply of water so that they can find the answers to the questions.

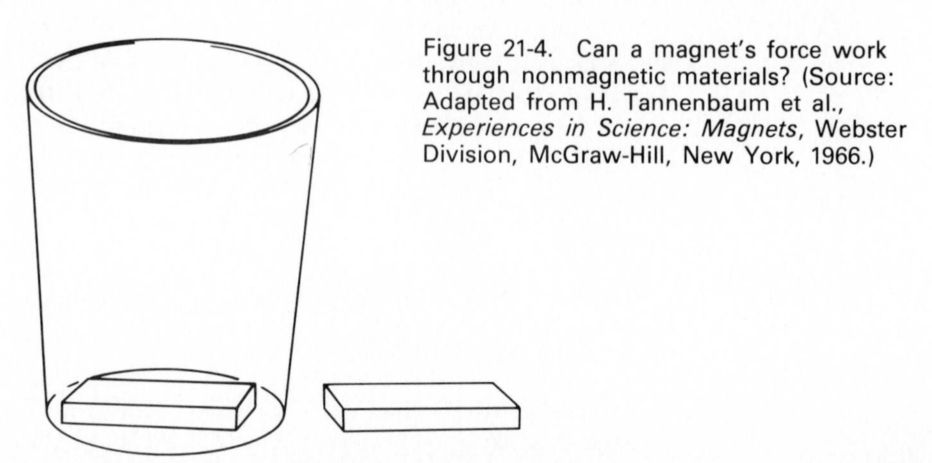

Figure 21-4. Can a magnet's force work through nonmagnetic materials? (Source: Adapted from H. Tannenbaum et al., *Experiences in Science: Magnets*, Webster Division, McGraw-Hill, New York, 1966.)

Fields of magnetic force can be revealed by using iron filings.

Then give your students pieces of cloth, plastic, paper, and other materials to see if the magnets will work through them. (*If the magnets are small and the material very thick—such as three or four layers of cardboard—the force may be so weak that it will be hard to detect.*)

CAN A MAGNETIC FIELD BE SEEN?

The answer is *no*! However, there is an activity which your students can do which will give them visual evidence of magnetic fields. Tell them to place a magnet beneath a piece of tagboard or light cardboard and sprinkle iron shavings onto the tagboard. The shavings will become arranged in a pattern around the magnet, and tapping the tagboard gently may make the arrangement become more clear.

To make a permanent record of a magnetic field, first put waxed paper on top of the tagboard. After the pattern is established, place another piece of waxed paper on top of the shavings. The waxed paper should then be heated by ironing it. These records of magnetic fields make nice wall hangings which the students may want to give to their parents.

Suggest to the students that they place two or more magnets beneath the tagboard to observe how the magnetic fields of the magnets interact. What happens if like poles are close to one another? Unlike poles? What if two magnets are placed side by side? Could a triangle-shaped pattern be made?

MAKE YOUR OWN MAGNET

Give each student a magnet, a large nail, and several small magnetic items (steel shavings, paper clips, or pins). Ask the students to find a magnetic way to pick up the small items without touching them with the magnet. After several minutes of work, ask students who have been successful to describe the method they used.

Tell the students that they can make a magnet of the nail by holding the head (or point) of the nail in one hand and stroking it with one pole (only) of the magnet. (See Figure 21-5.) Show them how to place the pole on the nail near the hand and with one stroke run the magnet off the other end of the nail, then again place the same pole on the nail at the same starting point and repeat the process. Stroking the nail 20 to 30 times should magnetize it. After they have made their magnet, let them pick up the small items using their magnetized nails only. Ask them if they think their nails would become stronger magnets if they stroked them 100 times. Tell them to try it if they would like to. How can they find out whether or not the magnets become stronger?

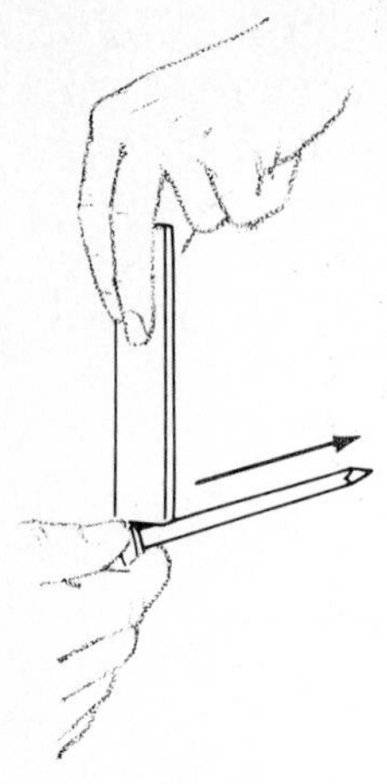

Figure 21-5. Making a magnet.

⋗ MAKE A COMPASS

For this activity, try to obtain several large bar magnets; small magnets may also be used. Make string holders for the magnets as shown in Figure 21-6. Take the class outdoors away from buildings, and have the students hang their magnets in the holders from trees or other supports. After the magnets have stopped moving, have the students hold out their arms and line them up in the same direction as the orientation of the bar magnets. (*They should show that almost all the magnets are lined up in the same general direction.*) Compare the orientation of the bar magnets with magnetic north as determined by a compass. (*The orientations should be the same or nearly the same. Stronger magnets give better results, unless there are magnetic objects interfering.*)

⋗ BE A MAGNET

Have your students line up in pairs along a long rope, with the members of each pair on opposite sides of the rope. Tell the students on one side of the rope to pull on the rope in various directions—some in one direction along the length of the rope, some in the opposite direction, some at right angles to the rope, and others at various other angles to the rope. Each student on the other side of the rope is to pull in the exact opposite direction from his or her partner. It should quickly become apparent that the rope will do little moving with the students pulling in all directions.

Figure 21-6. Making a compass.

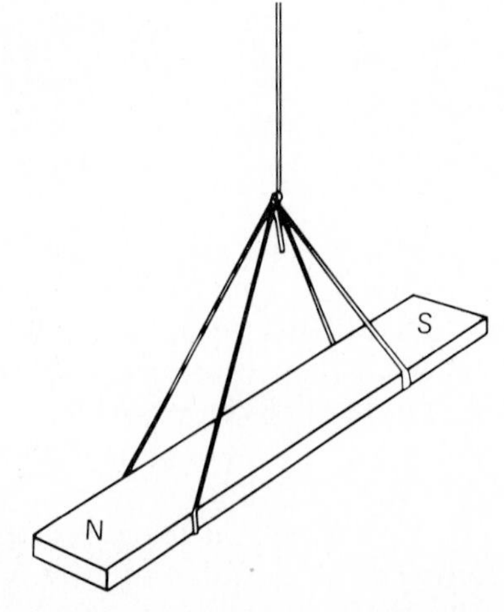

Tell the class that this part of the activity is analo... ways each small "magnetic force" in a piece of matter ...

Next have all the students on both sides of the rope pull ... direction. It should be obvious that now a great deal of co... movement is possible. Tell them that this part of the activity is a... gous to the manner in which the small "magnetic forces" in a piece ... matter are aligned when the matter is magnetized.

Children's literature

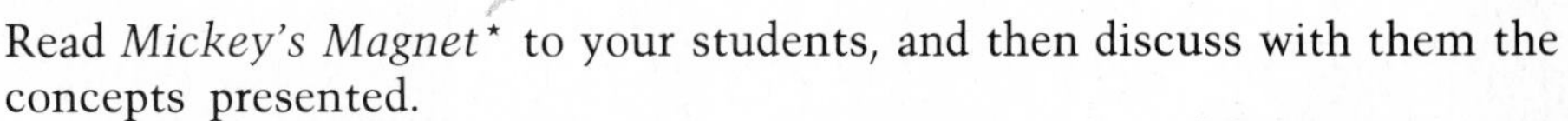

Read *Mickey's Magnet** to your students, and then discuss with them the concepts presented.

Find other books on magnets in your school library, and identify them for the students or place them in an interest center.

Activity cards

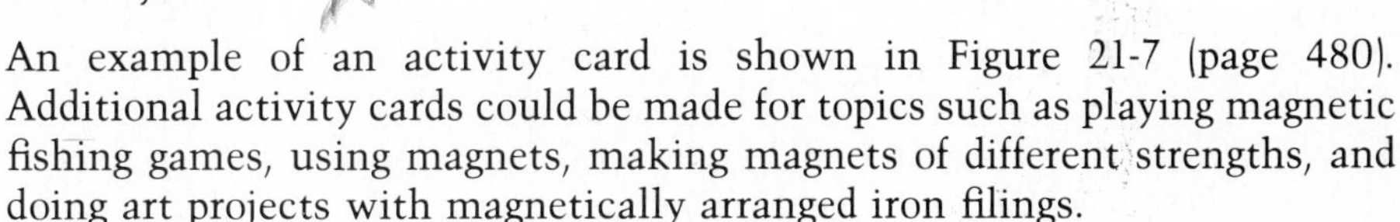

An example of an activity card is shown in Figure 21-7 (page 480). Additional activity cards could be made for topics such as playing magnetic fishing games, using magnets, making magnets of different strengths, and doing art projects with magnetically arranged iron filings.

Bulletin boards

WHICH MATERIALS ARE MAGNETIC?
Place several types of materials—some magnetic and some nonmagnetic, including nonmagnetic metals such as aluminum and copper—on a bulletin board as illustrated in Figure 21-8. Attach a magnet on a long string to the bulletin board, and tell the students they can touch the magnet to each of the materials to determine whether or not the magnet is attracted.

ELECTROMAGNETS
Post several pictures showing electromagnets doing work. The most useful examples for the primary grades are electromagnets which lift heavy loads, as in a scrap metal yard or in the metals industry. Explain to the class the basics of electromagnets; if you need to, do some research so that you can make a clear and accurate presentation.

Field trips

Few field trips are helpful in the study of magnetism. A brief trip to the school grounds or other nearby areas would allow students to determine whether materials not found in the classroom are attracted by a magnet. If there is an electromagnet in operation nearby, you may wish to schedule a trip to see it—but weigh the effort against the benefits before you make plans.

*Franklyn Branley and Eleanor Vaughan, *Mickey's Magnet*, Thomas Y. Crowell, New York, 1956.

FIND THINGS IN THE ROOM WHICH A MAGNET WILL STICK TO.

DOES IT STICK TO RED THINGS BETTER THAN BLUE ONES?

COULD YOU FIND A "NEEDLE" IN A HAYSTACK?

Figure 21-7. Activity card: A real sticker (above).

Figure 21-8. Bulletin board: Which materials are magnetic? (right).

ELECTRICITY AND MAGNETISM FOR INTERMEDIATE GRADES

Activities

Most of the activities in this section focus on electricity; but two of them illustrate the interrelationship of magnetism to electricity—moving a wire through a magnetic field to produce electricity and creating a magnet by using electricity.

Activities with electricity provide many opportunities to challenge your students to develop and use higher-level cognitive skills: especially application, analysis, and synthesis. Do not have the class do electrical experiments in a step-by-step, predetermined manner; instead, present the activities in a way that will require your students to question what they know and what they perceive, to draw together all they know about a problem before they can arrive at a solution. The students are especially likely to create electric circuits which do not appear in this book. When questions arise, have them carefully analyze the electrical flow through the circuit or circuits they are working with. A great source of help to you can be the Elementary Science Study unit *Batteries and Bulbs*, listed in "Building Your Own Library."

➢ STATIC ELECTRICITY—WITH A BALLOON*

The day before the activity, ask a few students to wear wool sweaters to school the next day. Then, to begin the activity, have your students inflate balloons and rub them against the wool sweaters. Next hold the balloons near small bits of paper towel or writing paper. The bits of paper will be attracted to the balloon and will stick to it, because the balloon has become electrically charged. (See Figure 21-9.)

Other things the class can do with the charged balloons are to place them against a wall (*they will stick to the wall*) and to move them carefully toward a very thin stream of water flowing from a faucet (*the stream will be bent toward the balloon*).

Explain to the students that you have chosen a dry day for this activity. If they are interested in how much effect high humidity would have on this activity, ask them how they could investigate the effect. Have them develop and explain plans for finding answers; and after they complete their investigations, have them report the results to the whole class. (*One way to investigate the effect of relative humidity on static electricity is to determine the relative humidity in the room—which may differ considerably from the weather report—on a day of high humidity and on a day of low humidity, and to perform the experiments described in this*

**Note*: Activities with static electricity do not work well when the humidity is high. If you teach in a location where the humidity is high, wait for a relatively dry day to attempt the static electricity activities—or you may have to omit them entirely. In either event, try the activities yourself first before attempting them with your class.

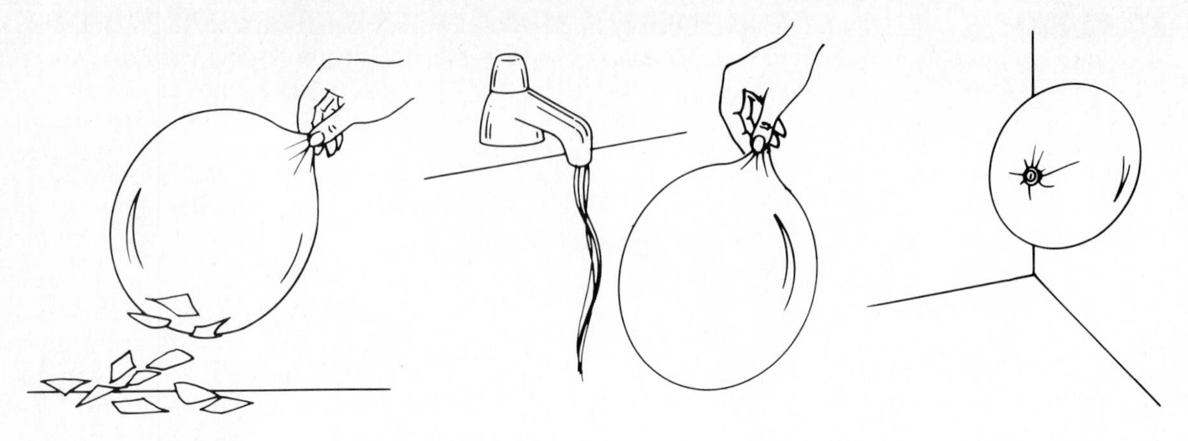

Figure 21-9. Testing for static electricity.

activity on both days and then compare the results. The results may not be sharply defined; but with a little practice, the students should be able to notice differences. Make sure the students recognize that they must be careful not to introduce unsuspected variables—that they must use the same kind of balloon for each trial, the same piece of wool, the same kind of paper, the same wall.)

⊳ STATIC ELECTRICITY—WITH A COMB

Tell the students that they can also generate static electricity by running a comb through their hair.* Then they can touch their combs to bits of paper or hold them near a thin stream of water to observe the effects of the static electricity, as in the above activity.

Suggest that they make additional investigations: Does the material from which the comb is made cause a difference in results? Does hair color make a difference? What about running the comb through their hair more or fewer times? Have students suggest other variables, and allow them to investigate the questions which interest them and report their findings to the class.

⊳ STATIC ELECTRICITY—A REAL SHOCKER

If you live in a dry climate, you and your students are probably already well aware that walking across a rug and then reaching for a metal doorknob can be a real shocker. This is another example of static electricity. Again, suggest ways your students can investigate this phenomenon: Does relative humidity make a difference? Type of rug? Kind of shoes? What happens with bare feet? Does walking four times as far on the rug make a difference? Ask the students to think of other questions and to investigate them. Then have them report their results to the class.

**Note*: Tell the students to bring their own clean combs to class for this activity. Caution them never to share their combs with other people, for reasons of cleanliness and health.

Explain to your students that they are working in much the same way as scientists work. Scientists find questions to investigate, perform investigations, and report the results to other scientists and to people in general. When they report on the results of their investigations, scientists often get feedback in the form of additional information and suggestions from others.

NOTE ON ACTIVITIES WITH CIRCUITS

Some of the activities with electricity which are most appealing to students include those which use wires, bulbs, switches, and batteries (or cells) to create electrical circuits. Many questions will arise. Some can be answered quickly, but others will require careful analysis and application of all the student's knowledge. A few may not be answerable within the limitations of the classroom. Point out to students, when this happens, that scientists do not always find their answers either.

Easily obtained (though not always inexpensive) materials are used in the circuit activities. The bell wire used can be obtained at most hardware stores or from the telephone company, which may be willing to give you short scraps or used wire, or to sell it to you inexpensively. Screw-based flashlight bulbs, D cells, and cell holders can also be obtained from local stores, or some science supply companies (see the Appendix) may sell them less expensively. You can make do without knife switches and bulb holders; but performing the activities is more convenient, less distracting, and less time-consuming with them. They may be available locally, or you may have to order them from a science supply company.

Simple materials—easily obtained at a local electronics supply house or from a science equipment supplier—can be used safely for many activities related to electricity.

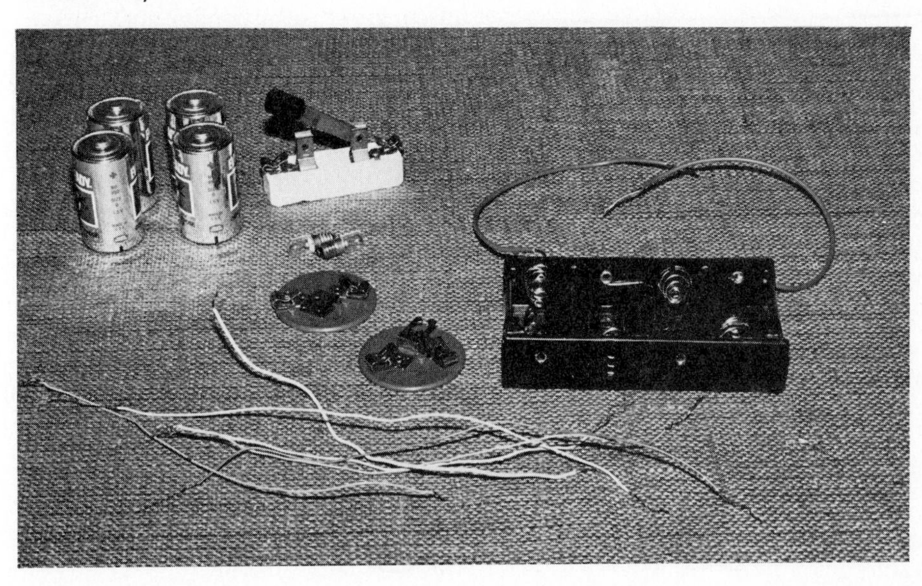

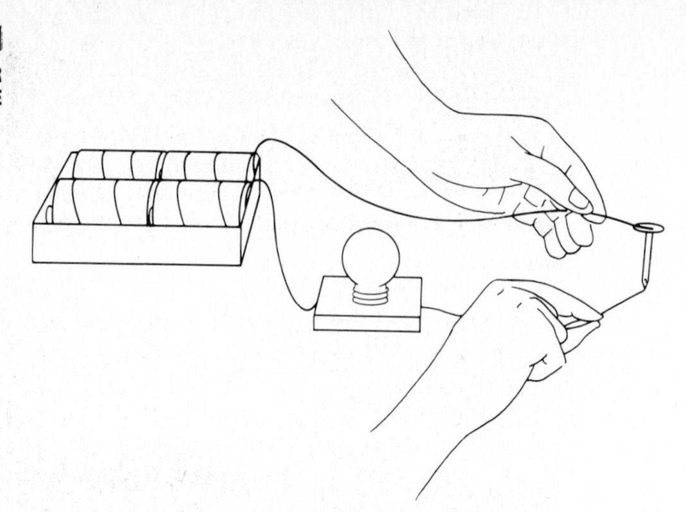

Figure 21-10. Testing for conduction and insulation.

▶ CONDUCTOR OR INSULATOR?
Introduce this activity by telling the class what conductors and insulators are and how they work. Then the students can use the setup shown in Figure 21-10 to test materials to determine whether they are conductors or insulators. The cell, light bulb, and wires are arranged so that they make a complete circuit except for the gap between the two wires. If the two wires are touched to the ends of a steel nail, which is a conductor, the circuit is completed and the bulb lights. Other materials can be tested in a similar fashion: conductors will complete the circuit and the bulb will light; but when insulators are touched to the wires, the bulb will not light. Give the students a variety of materials to be tested, and also allow them to test materials which they bring to school.

▶ THE FIRST CIRCUIT
Give each student or group of students a D cell, a piece of wire, and a bulb, and tell them that their job is to light the bulb—but do not tell them *how* to light the bulb. Even if it takes several days, let them figure it out for themselves. After some students or groups have succeeded, allow others to consult with them; but tell them that there is more than one way to light the bulb. (*The ways in which the bulb can be lit are shown in Figure 21-11.*) Tell the students that electricity flowing in the

Figure 21-11. Making a circuit.

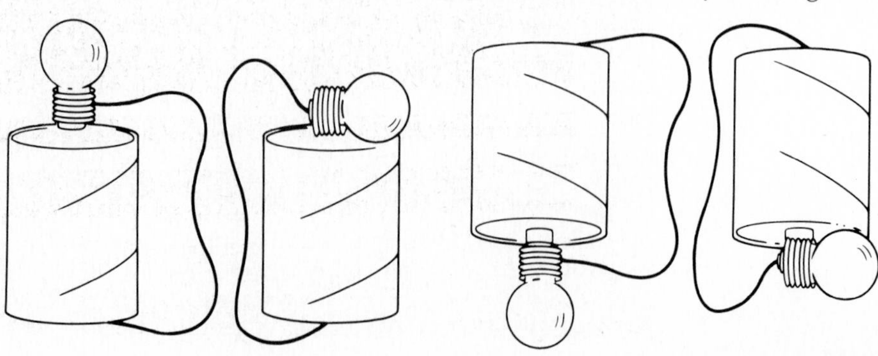

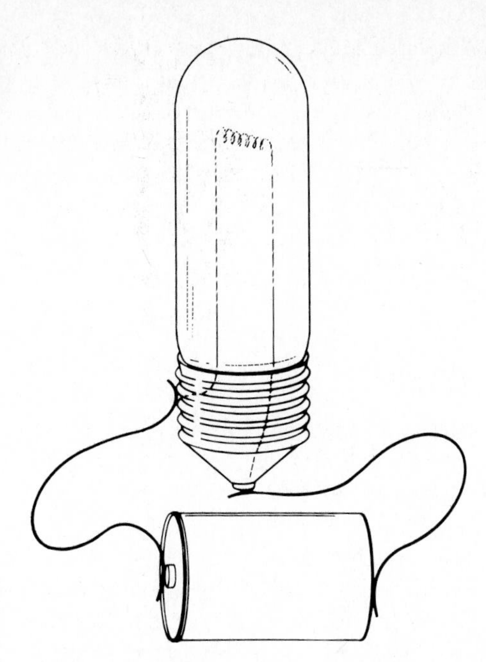

Figure 21-12. Understanding light bulbs.

circuit is what causes the bulb to light. Explain that the electricity heats up the filament, causing it to glow. Then ask the students to trace the path of the electricity from the battery through the bulb and back to the battery.*

➤ HOW DO BULBS FUNCTION?

Ask students to draw the inside of a bulb. (See Figure 21-12, above.) If they do not include the two wires which come from the ends of the filament, tell your students to add them to their drawings. Ask where the two wires go. The students will probably draw several different models. Tell them to analyze their models in view of what they know about where the bulb must be in contact with the battery and wire in order to be lighted. Carefully break open at least one flashlight bulb and several burned-out household bulbs to compare with the models.†

Ask the students to explain why a bulb does not "burn up" when it is lighted, since it becomes so hot that it glows. The students may respond that the bulb contains something special that prevents burning from occurring. Ask them how their explanations might be tested. (*One way is to carefully break away the glass on a flashlight bulb and then, after telling the students to watch closely, light the bulb—it will burn up. An intact bulb contains no oxygen.*)

One important thing that the students should learn about bulbs is that

**Note*: At this point stress to the students that they *must not* use electricity from any source other than cells. Especially point out the danger of using home and school electrical outlets.

†*Caution*: Break the bulbs yourself, and do not allow the students to touch the bulbs or the broken glass.

the reason they light is that the filament has a *high resistance* to the flow of electricity. When electricity flows through the filament, the filament becomes hot and begins to glow because of its high resistance. Tell the class to remember that the filament has a high resistance to the flow of electricity so that they will be able to understand several circuit activities they will be doing later.

➤ INTRODUCING SWITCHES

In this activity the students will begin to focus their attention upon various kinds of electric circuits. Not only is circuit building an interesting activity for students, but it also gives them many surprises and causes them to think of many questions. By asking probing questions which cause students to analyze the problems with which they are confronted, the teacher can help them draw together their information and concepts to produce explanations. Frequently ask the students to make predictions and check them out.

Have the students work individually or in small groups to build circuits (paths for electricity) that will use all the equipment and light the bulbs. The materials required are several pieces of wire 20 to 30 centimeters (8 to 12 inches) long, a battery holder and four cells, a bulb holder and bulb, and a knife switch. The three most common circuits which will be constructed are shown in Figure 21-13. Circuit A is the expected arrangement, and circuits B and C provide some unique learning opportunities.

In circuit A when the knife switch is open the bulb does not light, but when it is closed the bulb lights. Ask the students to trace the path of the electricity as it leaves the battery, passes through the bulb, passes

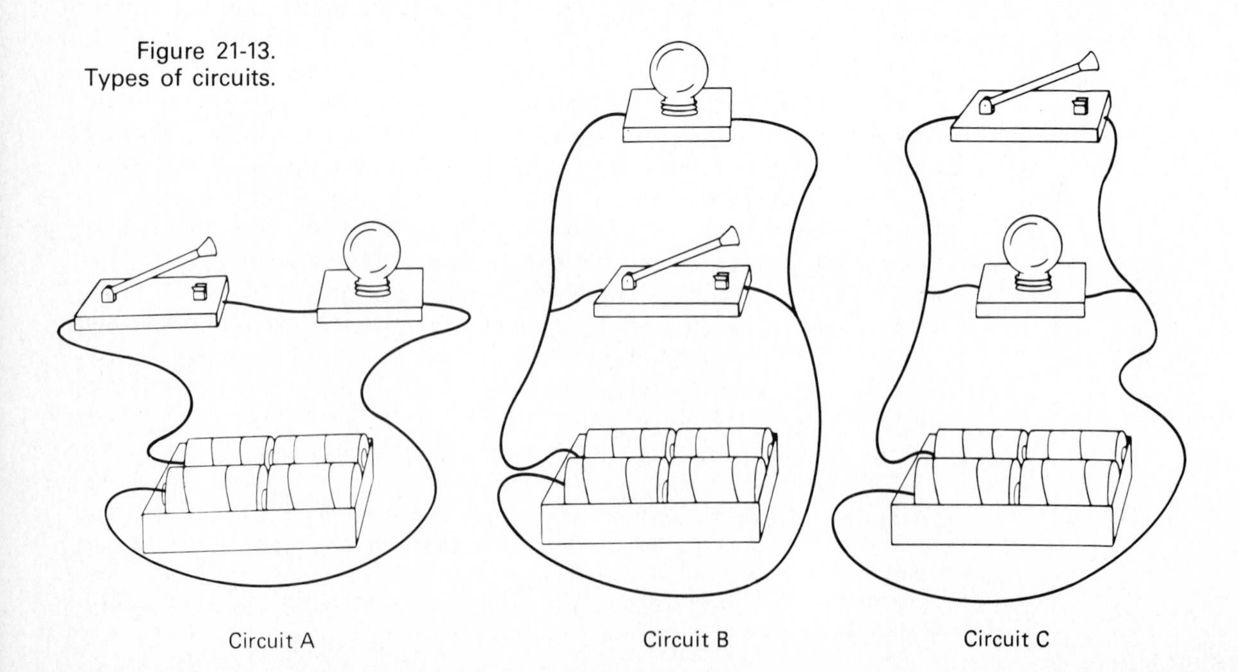

Figure 21-13. Types of circuits.

through the switch, and finally arrives at the other end of the battery. Stress that there must be a complete circuit in order for the electricity to flow.

If no student or group of students constructs circuit B, construct it yourself. In this circuit the bulb lights when the switch is open but does not light when the switch is closed. Have the students trace the path of the electricity when the bulb is lighted. Ask the students what happens when the switch is closed. (*There is another possible path for the electricity.*) Tell the students that there is electricity flowing from the battery, and ask them where it goes. (*It passes through the switch and back to the battery; in fact, there probably is some electricity flowing through the bulb, but not enough to cause it to light.*) Ask the students why the electricity goes through the switch rather than lighting the bulb. (*The students will probably indicate that it is a shorter path through the switch.*)

If necessary, also construct circuit C yourself. Before closing the switch, ask the students to predict what will happen and to explain why they think it will happen. (*The students will probably be uncertain, but several may suggest that the bulb will stay lighted, since it is closer to the cells.*) Close the switch. (*The bulb goes out.*) Now ask the students for explanations. (*The bulb is a source of high resistance while the wire has low resistance; thus the electricity flows through the wire rather than the bulb. Electricity flows most easily through the path of least resistance. Again, there will be some electricity flowing through the bulb, but not enough to cause it to light.*) Ask the students what would happen if a wire connected the two terminals of the switch in circuit A and then the switch were opened. (*The light would continue to burn.*) Ask them for explanations. (*There would still be a complete circuit.*)

➤ SERIES CIRCUITS

Tell the class that a series circuit is an electrical path in which all the electricity must pass through all the electrical equipment. Ask the students to identify which of the circuits in the previous activity was a series circuit. (*Circuit A is a series circuit. Circuits B and C each have two possible paths through which electricity can flow, and thus not all the electricity flows through all the electrical equipment.*)

When you are ready to begin the activity, give each student or group the same equipment as for the activity "Introducing Switches," and add an extra bulb holder and bulb. Tell them to build a circuit like the one shown in Figure 21-14, which is similar to circuit A in the previous activity but has two bulbs. Ask the students to predict whether or not there will be any noticeable differences between this circuit and the previous one. Then tell them to light the circuits. Are there any differences? (*The bulbs do not burn as brightly.*) Ask for possible explanations. Before they arrive at a conclusion, ask the students what would happen in the circuit if a wire were connected across the two terminals of one bulb holder. After they offer several suggestions, have them try it. What happens? (*The bulb in the affected holder goes out and the other one*

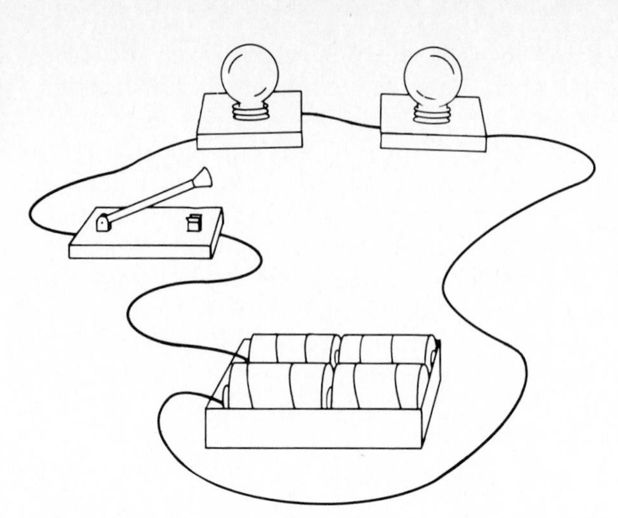

Figure 21-14.
A series circuit.

becomes brighter.) Ask for explanations: Why does the affected bulb go out? (*When the wire connects the two terminals of the bulb holder, there are two possible paths for the electricity: through the bulb or through the wire. The electricity flows through the wire, since it has a much lower resistance.*) Why does the other bulb become brighter? (*When one bulb—one source of high resistance—is eliminated, only one source of high resistance remains. Therefore, more electricity flows, and the increased electrical flow through the remaining bulb causes it to become brighter.*)

Ask the students to predict what will happen if another bulb is added to the two-bulb circuit, and then have them try it. (*Three bulbs burn even less brightly.*)

Ask if more electricity is flowing in the three-bulb circuit than in the one-bulb circuit. How could this be determined? Have students make as many suggestions as they can. (*Using an ammeter may be one suggestion. Even if you have an ammeter, ask them to describe other methods. One other way is to obtain fresh batteries, hook one—or a set of four—into the two circuits, and let them burn continuously to see which one is used up first. The circuit that went out first would be using the most electricity.*)

Here are some other questions which could be investigated by the students:

What would happen if one bulb were removed from its holder? (*All would go out.*) Why? (*There would no longer be a complete circuit.*)

Would it make a difference if the knife switch were in a different location? (*No.*) Why? (*It would still either complete or break the circuit.*)

What would happen if another set of four cells were added to the circuit?

Ask the students to identify an important characteristic of a series circuit. (*All the electricity passes through everything in the circuit. If the circuit is broken at any point, the entire flow of electricity stops.*)

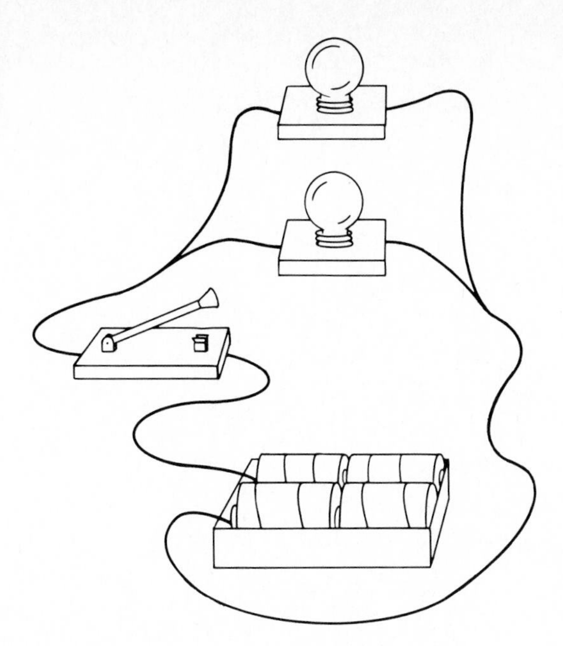

Figure 21-15.
A parallel circuit.

PARALLEL CIRCUITS

A parallel circuit is a circuit in which there is more than one possible electrical path. A portion of the electricity flows along each of the available paths.

For this activity each student or group will need several wires (20 to 30 centimeters or 8 to 12 inches long), a battery holder and four cells, three bulb holders and three bulbs, and one knife switch. Have the students construct a circuit similar to the one in Figure 21-15. Ask the students how bright they think the bulbs will be when the circuit is completed. Will the brightness of each bulb be similar to that in the one-bulb circuit, or will it be similar to that in the two-bulb circuit? (*It will be similar to the single bulb.*) Have the students explain. (*If either of the single paths is followed, there is only one bulb—one source of high resistance—along it.*) Ask how much electricity they would estimate is flowing in this new circuit as compared with the single-bulb series circuit. (*There should be approximately twice as much, since each of the two bulbs should require as much electricity as the single one to make them each be lighted as brightly.*) Ask the students how they could find out whether or not the two-bulb parallel circuit used more electricity than the one-bulb series circuit. (*One way would be to place fresh cells in each circuit and allow the bulbs to burn continuously until the batteries were used up.*)

Ask what would happen if a third bulb were added so that it was in another parallel circuit with the other two, as illustrated in Figure 21-16. (*It would be as bright as the other two.*)*

Here is a powerful activity which will cause students to examine all they have learned so that they can apply their knowledge to a real

*Note: As the cells begin to be used up, the brightness might vary. Try the activity with fresh batteries.

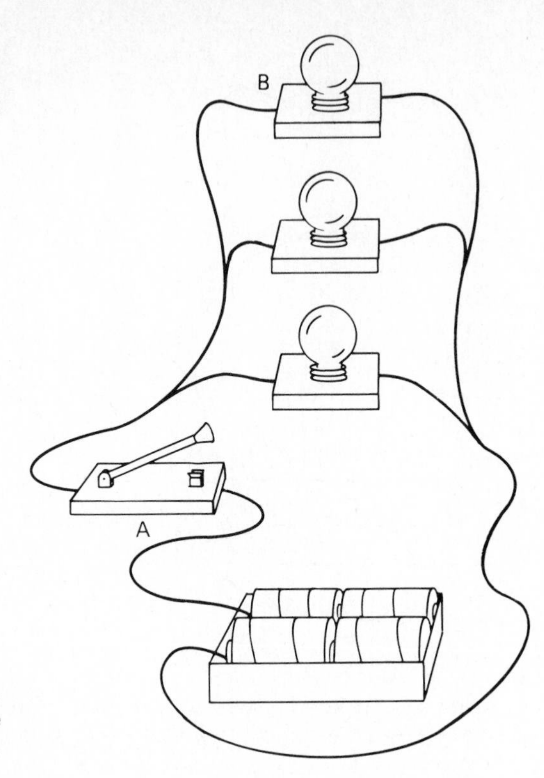

Figure 21-16.
Extending the parallel circuit.

problem. Ask what would happen if a wire were attached across the two terminals of one bulb of the three-bulb parallel circuit. The results should surprise almost all the students. They will probably say that the affected bulb will go out and the other two will remain lighted at the same brightness. Try it. (*All the lights go out.*) Ask the students to explain what happened. Suggest that they examine each possible route that electricity can take in this circuit. (*One possible route—the one in which the wire has been attached—consists entirely of wires with no bulbs, i.e., no sources of high resistance. The electricity follows that route.*)

Here are some more questions which could be investigated:

Would the knife switch serve the same function if it were located anywhere in the circuit? (*No.*) What if it were placed at the location indicated by the letter B in the circuit (Figure 21-16), rather than where it is, at location A? (*It would affect only the bulb near it.*)

What would happen if one of the bulbs were taken out of its holder? (*The others would stay lighted at the same brightness.*)

Ask the students to identify some of the important characteristics of parallel circuits. The most important characteristic is that not all the electricity passes through all parts of the circuit. Instead, there is more than one possible path for the electricity to take; and when one part of the circuit is broken, the others remain in operation.

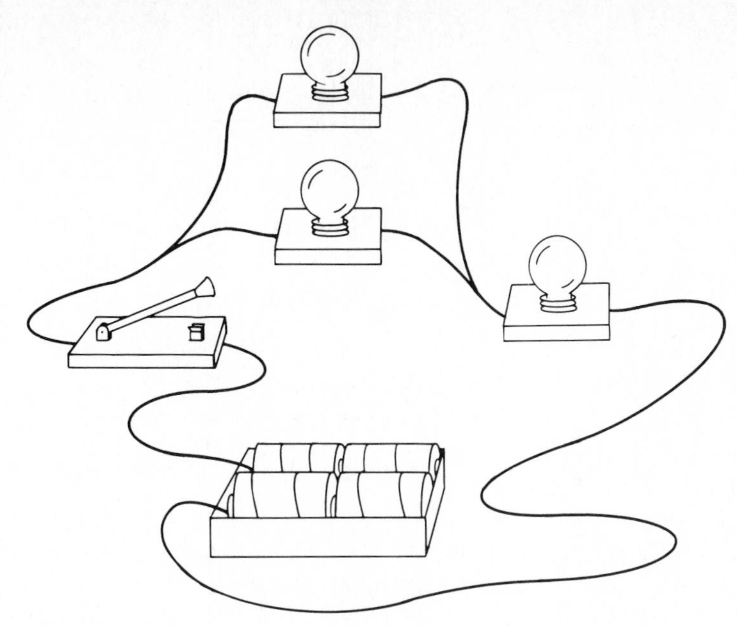

Figure 21-17. A combination of series and parallel circuits.

➢ COMBINATION PARALLEL AND SERIES CIRCUITS

Various combinations of parallel and series circuits can be constructed. One such circuit is shown in Figure 21-17. Many of the questions that can arise from such a combination are dealt with in the two previous activities. You may wish to allow your students to try building many unusual circuits. As questions arise, encourage them to carefully analyze each possible route that the electricity can follow. They will probably be able to answer their own questions; but if an individual student cannot answer a question, have others help.

Also encourage students to make simple circuit drawings like the ones shown in this chapter. Making the drawings will frequently help them answer their own questions.

➢ HOUSEHOLD CIRCUITS

This is not an activity as such, but rather an exercise which will allow students to apply some of the knowledge they have gained. Ask your students whether they think that the circuits in their homes are parallel, series, or combination circuits. (*Household circuits are parallel.*) What would be the implication if they were series circuits? (*Everything—television set, radio, clothes washer, clocks, stereo equipment, lights, etc.—would have to be operating, or none of it would work.*)

Explain to the class the purpose of circuit breakers or fuses: basically they are devices which interrupt (break) a circuit if too much electricity is flowing. Otherwise, the wires would become overheated and might start a fire. Would circuit breakers or fuses be in series or in parallel with the circuit they control? (*In series—or else electricity would continue to flow in the circuit.*)

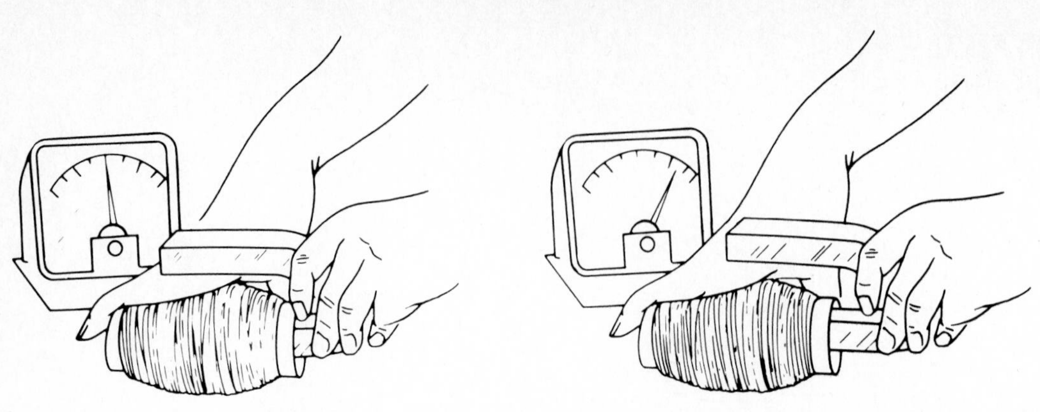

Figure 21-18. Generating electricity with a magnet. (Source: P. F. Brandwein et al., *Concepts in Science: Brown Level (Grade 6)*, Newton ed., Harcourt Brace Jovanovich, New York, 1975, p. 226, or Teachers ed., p. T-272.)

Ask the students if they have sets of Christmas tree lights in their homes. If so, can they remember whether the set continues to burn if one bulb burns out, or whether all go out when one bulb burns out? What kind of circuit is each of the types? (*If only one bulb goes out, the circuit is parallel; if all go out, the circuit is series.*) If possible, have a set of each on hand to demonstrate.*

➤ GENERATING ELECTRICITY WITH A MAGNET

One way to demonstrate the interrelationship between electricity and magnetism is to generate electricity with the use of a magnet. For this activity, you will need wire, a relatively powerful U-shaped magnet, a tube from a roll of paper towels, and a galvanometer (a meter which can detect a flow of electricity); see Figure 21-18, above. Wrap a wire around the tube several times, and connect the two ends of the wire to the galvanometer. Then move one part of the U magnet into and out of the tube. The galvanometer needle will swing back and forth as the magnet moves, indicating a flow of electricity.

Ask the students to predict what would happen if a longer wire were wrapped around the tube and the activity repeated. Try it. What would happen if the other end of the magnet were used? Try it. Could other factors affect the results?

➤ ELECTROMAGNET

A magnet can be produced by wrapping insulated wire around a nail and connecting the ends to a cell or battery of cells as shown in Figure 21-19.

**Caution*: Once again, warn the students not to experiment with household electricity. Tell them it is very powerful and can cause severe burns or even death. Remind them again that only people who are trained to work with that kind of electricity should do so.

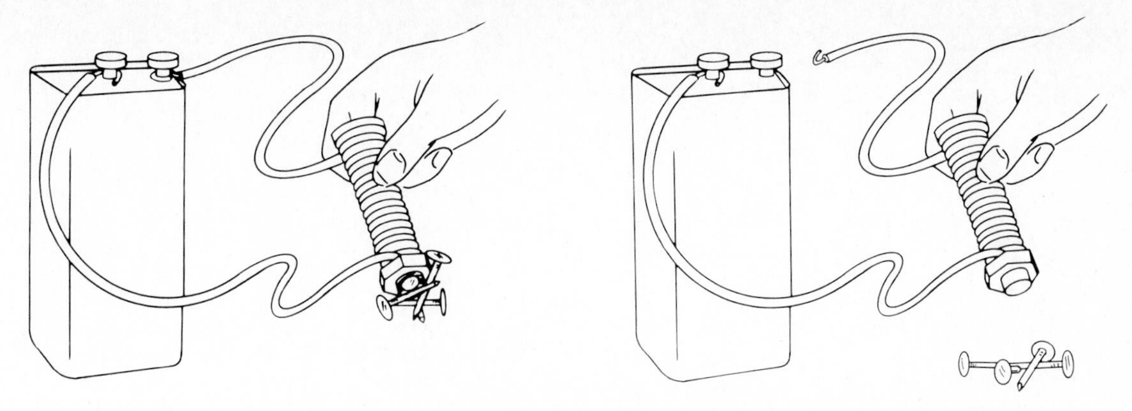

Figure 21-19. Making an electromagnet. (Source: P. F. Brandwin et al., *Concepts in Science: Brown Level (Grade 6)*, Newton ed., Harcourt Brace Jovanovich, New York, 1975, p. 231, or Teachers ed., p. T-277.)

Place the electromagnet into a box of paper clips. How many paper clips will it pick up? Detach the wire from one side of the electromagnet. What happens?

Ask the students to predict how the electromagnet could be made stronger. By wrapping more wire around the nail? By attaching more cells?*

What are the advantages of being able to "turn off" the magnet? How about in loading scrap metal into a railroad car? What would happen if a permanent magnet were used?

➤ NO ELECTRICITY!

Tell your students to relax and close their eyes. Then slowly read this guided fantasy aloud. Pause 5 to 15 seconds wherever a slash mark occurs.

We are going to spend a night without electricity. Picture yourself alone in your house. It is late on a winter night. / You are lying on the couch, cuddled up in a warm blanket watching your favorite television show. / Outside it is very cold and there is frost on the windows. The wind is blowing very hard, and the trees are swaying to and fro, and you can hear the limbs scrape against the window. / After a few minutes, you hear the sound of raindrops begin to hit against the windowpane. / The sound makes you drowsy, and you close your eyes and drift off to sleep. / Suddenly, you are awakened by a crash of thunder, and you realize that you are in total darkness. The television, which you had left on, is now off also. / You stand up and begin to feel your way across the room toward the kitchen. On the way you run into a footstool and almost fall. / You go to the pantry in the kitchen and get out some candles and matches. You light the

**Caution*: Do not exceed 6 cells. The wires could become hot if a large amount of electricity were carried.

candles and set them around the house. / You glance at the clock sitting on the desk, and notice that the face of the clock is dark. The hands of the clock are stopped at 11:30. / Since the electricity is out and all the heaters are electric, the house begins to feel cold. / You take a candle from the table and go to your closet to get a sweater. / Then, after several minutes of walking around doing nothing, you decide to do something to keep yourself busy until the electricity comes back on. / In trying to think of something to do, though, you run into problems. Nearly all the things that you do around the house to entertain yourself run on electricity. The television will not work, and you can't play records or listen to the radio. / You would like to make some cookies or hot chocolate, but the stove is also electric. / Finally, you decide just to sit by candlelight and read a book, the way Abraham Lincoln did as a boy. / You soon realize it is difficult to read by the flickering light of a candle. / After what seems like a lifetime, the electricity finally comes back on. You breathe a sigh of relief and realize how lucky you are to have electricity. / You find it hard to believe that people were ever able to live without it. You now realize how important electricity is to you and how many things you use depend on it. /*

Children's literature

Have your students read *Wonder of Electricity*† or read selections from it to the class, and then discuss with them the concepts presented.

Find other books on electricity in the school library, and identify them for the students or place them in an interest center.

Activity cards

An example of an activity card is shown in Figure 21-20. Many activity cards could be made by inventing slight variations on the activities in this chapter. Designing electric circuits is an especially rich area for investigation. You might include one activity card challenging your students to make activity cards for new circuits. Additional activity cards could be made for topics such as making an electric motor, making a radio, working with fuses and circuit breakers, reading home electric meters, determining how much electricity selected household devices use, studying lightning, finding out how electricity is produced for the local community, and developing safe guidelines for use of electricity in the home.

Bulletin boards

HOW IS ELECTRICITY PRODUCED?
Make a bulletin board showing how electric energy is produced by burning coal (or however it is produced in your area).

*Written by Nancy Jo Cooper.
†Hyman Ruchlis, *Wonder of Electricity*, Harper and Row, New York, 1965.

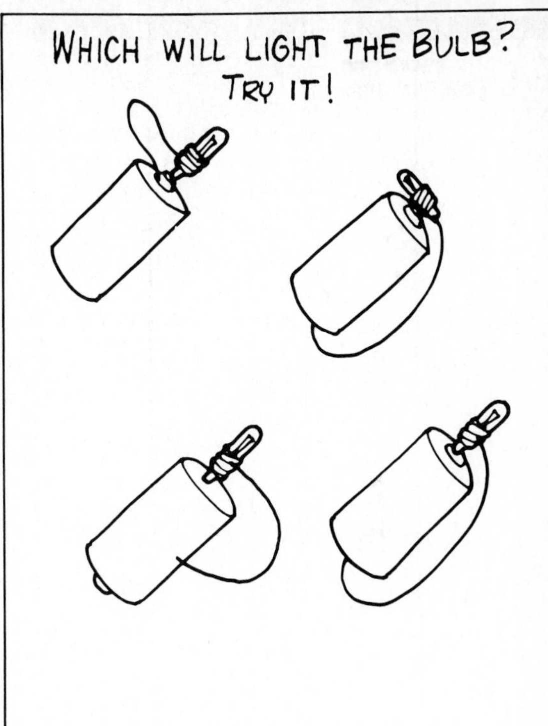

Figure 21-20. Activity card: Light it.

STATIC ELECTRICITY
Attach materials such as combs, balloons, small pieces of paper, and a piece of wool to a bulletin board. Post directions on how to generate static electricity.

Field trips

Electricity does not offer many possibilities for field trips on the school grounds; but one thing the students could do is look around and list evidence of use of electricity—electric lines, automobile headlights, outdoor lamps, and so forth.

A very useful trip could be made to the local electric utility company. Students might be able to see generators in action, and you could ask a company representative to give an illustrated talk on how electricity is generated for local use. Safe use of electricity and conservation of energy are other topics which might be presented.

A visit to a local department store could be very enlightening. Ask the class to make a list of all the devices and equipment in the store which use electricity. The list should be impressive.

RESOURCES

Sources of assistance

FREE AND INEXPENSIVE MATERIALS
Here are some of the best sources of free and inexpensive materials for use in teaching magnetism and electricity:

1. *American Museum of Atomic Energy*.* Folders of experiments dealing with several types of energy, including electric energy.
2. *Thomas Alva Edison Foundation*. A great number of teaching materials on electricity. Write for a list of the available aids.
3. *Edison Electric Institute*. Many materials, including comic-format materials on electricity.
4. *Energy Research and Development Administration*. Especially helpful on energy conservation materials.
5. *General Electric Company*. Many aids, including lesson plans and career information.

LOCAL RESOURCES
Local resources which can help in teaching magnetism and electricity include:

1. An educational consultant from the electric utility company.
2. An electrical repair worker who is willing to show students the delicate and intricate nature of equipment like radios and televisions.
3. A representative of a microcomputer company who could demonstrate some of the capabilities of small computers.
4. Representatives of industry or an organization who are familiar with electric automobiles or with wind, water, or solar generation of electricity.

ADDITIONAL RESOURCES
1. Magazines such as *National Geographic World*. In addition, almost all popular and news magazines carry articles on the production, use, and conservation of electric energy.
2. Several state and regional electric utility companies offer teaching materials and helpful general materials on electric energy.

Strengthening your background

1. Do the activities in this chapter—especially the circuits activities.
2. Take apart a flashlight cell to learn what it is like inside.

*The addresses of specific agencies, organizations, and businesses are listed in the Appendix.

3. Ask one of your students who has a special interest in magnetism or electricity to share some of her or his knowledge with you.
4. Locate the north and south poles on an inexpensive magnet of the type used to attach notes to a refrigerator or note board.
5. Learn how a circuit breaker works. Ask a friend who is knowledgeable about electricity to help you.
6. Examine an automobile battery. Why is it a battery rather than a single cell?

Getting ready ahead of time

1. Obtain, through your school, enough small, inexpensive bar magnets to carry out the activities in this chapter. You should have at least two for each student, plus a few spares. (See the Appendix for names and addresses of science supply companies.)
2. Obtain a variety of magnets other than bar magnets, including at least one strong U magnet.
3. Get a supply of iron filings from a local machine shop—or talk to the high school metal shop teacher.
4. Visit a local electrical contractor and collect scrap pieces of materials used in household wiring.
5. Obtain, through your school, the supplies needed for electric circuit activities. (Ask the telephone company for scraps of wire.)
6. Build a "shining light" box (see Chapter 7).
7. Make overhead transparencies of the circuits you will want your students to construct.
8. Check the school library for books on magnetism and electricity, and give the librarian a list of supplementary books which you feel your students will need. Ask that at least some of them be added to the library collection.

Building your own library

Blecha, Milo, Franklin Fish, and Joyce Holly: *Exploring Matter and Energy*, Laidlaw, River Forest, Ill., 1980.

Brandwein, Paul, Daniel Brovery, Arthur Breenstone, and Warren Yasso: *Energy: A Physical Science*, Harcourt Brace Jovanovich, New York, 1975.

Elementary Science Study (ESS): *Batteries and Bulbs II*, Webster, McGraw-Hill, Manchester, Mo., 1971.

———: *Teacher's Guide for Batteries and Bulbs*, Webster, McGraw-Hill, Manchester, Mo., 1974.

Miller, Franklin, Jr., Thomas Dillon, and Malcolm Smith: *Concepts in Physics*, Harcourt Brace Jovanovich, New York, 1974.

Oxenhorn, Joseph, and Michael Idelson: *Pathways in Science: The Forces of Nature*, Globe Book, New York, 1975.

Press, Jans Jurgen: *Science Projects for Young People*, Van Nostrand, Princeton, N.J., 1971.

Sootin, Harry: *Experiments with Electric Currents*, Norton, New York, 1969.
Wilson, Mitchell: *Energy*, Life Science Library, Time-Life, New York, 1963.

Building a classroom library

Adler, Irving: *Magnets*, John Day, New York, 1966.
Amery, Heather, and Angela Littler: *The Know How Book of Experiments with Batterys and Magnets*, Sterling, New York, 1976.
Asimov, Isaac: *How Did We Find Out about Electricity?* Walker, New York, 1973.
Beeler, Nelson, and Franklyn Branley: *Experiments with Electricity*, Thomas Y. Crowell, New York, 1949.
Branley, Franklyn, and Eleanor Vaughan: *Mickey's Magnet*, Thomas Y. Crowell, New York, 1956.
Epstein, Samuel, and Beryl Epstein: *The First Book of Electricity*, F. Watts, New York, 1977.
Feravolo, Rocco: *Electricity*, Garrad, Champaign, Ill., 1960.
Keen, Martin: *The How and Why Book of Magnets and Magnetism*, Grosset and Dunlap, New York, 1969.
Kirkpatrick, Rena: *Look at Magnets*, Raintree, Milwaukee, Wis., 1978.
Knight, David: *Let's Find Out about Magnets*, F. Watts, New York, 1967.
Math, Irwin: *Wires and Watts: Understanding and Using Electricity*, Scribner, New York, 1981.
Pine, Tillie, and Joseph Levine: *Electricity and How We Use It*, McGraw-Hill, New York, 1962.
———: *Magnets and How We Use Them*, 9th ed., McGraw-Hill, New York, 1973.
Podendorf, Illa: *True Book of Magnets and Electricity*, Children's Press, Chicago, Ill., 1971.
Ruchlis, Hyman: *Wonder of Electricity*, Harper and Row, New York, 1965.
Shapp, Martha, and Charles Shapp: *Let's Find Out about What Electricity Does*, rev. ed., F. Watts, New York, 1975.
Sootin, H.: *Experiments with Electric Current*, Grosset and Dunlap, New York, 1969.
Veglahn, Nancy: *Coils, Magnets, and Rings: Michael Faraday's World*, Coward-McCann, New York, 1976.
Wade, Harlan: *A Book about Electricity*, Raintree, Milwaukee, Wis., 1979.

CHAPTER 22

MACHINES

IMPORTANT CONCEPTS

Machines enable us to do work that might otherwise be impossible.

Six simple machines are the lever, wedge, screw, inclined plane, wheel and axle, and pulley. It could be said that there are only two simple machines—the lever and the inclined plane—and that special applications produce the other four.

Simple machines can be used in everyday tasks.

Simple machines can be used to change the direction of a force.

Simple machines can be used to increase force or speed.

An increase in force is required to increase speed.

An increase in distance is required to increase force.

Work is done when a force is used to move matter over a distance.

INTRODUCTION: STUDYING MACHINES

Children should become familiar with the basic uses of simple machines. There are many practical applications of such knowledge in everyday life. Opening a can or moving a heavy object can become much more manageable and safe when simple machines are used. In teaching simple machines, you should focus on basic concepts and practical applications. Many interesting questions can be posed and investigated. Several questions can be answered by relatively simple activities—questions such as how many times a person's effort is multiplied by use of a simple machine, or how much distance a simple machine's effort must move through to move an object a desired distance.

MACHINES FOR PRIMARY GRADES

Activities

➤ KINDS OF MACHINES

Have your students cut out magazine pictures of machines which help people do work. Ask some of these questions: What would the world be like without machines? What machines have you used today? What would you have done if those machines did not exist? (*Most of the machines named will be quite complex, and to analyze them in terms of what simple machines were represented would be difficult. Just stress the concept that machines do help us accomplish work.*)

➤ BALANCING PEOPLE

Make a balance or teeter-totter out of a 5- by 20-centimeter (2- by 8-inch) plank which is 2½ to 3 meters (8 to 10 feet) long and a short piece of 5-centimeter (2-inch) pipe, as shown in Figure 22-1. Stand on the balance

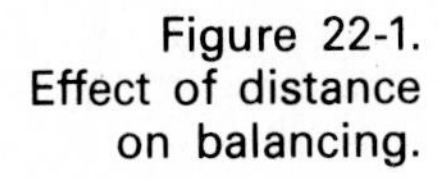

Figure 22-1. Effect of distance on balancing.

and ask if any students think they could balance you. Have them try it. (If the students are much lighter than you, stand near the pipe.) Then have students balance each other. After they have done this a few times, ask them to predict how far from the pipe each of two persons must stand to be balanced. Repeat the procedure several times until the students become accurate in making predictions. What do they observe about balancing a heavy person on one end against a lighter person on the other end? (*The lighter person must be farther from where the plank rests on the pipe than the heavy person is.*)

If the students become quite good at predicting where people must stand on the balance, or if some students want to do more work with the balance, obtain a few bricks and have them do balancing by placing different numbers of bricks at different distances from the pipe, with the board centered on the pipe. Can one brick balance six?

➢ WHEELS MAKE PULLING EASIER

Place a nail in a piece of board as shown in Figure 22-2. Hook a rubber band over the nail and have the students pull the board over a rough surface such as a rug. Make sure that they notice how much the rubber band stretches as they pull the board. Next place some wooden dowels (or round pencils) under the board. Again have the students notice how far the rubber band stretches. Ask them to compare the amount of stretching with and without the dowels. (*The rubber band should stretch much less with dowels under the board, indicating that much less force is required to move the board with dowels.*) Tell the students that the dowels are simple wheels, and ask them to name other examples of wheels which they see frequently.

➢ SIMPLE MACHINES USED AT HOME

Discuss simple machines in class, and then ask the students to name simple machines which they use at home. Make a list on the board, and ask the students to look around at home for more simple machines to add to the list.

Show the class a long and a short screwdriver and an empty, clean paint can with a press-on lid. Ask the students if they would be able to open a full can of paint (and not spill any paint) without using a tool.

Figure 22-2. Effect of wheels on pulling.

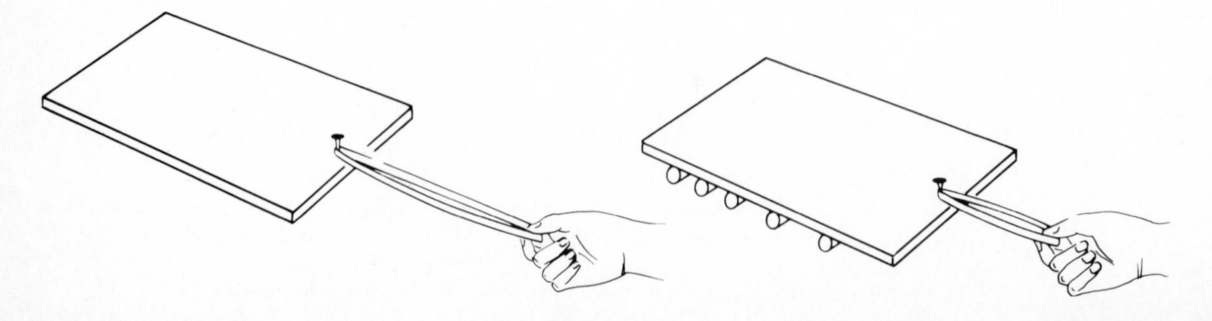

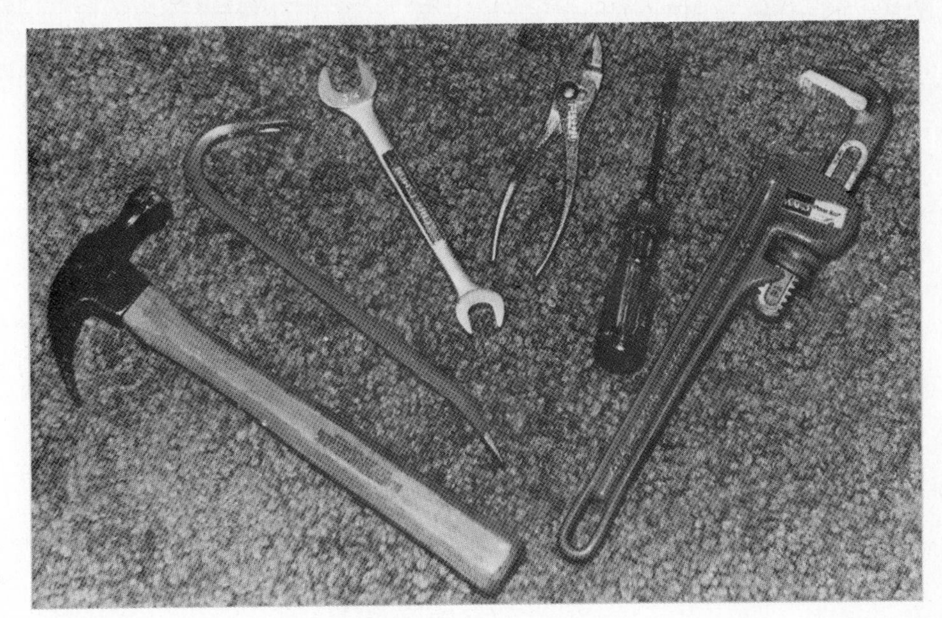

These simple machines, which are found in many homes, are types of levers.

Then ask if they could do it with a screwdriver. They would use a screwdriver like what kind of simple tool to open the can? (*Lever.*) Let the students practice opening the can with the screwdrivers. How does using a short screwdriver compare with using a long one?

➢ OPENING A CAN MORE EASILY

Bring several clean, empty metal cans to school.* Have students use a can opener of the type which produces triangular punctures to make openings in the intact ends of the cans. Then tape or tie a screwdriver or sturdy stick securely to the opener and have them make other holes with the longer-handled can opener. Is it easier or harder to make holes using the opener with the longer handle? What if the handle were even longer?

➢ MAKING WORK EASIER

Ask students to pretend that they have been asked to move a heavy weight. Then have them either write down or tell the class how they successfully moved that weight using only simple machines.

Children's literature

Read *A Book about the Lever*† to your students or have them read it, and then discuss with them the concepts presented.

Find other books on machines in the school library, and identify them for the students or place them in an interest center.

**Note*: These should be cans that you have opened at one end carefully at home, using a rotary-type can opener which does not leave jagged edges on the cans.

†Harlan Wade, *A Book about the Lever*, Raintree, Milwaukee, Wis., 1979.

Figure 22-3. Activity card: Simple machines.

Activity cards

A sample activity card is shown in Figure 22-3. Other activity cards could be made for topics such as finding pictures of examples of each type of simple machine and using simple machines in everyday tasks which students do.

Bulletin boards

KINDS OF SIMPLE MACHINES
Post a colored drawing of each type of simple machine on a bulletin board. (See Figure 22-4.) In a small box nearby place samples of simple machines (e.g., a screwdriver, wheels from a toy, a screw) along with colored pieces of paper, and ask students to identify what type of machine each sample is by attaching paper of the appropriate colors to the samples.

USING MACHINES
Place several pictures of people using machines on the bulletin board. Ask students to add pictures or drawings which they cut out or make.

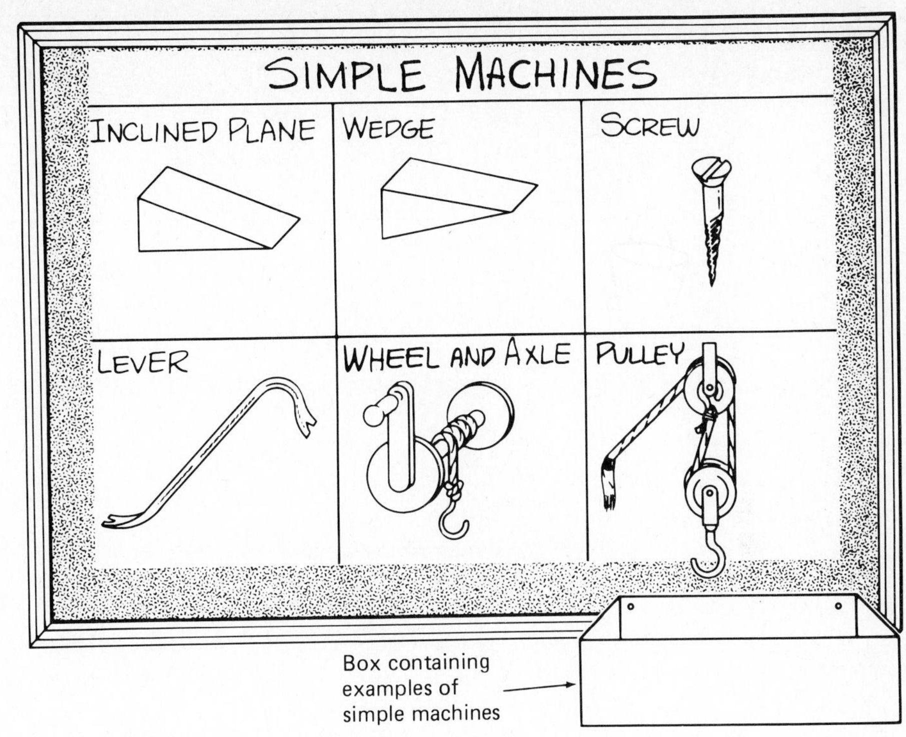

Figure 22-4. Bulletin board: Simple machines.

Field trips

Take a walk around the school grounds and have your students identify examples of simple machines. (*Stairways are inclined planes; it is much easier to go up several small steps than to go straight up the total height. Driveways are often inclined planes. Automobiles and wagons have wheels. A teeter-totter is a lever. Shovels may be used as levers.*)

MACHINES FOR INTERMEDIATE GRADES

Activities

➢ INCLINED PLANES

To help students understand the advantage of using inclined planes, have them place a weight (sand, blocks of wood, or other easily available materials) into a toy truck or other carrier with wheels. Then attach a spring scale as shown in Figure 22-5 and pull the carrier up inclined planes of different degrees of steepness. Have the students record the readings on the scale for each trial. Do gentle slopes require more or less force than steep ones? (*Less.*) Although less force is required for gently sloped inclined planes, what is true about the distance the carrier must be moved to gain the same height as on a steep inclined plane? (*The distance is*

Figure 22-5.
Inclined planes.

greater—once again illustrating that while a simple machine may reduce the amount of force that must be applied, it requires that the force be applied over a greater distance.)

➢ SPLITTING WOOD WITH WEDGES

Obtain a few "logs" about 8 to 15 centimeters (3 to 6 inches) in diameter, and ask your students to split them. Have them decide how the job can best be done. At some point either they or you should suggest using a

Wood-splitting wedges make the job of breaking up large pieces of wood much easier.

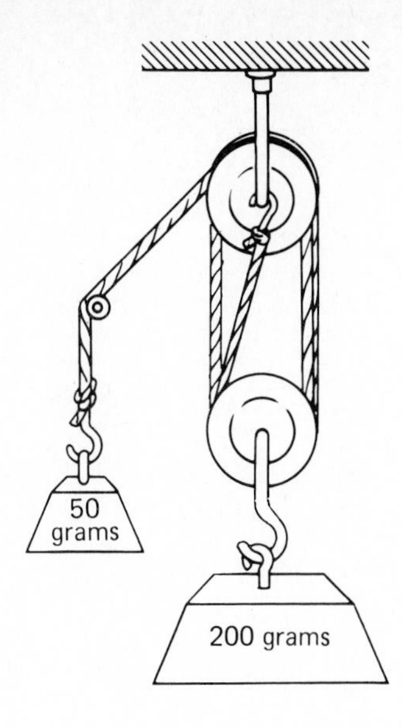

Figure 22-6. The pulley is a wheel and axle.

wedge and hammer. Obtain a wood-splitting wedge if possible; otherwise use a wood chisel and a small log. Use a normal-sized hammer for the activity.*

Make sure that your students understand that wood splitting is made easier with the help of a wedge, and that they have no doubts. If more than one wedge is available, have them determine whether sharp, narrow wedges or blunt, wide wedges are easier to pound into the wood. (*Narrow wedges are easier because they can be inserted with less force, but they must be pounded in farther to separate the wood as much as a blunt wedge would.*)

➤ PULLEYS

Obtain a set of pulleys from a science supply company.† Hang one as shown in Figure 22-6 (above), where one weight is balancing another weight 4 times as heavy. Challenge the students to design configurations with other weight-balancing ratios. (*The weight ratio necessary for balancing is determined by the number of movable strings which support the movable pulley assembly.*)

Also, have the students determine how far the smaller weight must be moved to move the larger weight 10 centimeters (4 inches). (*If the heavier weight is 4 times the weight of the lighter one, the lighter one will have to be moved 4 times as far—40 centimeters or 16 inches.*)

**Caution*: Warn the students to be very careful so that they do not injure themselves or others—and supervise them closely during this activity.

†They are relatively inexpensive. A list of supply companies is found in the Appendix.

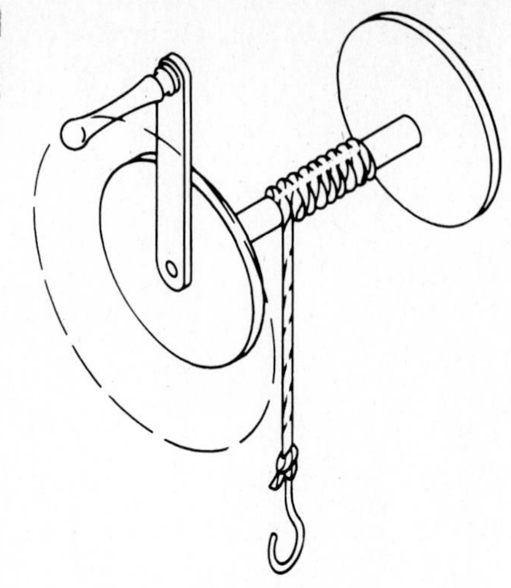

Figure 22-7.
The windlass is a wheel and axle.

WHEEL AND AXLE

The term "wheel and axle" is immediately reminiscent of the wheel-and-axle assemblies on automobiles and trucks—and indeed, these assemblies can function as simple machines. However, in demonstrating for elementary school classes the advantages of the wheel and axle as a simple machine, the windlass (a device which at first does not look like a wheel and axle) is much more fruitful. It may be helpful to think of the "lever" of the device as just one part of the wheel, as indicated in Figure 22-7.

Make or buy a simple windlass, and have your students attach a pail of sand to the rope and lift it with the wheel and axle. Then have them put a much longer handle on the windlass and again lift the pail of sand. Which is easier to use, a long or a short handle? (*A long handle makes the task easier.*) Ask the students to compare the distances that they had to move the long and short handles. (*Their hands had to travel a greater distance with the long handle.*) Ask the students to name any similar devices which they have used or have seen used.

LEVERS

One way to study the lever is by using a simply made balance. Drive a long nail through the middle of a meter stick (or yardstick) as shown in Figure 22-8, and support the nail on two cans filled with sand (or any other suitable support). If the meter stick does not balance, add pieces of masking tape to the "high" end.

Give the students paper clips which have been partially unbent to use as supports and washers to use as weights, and challenge them to try many combinations to cause the system to balance. Pose questions such as "If two washers were placed 20 centimeters (8 inches) from the nail, where would one washer have to be placed on the other side to make the

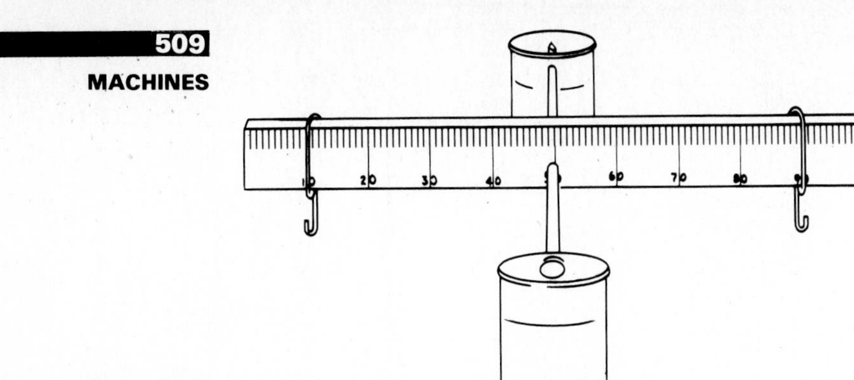

Figure 22-8.
Studying the lever.

system balance?" After students have acquired some skill in predicting where the weights must be located to cause the system to balance, ask them if they can state a rule for determining the locations of weights in a balancing system. (*The number of weights on one side of the nail multiplied by the distance from the nail must be equal to the number of weights on the other side multiplied by the distance from the nail.*) If you have to tell students the rule, have them verify it.

To demonstrate a practical everyday use of a lever, have students use a wrecking bar (sometimes called a "crowbar") to pull a large nail from a board. For comparison, have them pull out another nail the same size from the same board by using a pair of pliers and by pulling directly on the nail.

▶ THE BICYCLE AS A MACHINE

Have one of the students bring a 10-speed bicycle to school. What examples of simple machines can the class find on the bicycle? (*Levers for the brakes and possibly for the shifting aparatus. Wheels and axles for the pedal-to-rear-wheel drive chain.*)

When the bicycle is traveling fastest in the highest gear, on which size sprocket (wheel) on the rear wheel is the chain located? (*The smallest one.*) Why is the wheel moving fastest then? (*The distance that the chain has to move over the wheel to cause one revolution of the wheel is the least using that sprocket. The force being applied at that time is relatively great.*)

▶ A CONSTRUCTION PROBLEM

Have the students imagine that they are living back in the time of cave people. They have decided to build homes in a cliff using rather heavy materials. How can they lift the materials to the construction site? What kind of a system can they develop which will both lift them up to their homes and keep out unfriendly people or animals?

Figure 22-9. Activity card: Primitive tools.

Children's literature

Read parts of *Machine Tools** to your students or have them read the book, and then discuss with them the concepts presented.

Find other books on machines in the school library, and identify them for the students or place them in an interest center.

Activity cards

An example of an activity card is shown in Figure 22-9. Additional activity cards could be made for topics such as locating examples of simple machines in the home workshop or kitchen, building a block-and-tackle system for lifting weights, finding examples of machines which multiply distance, collecting pictures or making drawings of complex machines which people enjoy using, and using an erector set or Tinker-Toy set to build machines.

*H. S. Zim, and J. R. Shelley, *Machine Tools*, Morrow, New York, 1960.

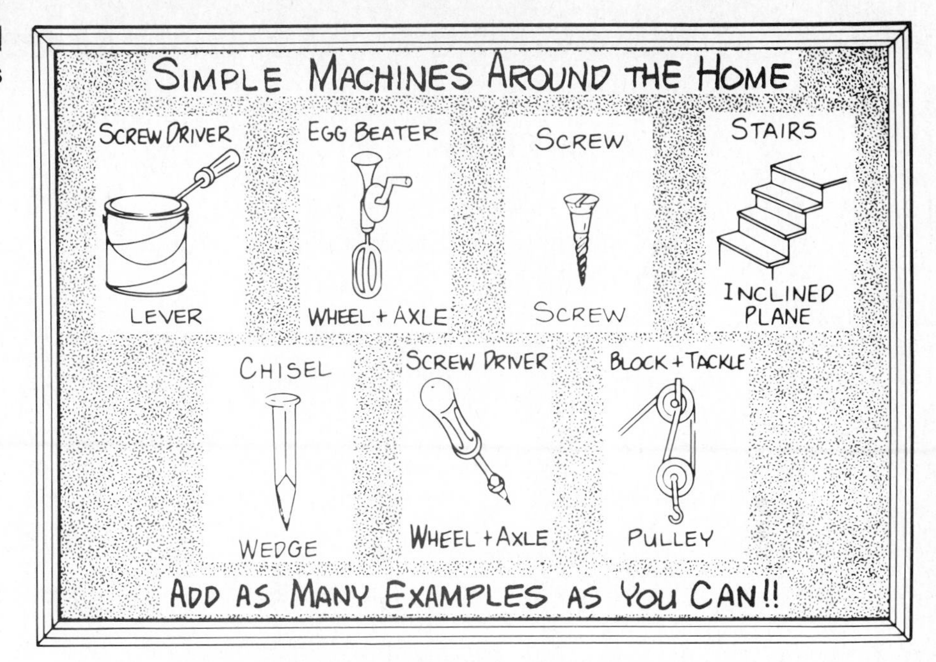

Figure 22-10. Bulletin board: Simple machines around the home.

Bulletin boards

USEFUL SIMPLE MACHINES
Have students draw pictures of simple machines which are commonly used around the home and post them on a bulletin board such as the one illustrated in Figure 22-10.

MACHINES MAKE LIFE ENJOYABLE
Post a series of pictures of some complex machine which adds convenience and enjoyment to life—for example, the airplane. Every few days the bulletin board could be changed to focus on another type of machine: e.g., the railway train, the automobile, the bicycle, the derrick (and other construction machines).

Field trips

Take the class to visit a repair garage or a service station, and ask a mechanic to demonstrate several of the tools used in working on automobiles or trucks. Tell the students to note especially the block and tackle (pulley) which lifts the engine from a vehicle, screwdrivers, pliers, wrenches, and pry bars. Ask the students to identify the simple machine on which each tool is based.

RESOURCES

Sources of assistance

FREE AND INEXPENSIVE MATERIALS

Here are some good sources of free and inexpensive materials for use in teaching about machines:

1. *Automobile Manufacturers Association.** Posters and bulletin-board ideas on automobiles and trucks.
2. *National Air and Space Museum.* Publications on the history of flight.
3. *Association of American Railroads.* Teaching activities for several curriculum areas related to railroad transportation.
4. *United Airlines.* A set of pictures of airplanes from past to present.

LOCAL RESOURCES

Local resources which can help in teaching about machines include:

1. A representative from an association of engineers.
2. Persons who work with tools—carpenters, mechanics, plumbers.

ADDITIONAL RESOURCES

1. Magazines such as *SciQuest* and *Nature.* (See especially articles relating to tools of people in primitive cultures.)
2. Catalogs which picture tools.

Strengthening your background

1. Try to identify at least one example of each simple machine in your house. (All except the pulley and possibly the wedge should be relatively easy to find.)
2. Obtain a book on ancient people and identify some of the earliest tools used by humankind.
3. Read *Simple Machines and How We Use Them* by Tillie Pine and Joseph Levine (see "Building a Classroom Library").

Getting ready ahead of time

1. Make a collection including at least one example of each simple machine which you can use in your classroom.
2. Construct five simple balances, such as the one used in the "Levers" activity, for use in your classroom.

*The addresses of specific agencies, organizations, and businesses are listed in the Appendix.

3. Write a script for a guided-imagery activity on the topic of machines.
4. Order the materials listed in "Free and Inexpensive Materials."
5. Obtain—or have your school order—a set of pulleys with which your students can do the "Pulleys" activity.

Building your own library

Barker, Eric, and W. Millard: *Machines and Energy*, ARCO, 1972.
Blackwood, Paul: *Push and Pull: The Story of Energy*, McGraw-Hill, New York, n.d.
Blecha, Milo, et al.: *Exploring Matter and Energy*, Laidlaw, River Forest, Ill., 1980.
Brandwein, Paul: *Energy: A Physical Science*, Harcourt Brace Jovanovich, New York, 1975.
Elementary Science Study (ESS): *Primary Balancing*, Webster, McGraw-Hill, Manchester, Mo., 1969.
Ramsey, William, et al.: *Physical Science*, Holt, New York, 1978.

Building a classroom library

Adkins, Jan: *Heavy Equipment*, Scribner, New York, 1980.
Bitter, Gary C.: *Exploring with Computers*, Messner, New York, 1981.
Buehr, Walter: *First Book of Machines*, F. Watts, New York, 1962.
Corbett, Scott: *Home Computers: A Simple and Informative Guide*, Atlantic Monthly, Little, Brown, Boston, 1980.
Darby, Gene: *What Is a Simple Machine?* Benefic Press, Chicago, Ill., 1961.
Englebardt, Stanley L.: *Miracle Chip: The Microelectronic Revolution*, Lothrop, New York, 1979.
Hellman, Hal: *Energy and Inertia*, M. Evans, Philadelphia, Pa., 1970.
Hellman, Harold: *The Lever and the Pulley*, Lippincott, Philadelphia, 1971.
James, Elizabeth, and Carol Barkin: *The Simple Facts of Simple Machines*, Lothrop, New York, 1975.
Lewis, Bruce: *Meet the Computer*, Dodd, Mead, New York, 1977.
Liberty, Gene: *The First Book of Tools*, F. Watts, New York, 1960.
Meyer, J. S.: *Machines*, World Publishing, Cleveland, 1972.
Miller, Lisa: *Levers*, Coward-McCann, New York, 1967.
———: *Wheels*, Coward-McCann, New York, 1965.
Milton, Joyce: *Here Come the Robots*, Hastings House, New York, 1981.
Pine, Tillie S., and Joseph Levine: *Simple Machines and How We Use Them*, McGraw-Hill, New York, 1965.
Rockwell, Anne F., and Harlow Rockwell: *Machines*, Macmillan, New York, 1972.
Schneider, H.: *Everyday Machines and How They Work*, McGraw-Hill, New York, 1950.
Shapp, Charles, and Martha Shapp: *Let's Find Out about Wheels*, F. Watts, New York.

Wade, Harlan: *A Book about the Lever*, Raintree, Milwaukee, Wis., 1979.
Wilson, Mike and Robin Scagell: *Jet Journal*, Viking, New York, 1978.
Zim, H. S., and J. R. Skelly: *Machine Tools*, Morrow, New York, 1960.
Zisfein, Melvin B.: *Flight: A Panorama of Aviation*, Pantheon, New York, 1981.

APPENDIX

SOURCES OF FREE AND INEXPENSIVE TEACHING MATERIALS

American Dental Association
211 East Chicago Avenue
Chicago, IL 60611

American Forest Institute
1619 Massachusetts Avenue, N.W.
Washington, DC 20036

American Gas Association, Inc.
Educational Services
1515 Witson Boulevard
Arlington, VA 22209

American Humane Association
P.O. Box 1266
Denver, CO 80201

American Institute of Aeronautics and Astronautics
1290 Avenue of the Americas
New York, NY 10019

American Institute of Baking
1213 Baker's Way
Manhattan, KS 66502

American Medical Association
535 North Dearborn Street
Chicago, IL 60610

American Museum of Atomic Energy
Oak Ridge Associated Universities
P.O. Box 177
Oak Ridge, TN 37830

American Petroleum Institute
1271 Avenue of the Americas
New York, NY 10020

Animal Protection Institute of America
5894 South Land Park Drive
P.O. Box 22505
Sacramento, CA 95822

Association of American Railroads
Office of Information and Public Affairs
1920 L Street, N.W.
Washington, DC 20036

Astrophysical Observatory
60 Garden Street
Cambridge, MA 02138

Automobile Manufacturers Association
300 New Center Building
Detroit, MI 48202

California Raisin Advisory Board
P.O. Box 5335
Fresno, CA 93755

Chevron Chemical Company
P.O. Box 3744
San Francisco, CA 94119

Council on Family Health
633 Third Avenue
New York, NY 10017

Cruise Lines International Association
17 Battery Place
New York, NY 10004

DARES
c/o Helen Georges
407 Concordia Drive
Tempe, AZ 85282

Del Monte
Box 9075
Clinton, IA 52732

Department of Citrus
State of Florida
Florida Citrus Commission
P.O. Box 148
Lakeland, FL 33802

Edison Electric Institute
1111 19th Street, N.W.
Washington, DC 20036

Energy Research and Development Administration (ERDA)
Office of Public Affairs
Washington, DC 20545

Eros Data Center
Sioux Falls, SD 57198

Estes Industries
Education Director
Penrose, CA 81240

Garden Clubs of America
598 Madison Avenue
New York, NY 10022

General Electric Company
Educational Communications
Fairfield, CT 06431

Green Giant Company
LeSeur, MN 56058

Johnson & Johnson Products, Inc.
Consumer Services
501 George Street
New Brunswick, NJ 08903

McDonald Observatory
RLM 15.308
University of Texas
Department of Astronomy
Austin, TX 78712

Metaframe Corporation
Elmwood Park, NJ 07407

National Aeronautics and Space Administration (NASA)
Washington, DC 20546

National Air and Space Museum
Smithsonian Institute
Washington, DC 20560

National Coal Association
Education Division–Coal Building
1130 17th Street, N.W.
Washington, DC 20036

National Dairy Council
6300 River Road
Rosemont, IL 60018

National Oceanic and Atmospheric Administration
Office of Coastal Zone Management
3300 Whitehaven Street, N.W.
Washington, D.C. 20235

National Wildlife Federation
1413 16th Street, N.W.
Washington, DC 20036

Nektonics
1015 East 35th Street
Brooklyn, NY 11210

Pendleton Woolen Mills
218 S.W. Jefferson
Portland, OR 97201

Rock of Ages Corporation
Education Department
P.O. Box 482
Barre, VT 05641

Sea Grant College Program
Texas A&M University
College Station, TX 77843

Shell Oil Company
Public Relations Department
P.O. Box 2463
Houston, TX 77001

Soil Conservation Society of America
7515 N.W. Ankeny Road
Ankeny, IA 50021

Superintendent of Documents
Washington, DC 20402

Survival Education Association
9035 Golden Given
Tacoma, WA 98445

Taylor Instrument Company
Consumer Products Division
Advertising Department
Arden, NC 28704

Thomas Alva Edison Foundation
143 Cambridge Office Plaza
18280 West Ten Mile Road
Southfield, MI 48075

Tuna Research Foundation
1101 17th Street, N.W.
Washington, D.C. 20036

United Airlines
P.O. Box 66100
Chicago, IL 60666

U.S. Department of Agriculture
Forest Service
P.O. Box 1963
Washington, DC 20013

U.S. Department of Agriculture
Office of Information
Washington, D.C. 20250

U.S. Department of Agriculture
Soil Conservation Service
Washington, DC 20250

U.S. Department of Commerce
610 Hardesty Street
Kansas City, MO 64124

U.S. Department of Commerce
National Oceanic and Atmospheric Administration
Central Logistics Supply Center
610 Hardesty Street
Kansas City, MO 64124

U.S. Department of Commerce
National Weather Service
610 Hardesty Street
Kansas City, MO 64124

U.S. Department of Energy
P.O. Box 62
Oak Ridge, TN 37830

U.S. Department of the Interior
Bureau of Mines
4800 Forbes Avenue
Pittsburgh, PA 15213

U.S. Department of the Interior
Fish and Wildlife Service
Bureau of Sport Fisheries and Wildlife
Washington, DC 20240

U.S. Department of the Interior
Geological Survey
Reston, VA 22092

U.S. Environmental Protection Agency
401 M Street, S.W.
Washington, D.C. 20460

University of Rhode Island
Marine Advisory Service
Publications Unit
P. O. Box 56
Narragansett, RI 02882

SOURCES OF SCIENCE MATERIALS AND EQUIPMENT

Bausch & Lomb (Microscopes)
P.O. Box 450
Rochester, NY 14692

Carolina Biological Supply Co. (Life science materials)
2700 York Road
Burlington, NC 27215

Delta Education, Inc. (General elementary education)
P.O. Box M
Nashua, NH 03061

Earth Science Materials, Inc. (Earth science)
P.O. Box 69
Florence, CO 81226

Edmund Scientific Co.
101 East Gloucester Pike
Barrington, NJ 08033

Estes Industries (Rockets)
Department 27
Penrose, CO 81240

Fisher Scientific Co. (General)
4901 West Le Moyne Street
Chicago, IL 60651

Jewel Industries, Inc. (Aquariums, cages, terrariums)
5005 West Armitage Avenue
Chicago, IL 60639

NASCO (General)
901 Janesville Avenue
Fort Atkinson, WI 53538

Ohaus Scale Corporation (Balances, scales)
29 Hanover Road
Florham Park, NJ 07932

Sargent-Welch Scientific Co. (General)
7300 North Linder Avenue
Skokie, IL 60077

Ward's Natural Science Establishment, Inc. (General)
P.O. Box 1712
Rochester, NY 14603

Webster Division (Tuning forks)
McGraw-Hill Book Company
Manchester Road
Manchester, MO 63011

INDEXES

NAME INDEX

SUBJECT INDEX*

**Note:* Page numbers printed in *italic* in the Subject Index indicate activities. Readers are also referred to the Contents, in which activities are listed.